Molecular Mechanisms of Parasite Invasion

Subcellular Biochemistry
Volume 47

SUBCELLULAR BIOCHEMISTRY

Recent Volumes in this Series

Volume 34 **Fusion of Biological Membranes and Related Problems**
Edited by Herwig Hilderson and Stefan Fuller

Volume 35 **Enzyme-Catalyzed Electron and Radical Transfer**
Andreas Holzenburg and Nigel S. Scrutton

Volume 36 **Phosphilipid Metabolism in Apoptosis**
Edited by Peter J. Quinn and Valerian E. Kagan

Volume 37 **Membrane Dynamics and Domains**
Edited by Peter J. Quinn

Volume 38 **Alzheimer's Disease: Cellular and Molecular Aspects of Amyloid beta**
Edited by Robin Harris and Falk Fahrenholz

Volume 39 **Biology of Inositols and Phosphoinositides**
Edited by A. Lahiri Majumder and B.B. Biswas

Volume 40 **Reviews and Protocols in DT40 Research**
Edited by Jean-Marie Buerstedde and Shunichi Takeda

Volume 41 **Chromatin and Disease**
Edited by Tapas K. Kundu and Dipak Dasgupta

Volume 42 **Inflammation in the Pathogenesis of Chronic Diseases**
Edited by Randall E. Harris

Volume 43 **Subcellular Proteomics**
Edited by Eric Bertrand and Michel Faupel

Volume 44 **Peroxiredoxin Systems**
Edited by Leopold Flohé and J. Robin Harris

Volume 45 **Calcium Signalling and Disease**
Edited by Ernesto Carafoli and Marisa Brini

Volume 46 **Creatine and Creatine Kinase in Health and Disease**
Edited by Gajja S. Salomons and Markus Wyss

Volume 47 **Molecular Mechanisms of Parasite Invasion**
Edited by Barbara A. Burleigh and Dominique Soldati-Favre

Molecular Mechanisms of Parasite Invasion

Subcellular Biochemistry Volume 47

Edited by

Barbara A. Burleigh, PhD

Department of Immunology and Infectious Diseases, Harvard School of Public Health, Boston, Massachusetts, USA

Dominique Soldati-Favre, PhD

Department of Biological Sciences, Imperial College, London, UK
and
Département de Microbiologie et Medecine Moléculaire, Université de Genève, Genève, Switzerland

Springer

LANDES BIOSCIENCE

A C.I.P catalog record for this book is available from the Library of Congress.

ISBN: 978-0-387-78266-9

Published by:

Landes Bioscience, 1002 West Avenue, 2nd Floor, Austin, Texas 78701, USA
Phone: 512/ 637 6050; FAX: 512/ 637 6079
www.landesbioscience.com

and

Springer Science+Business Media, LLC, 233 Spring Street, New York, New York 10013, USA
www.springer.com

DEDICATION

The editors have dedicated this volume to Lewis Tilney because of his key contributions to our understanding of the cytoskeleton in general and for elucidating the roles of the cytoskeleton in pathogen-host cell interactions in particular. It is difficult to summarize those contributions in a paragraph or two. It is even more daunting to convey to those who have not experienced it, Lew's unique vision, style and the irrepressible zest with which he pursued the myriad of model systems he helped to establish. Working with Lew has always been a delight and an adventure. In the lab, his comments and/or observations were constantly punctuated with his favorite adjectives—"Terrific!!!"or "Fantastic!!"—and my personal favorite, "*Formidable!!*" (first heard while working with him in France). His enthusiasm for life and science was and is irresistible.

The scope of Lew's contributions is particularly impressive given that throughout his career he had only a small handful of students and postdocs. He was, however, adept at forging collaborations, some of which spanned decades—and no collaborator was more central to his efforts than his wonderful late wife, Molly. It is also important to note that his efforts were immeasurably aided by the two gifted EM techs with whom he worked, first Doris Bush and later Pat Connelly. His hands and keen eye at the electron microscope (an antique Philips 200 he maintained by collecting a stable of abandoned scopes for parts) were always central to his studies. Indeed he continued to be first author on many of his papers throughout his career. The other hallmark of Lew's career was his knack for picking oddball cell types with "extreme" cytoskeletal arrays from which basic mechanisms of organization and function were relatively easy to parse out.

When I joined Lew's lab as an undergraduate in 1970, his research dealt primarily with the role of microtubule assembly/disassembly in changes in cell shape—highlighted by studies on his first and perhaps most beautiful model system, the protozoan *Actinosphaerium*. After a summer project as a student in the 1970 physiology course at the Marine Biological Laboratory (MBL), Lew's focus began to shift to the actin cytoskeleton. Some of his early contributions include the identification of microfilaments in a vertebrate nonmuscle cell as actin; the establishment of the intestinal brush border as a model system for investigating functional organization of the actin cytoskeleton; the identification of an oligomeric form of actin as a key component of the erythrocyte membrane skeleton. His decade-plus series of studies on the acrosomal reaction of echinoderm sperm resulted in a number of seminal findings that provided the foundation for understanding the role of actin assembly in cell movement. These include the first demonstration that actin polymerization can power the formation/extension of an asymmetric cell process; the characterization of a nonfilamentous, storage form of actin, profilamentous actin (from which the term profilin is derived); and the first experimental evidence for the role of profilin in promoting preferential actin monomer addition to the plus/barbed end of the growing actin filament. Of course the crowning achievement of his studies on actin-assembly driven cellular processes are those with his collaborator Portnoy and others, on the elucidation of how *Listeria* recruits a comet tail of actin filaments whose assembly powers the intracellular movement and intercellular infection of this bacterial pathogen.

One common element driving Lew's research effort throughout his career was an intense curiosity about the basis for biological pattern formation, particularly with regard to exquisitely ordered cytoskeletal arrays. Lew, together with his collaborators (most notably DeRosier), through studies on the structural organization of the actin bundle that comprises the acrosomal process of *Limulus* sperm and later the actin bundle within the stereo cilium of the cochlear hair cell, established the guiding principles for how filaments can be packed into highly ordered bundles by various combinations of filament cross linking proteins. Ironically, in his latest series of studies, Lew (in collaboration with Guild and others) turned away from the oddball to the classic "genetic" system *Drosophila* to investigate the molecular basis for actin bundle formation—primarily with respect to the actin bundles involved in bristle morphogenesis. Lew's analytical eye and ability to develop improved fixation techniques for visualization of actin in the fly allowed him to extract unique insights into the molecular basis of actin bundle formation and the role of these bundles in morphogenesis.

Lew also made key observations directly relevant to the topics covered in this volume. He, together with his collaborators (including Shaw and Roos) published more than a dozen papers detailing key aspects of the cell biology of intracellular parasites including *Toxoplasma*—perhaps the most notable of which was the discovery that *Toxoplasma* has an algal symbiont-derived plastid. One of my most memorable "Tilney moments" occurred in the mid 1990s when I ran into Lew in the MBL library at Woods Hole. He was, as usual, at a table surrounded by dozens of micrographs. As I entered the reading room he looked up at me, held up one of the micrographs and exclaimed—"Moose, look at this—*Toxoplasma* is a damned plant!!" A keen eye, indeed.

Dr. Mark Mooseker
Department of Molecular, Cellular and Developmental Biology
Yale University
New Haven, Connecticut, USA

ABOUT THE EDITORS...

BARBARA BURLEIGH is an Associate Professor of Immunology and Infectious Diseases at the Harvard School of Public Health in Boston, MA. Her main research interests include determination of the molecular and cellular requirements for establishment of intracellular infection in mammalian host cells by the intracellular pathogen *Trypanosoma cruzi*, with an emphasis on signal transduction and regulation of host cell gene expression. Dr. Burleigh received her PhD from the University of Western Ontario in London, Ontario and carried out post-doctoral research at Yale University. She currently serves as an Associate Editor of *PLoS Pathogens*.

DOMINIQUE SOLDATI-FAVRE is an Associate Professor at the Faculty of Medicine of the University of Geneva, in the Department of Microbiology and Molecular Medicine. Her main research interest is to understand how an eukaryotic cell can penetrate, survive and replicate within another eukaryotic cell, exploiting *Toxoplasma gondii* as a model organism, which is conveniently amenable to genetic manipulation. She is studying the molecular mechanisms involved in motility, host cell attachment and invasion and the strategies allowing intracellular survival. She received her PhD from the Department of Molecular Biology of the University of Zurich and did postdoctoral training with Prof. John Boothroyd at the Department of Microbiology and Immunology at the Stanford School of Medicine.

TABLE OF CONTENTS

SECTION IV: ACTIN DYNAMICS: HOST AND PARASITE

SECTION V: PROTEASES

SECTION VI: THE PARASITOPHOROUS VACUOLE

SECTION VII: PENETRATING BIOLOGICAL BARRIERS

LIST OF CONTRIBUTORS

Maria Júlia M. Alves
Departamento de Bioquímica, Instituto de Química, Universidade de São Paulo, São Paulo, Brazil

Walter K. Andreoli
Disciplina de Parasitologia, Departamento de Microbiologia, Imunologia e Parasitologia, UNIFESP, Escola Paulista de Medicina, São Paulo, Brazil

Kerstin Baer
Department of Medical Parasitology, New York University School of Medicine, New York, New York, USA

Antonio Barragan
Swedish Institute for Infectious Disease Control, and Center for Infectious Medicine, Karolinska Institutet, Stockholm, Sweden

Michael J. Blackman
Division of Parasitology, National Institute for Medical Research, Mill Hill, London, UK

Barbara A. Burleigh
Department of Immunology and Infectious Diseases, Harvard School of Public Health, Boston, Massachusetts, USA

Vern B. Carruthers
Department of Microbiology and Immunology, University of Michigan School of Medicine, Ann Arbor, Michigan, USA

Xian-Ming Chen
Miles and Shirley Fiterman Center for Digestive Diseases, Mayo Clinic, Rochester, Minnesota, USA

Eduardo N. Chini
Department of Anesthesiology, Mayo Medical School, Mayo Clinic and Foundation, Rochester, Minnesota, USA

Walter Colli
Departamento de Bioquímica, Instituto de Química, Universidade de São Paulo, São Paulo, Brazil

Mauro Cortez
Departamento de Microbiologia, Imunologia e Parasitologia, Escola Paulista de Medicina, Universidade de São Paulo, São Paulo, Brazil

Claudio V. da Silva
Disciplina de Parasitologia, Departamento de Microbiologia, Imunologia e Parasitologia, UNIFESP, Escola Paulista de Medicina, São Paulo, Brazil

Solange da Silva
Disciplina de Parasitologia, Departamento de Microbiologia, Imunologia e Parasitologia, UNIFESP, Escola Paulista de Medicina, São Paulo, Brazil

Frédéric Delbac
Equipe Parasitologie Moléculaire et Cellulaire, LBP, UMR CNRS 6023, Université Blaise Pascal, Aubière, France

Albert Descoteaux
INRS-Institut Armand-Frappier, Laval, Quebec, Canada

Tiffany DeSimone
Department of Immunology and Infectious Diseases, Harvard School of Public Health, Boston, Massachusetts, USA

Timothy J. Dowse
Department of Biological Sciences, Imperial College, London, UK

Manoj T. Duraisingh
Department of Immunology and Infectious Diseases, Harvard School of Public Health, Boston, Massachusetts, USA

Adriana B. Fernandes
Disciplina de Parasitologia, Departamento de Microbiologia, Imunologia e Parasitologia, UNIFESP, Escola Paulista de Medicina, São Paulo, Brazil

Maria Cecília D.C. Fernandes
Disciplina de Parasitologia, Departamento de Microbiologia, Imunologia e Parasitologia, UNIFESP, Escola Paulista de Medicina, São Paulo, Brazil

Ute Frevert
Department of Medical Parasitology, New York University School of Medicine, New York, New York, USA

Niclas Hitziger
Swedish Institute for Infectious Disease Control, and Center for Infectious Medicine, Karolinska Institutet, Stockholm, Sweden

Cameron Jennings
Department of Immunology and Infectious Diseases, Harvard School of Public Health, Boston, Massachusetts, USA

Christian Klotz
Department of Medical Parasitology, New York University School of Medicine, New York, New York, USA

Konstantinos Koussis
Division of Parasitology, National Institute for Medical Research, Mill Hill, London, UK

Carolina L'Abbate
Disciplina de Parasitologia, Departamento de Microbiologia, Imunologia e Parasitologia, UNIFESP, Escola Paulista de Medicina, São Paulo, Brazil

Nicholas F. LaRusso
Miles and Shirley Fiterman Center for Digestive Diseases, Mayo Clinic, Rochester, Minnesota, USA

Ana Paula C.A. Lima
Instituto de Biofisica Carlos Chagas Filho, UFRJ, Rio de Janeiro, Brazil

Robert Lodge
INRS-Institut Armand Frappier, Laval, Quebec, Canada

Kai Matuschewski
Department of Parasitology, Heidelberg University School of Medicine, Heidelberg, Germany

Jeffrey Mital
Department of Microbiology and Molecular Genetics, University of Vermont, Burlington, Vermont, USA

Silvia N. Moreno
Department of Cellular Biology, and Center for Tropical and Emerging Global Diseases, University of Georgia, Athens, Giorgia, USA

Renato A. Mortara
Disciplina de Parasitologia, Departamento de Microbiologia, Imunologia e Parasitologia, UNIFESP, Escola Paulista de Medicina, São Paulo, Brazil

G. Adam Mott
Department of Immunology and Infectious Diseases, Harvard School of Public Health, Boston, Massachusetts, USA

Kisaburo Nagamune
Department of Molecular Microbiology, Washington University School of Medicine, St. Louis, Missouri, USA

Steven P. O'Hara
Miles and Shirley Fiterman Center for Digestive Diseases, Mayo Clinic, Rochester, Minnesota, USA

William A. Petri, Jr.
Division of Infectious Diseases and International Health, University of Virginia Health System, Charlottesville, Virginia, USA

Valérie Polonais
Department of Microbiology and Molecular Medicine, University of Geneva, Geneva, Switzerland

Philippe Refour
Department of Immunology and Infectious Diseases, Harvard School of Public Health, Boston, Massachusetts, USA

Julio Scharfstein
Instituto de Biofisica Carlos Chagas Filho, UFRJ, Rio de Janeiro, Brazil

Herwig Schüler
Max Delbrück Center for Molecular Medicine, Berlin, Germany

L. David Sibley
Department of Molecular Microbiology, Washington University School of Medicine, St. Louis, Missouri, USA

Anthony P. Sinai
Department of Microbiology, Immunology and Molecular Genetics, University of Kentucky College of Medicine, Lexington, Kentucky, USA

Aaron J. Small
Miles and Shirley Fiterman Center for Digestive Diseases, Mayo Clinic, Rochester, Minnesota, USA

Shahram Solaymani-Mohammadi
Department of Medicine—Infectious Diseases, University of Virginia, Charlottesville, Virginia, USA

Dominique Soldati-Favre
Department of Biological Sciences, Imperial College, London, UK
and
Département de Microbiologie et Medecine Moléculaire, Université de Genève, Genève, Switzerland

Fiona M. Tomley
Division of Microbiology, Institute for Animal Health, Compton Newbury, Berkshire, UK

Ivan Usynin
Department of Medical Parasitology, New York University School of Medicine, New York, New York, USA

Gary E. Ward
Department of Microbiology and Molecular Genetics, University of Vermont, Burlington, Vermont, USA

Chenwei Wu
Department of Immunology and Infectious Diseases, Harvard School of Public Health, Boston, Massachusetts, USA

Nobuko Yoshida
Departamento de Microbiologia, Imunologia e Parasitologia, Escola Paulista de Medicina, Universidade de São Paulo, São Paulo, Brazil

PREFACE

In biological terms, a **parasite** can be defined as "*an organism that grows, feeds and is sheltered on, or in, a different organism while contributing nothing to the survival of the host*."[1] This relationship is particularly intimate for parasitic protozoa (and single-celled fungi such as the Microsporidia), many of which not only penetrate tissue barriers, but gain entry into host cells. The entry mechanisms are as diverse as the organisms employing them and are often critical components of their pathogenic profile. All of the parasitic organisms highlighted in this book represent medically important human pathogens that contribute significantly to the global burden of disease. As such there is intense interest in understanding the molecular basis of infection by these pathogens—not only with regard to their clinical relevance but also the fascinating biology they reveal.

For most of the parasites discussed here (*Plasmodium falciparum*, *Toxoplasma gondii*, *Cryptosporidium parvum*, *Trypanosoma cruzi* and *Leishmania spp.*), the ability to penetrate biological barriers and/or to establish intracellular residence is critical to survival of the pathogen in the mammalian hosts. For other parasites, such as *Entamoeba*, a tissue invasive phenotype is a key virulence determinant. In the ensuing 18 chapters, select members of this diverse set of protozoan parasites, as well as some examples of the extremely reduced fungal parasites classified as Microsporidia, are discussed within the context of the fascinating molecular strategies employed by these organisms to migrate across biological barriers and to establish residence within target host cells.

Given the emphasis of this book on the Apicomplexa (including *Plasmodium*, *Toxoplasma* and *Cryptosporidium*) within which the biology of invasion and intracellular establishment are best studied at the molecular level, the first chapter has been dedicated to a comprehensive overview of the array of experimental approaches that are routinely used to study this group of parasites. The remaining chapters have been arranged according to subtopic to facilitate comparisons of the molecular and cellular mechanisms governing invasion (entry), the molecular determinants of these processes and the establishment of the replication-permissive niches exploited by these diverse pathogens. What emerges is a diverse tapestry of mechanisms, molecular determinants and cellular responses that share both common patterns as well as patterns unique to the specific pathogen-host interaction. With this state-of-the-art collection we hope to highlight the recent advances while paving the way for exciting discoveries in the future.

Barbara A. Burleigh, PhD
Anthony P. Sinai, PhD

1. American Heritage® Dictionary of the English Language

CHAPTER 1

Current and Emerging Approaches to Studying Invasion in Apicomplexan Parasites

Jeffrey Mital and Gary E. Ward*

Abstract

In this chapter, we outline the tools and techniques available to study the process of host cell invasion by apicomplexan parasites and we provide specific examples of how these methods have been used to further our understanding of apicomplexan invasive mechanisms. Throughout the chapter we focus our discussion on *Toxoplasma gondii,* because *T. gondii* is the most experimentally accessible model organism for studying apicomplexan invasion (discussed further in the section, "*Toxoplasma as a Model Apicomplexan*") and more is known about invasion in *T. gondii* than in any other apicomplexan.

Host Cell Invasion by Apicomplexan Parasites

Host cell invasion is a complex, multi-step process that is relatively conserved among apicomplexans (see Fig. 1). *T. gondii* invasion begins with movement of the parasite over the surface of a host cell, driven by an actin-myosin-based motor within the parasite. This gliding motility is dependent on a solid substrate, and occurs in the absence of flagella, pseudopodia, or other traditional locomotive organelles (reviewed in ref. 2; see also Matuschewski, this volume). As the parasite glides over host cells, a trail of surface proteins and lipids is deposited in the parasite's wake.[3] The conoid, a thimble-shaped cytoskeletal structure comprised of tubulin assembled into spiral ribbons,[4] repeatedly protrudes and retracts from the apical tip of the gliding parasite (reviewed in ref. 5). Initial attachment of the parasite to the host cell is coupled to the discharge of the micronemes, apical secretory organelles that release adhesins onto the parasite surface[6,7] (see also Carruthers, this volume). A more intimate apical interaction then develops between the parasite and host cell[8] and a second set of apical organelles, the rhoptries, discharge. Secretion from the micronemes and the rhoptries is tightly controlled, and there is some evidence that secretion from the bulb and neck regions of the rhoptries are distinct, regulated events,[8] although the relative order of these various secretory events is not entirely clear. Very early in the process of invasion, perhaps when the micronemes and/or rhoptries are first discharged, a transient change in host cell plasma membrane conductance can be detected,[9] suggesting a temporary, localized breach in host plasma membrane integrity. A third set of secretory organelles, the dense granules, are constitutively discharged, and are thought to function primarily in establishing and modifying the parasite's intracellular niche after invasion (reviewed in ref. 1).

Active penetration of the host cell begins at the apical tip of the parasite and is thought to be dependent upon the same actin-myosin machinery that powers gliding motility, since disruption

*Corresponding Author: Gary E. Ward—Department of Microbiology and Molecular Genetics, University of Vermont, Burlington, VT 05405, USA. Email: gary.ward@uvm.edu

Molecular Mechanisms of Parasite Invasion, edited by Barbara A. Burleigh and Dominique Soldati-Favre.

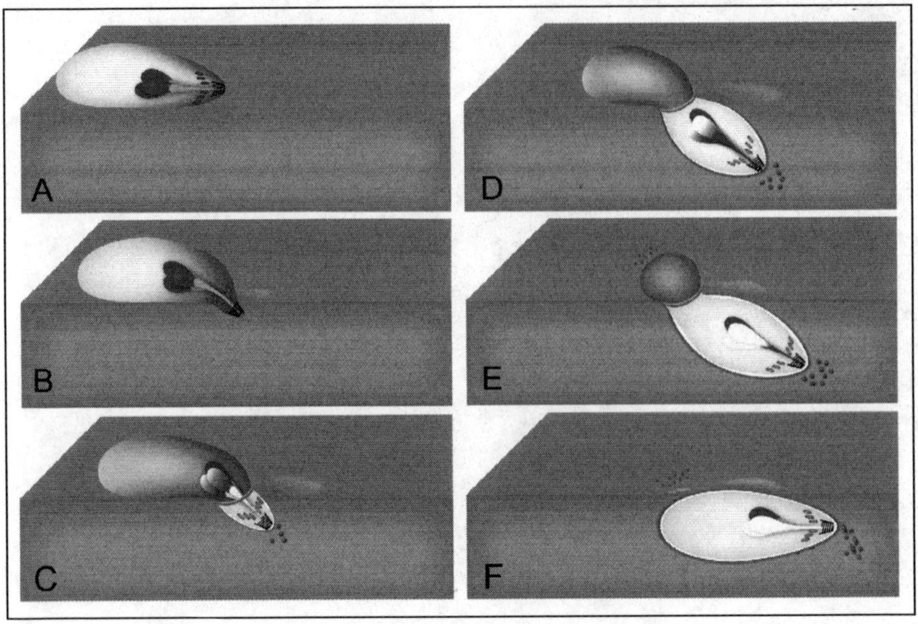

Figure 1. Overview of host cell invasion by *T. gondii* tachyzoites. A) A tachyzoite glides over the surface of a host cell, powered by its actin-myosin-based motor. The thimble-shaped conoid at the apical tip of the parasite repeatedly protrudes and retracts. Adhesins (green) are found both intracellularly, within small tubular micronemes, and on the parasite surface, concentrated towards the apical end. B) The conoid is extended when the parasite begins to invade, bringing the apical tip of the parasite into intimate contact with the host cell. Microneme secretion increases, delivering more proteins to the parasite surface. An invagination of the host cell membrane begins to develop. C) Active penetration of the host cell begins, driven by the same actin-myosin motor that drives gliding motility. Rhoptry neck proteins (light blue) are secreted and associate with the microneme protein TgAMA1 to form the moving junction, through which the parasite enters the host cell. Most microneme proteins are capped, accumulating posteriorly; they are thought to be kept out of the parasitophorous vacuole by sieving action of the moving junction. Rhoptry bulb proteins (dark blue) begin to be secreted into the host cell. D,E) Microneme proteins continue to be capped towards the posterior end, where they are cleaved within their transmembrane domains. Rhoptry secretion continues, with rhoptry proteins contributing to the formation of the parasitophorous vacuole. The parasite squeezes through the moving junction, visibly constricting at the point of entry, into the deepening vacuole. F) Microneme proteins are shed from the posterior end of the parasite. Invasion is complete when the parasitophorous vacuole membrane pinches closed. A patch of protein originally derived from the rhoptry necks is found transiently at the site of entry, associated with the host plasma membrane. Note that dense granules and dense granule secretion[1] are not illustrated in this diagram, for simplicity.

of a single myosin gene, *MyoA*, disrupts both processes[10] and pharmacological agents that disrupt gliding motility also disrupt invasion.[11] Due to the duality of function of the actin-myosin motor, it is difficult to determine whether gliding motility is required for invasion; however, video-microscopic analysis of *T. gondii* suggests that tachyzoites may be probing, perhaps by way of conoid extension, for some specific receptor or membrane microdomain as they glide over the host cell surface before initiating invasion. Both gliding motility and invasion involve receptor-ligand interactions (see Matuschewski, this volume; Carruthers, this volume), resulting in anterior to posterior capping of surface proteins, followed by their protease-mediated shedding[7,12] (see also Dowse, this volume). Capping and shedding are thought to be responsible for the trails of surface proteins deposited at the posterior end of gliding parasites.

During penetration, a zone of tight interaction between the host and parasite plasma membranes forms at the site of parasite entry. This "moving junction" translocates from the anterior to posterior end of the parasite concurrent with penetration. Proteins secreted from the necks of the rhoptries, physically complexed with at least one microneme protein, were recently identified as components of the moving junction.[13,14] As the tachyzoite penetrates into the host cell it becomes surrounded by a parasitophorous vacuole membrane (PVM), derived from host cell plasma membrane lipids.[9] Host plasma membrane proteins are largely excluded from the PVM as it forms, creating a compartment that does not fuse with the endolysosomal system of the host cell[15,16] (see also Sinai, this volume). The entire process of invasion takes approximately 20-25 seconds.[17] Shortly after invasion, the PV relocates to a position adjacent to the host cell nucleus and becomes tightly associated with host cell mitochondria and endoplasmic reticulum.[18]

While the various steps in invasion, described above for *T. gondii,* are broadly conserved among apicomplexan parasites,[19-21] several interesting and potentially informative variations are observed in other genera. For example, the conoid appears to be absent from *Plasmodium* and *Babesia*[22] (reviewed in ref. 5). Parasite reorientation, to bring the apical end in contact with the host cell, is more pronounced in other apicomplexans such as *Plasmodium*[23] and *Cryptosporidium*[21] than it is in *T. gondii*. An entirely different invasive mechanism, which does not involve formation of a PVM, is employed by *Plasmodium* sporozoites as they migrate through host cells en route to their preferred site of replication (see Frevert, this volume). *Theileria* sporozoites are perhaps the most unusual of all the apicomplexans in terms of their invasive mechanisms. The invasive stages of *Theileria* are not motile, have no recognizable conoid or micronemes, and do not apically reorient prior to invasion. Internalization appears to be a passive process on the part of the parasite, with no role for parasite actin. However, other features of *Theileria* invasion are similar to those of other apicomplexans, including protease-mediated shedding of surface proteins during internalization, and formation of a PVM from the host cell plasma membrane (reviewed in ref. 24).

Assaying Host Cell Invasion

Given the importance of host cell invasion in the life cycle and pathogenesis of apicomplexan parasites, there is much interest in studying the mechanisms underlying invasion. Many different means of quantitatively assaying invasion have been developed, particularly for *T. gondii*, as described below.

Plaque Assay

Parasites are added to host cell monolayers and the resulting number of plaques, formed by successive rounds of host cell invasion, lysis, and invasion of surrounding cells by the released daughter parasites, is determined visually.[25] Plaque visualization can be facilitated by staining with Giemsa or crystal violet. While the plaque assay is simple to perform and requires no special equipment, it does not separate the process of invasion from the rest of the parasite's intracellular life cycle. A decrease in plaque formation can result from a defect in invasion, but also from defects in intracellular growth, parasite replication, host cell lysis and parasite egress, or any combination of these processes.

Vacuole Assay

The number of mature parasitophorous vacuoles present 24 hours after invasion can be determined by phase microscopy and used as a measure of invasion.[10] Caution must be exercised when performing and interpreting this assay since overlapping or adjacent vacuoles can be difficult to count accurately and, like the plaque assay, it may confuse a growth or replication defect with an invasion defect.

^{3}H-Uracil Incorporation Assay

Since many host cells lack a uracil salvage pathway, addition of ^{3}H-uracil to an infected monolayer results in specific incorporation by parasites and can therefore be used to assay the number of intracellular parasites.[26,27] However, since both intracellular and extracellular parasites can incorporate ^{3}H-uracil (although extracellular parasites may have a shorter period of incorporation[28]) it is critical to wash off or kill as many extracellular parasites as possible prior to assaying ^{3}H-uracil incorporation. Extracellular parasites can be killed with pyrrolidine dithiocarbamate (PDTC[29]), although this adds another potential source of variability to the assay. Like the plaque assay, it can also be difficult to definitively ascribe a decrease in ^{3}H-uracil incorporation to a defect in invasion, rather than some other post-invasion defect that alters ^{3}H-uracil uptake or incorporation. Prelabeling parasites with ^{3}H-uracil and measuring the amount of radioactivity associated with the monolayer after infection[30] can circumvent this problem. This approach is best suited for experiments that start with one population of parasites (e.g., comparing the effect of different pharmacological agents on the invasion of a single population of ^{3}H-labeled parasites); it is more problematic to compare two different populations, which may differentially incorporate exogenous uracil. Again, the consistency with which extracellular parasites can be removed or killed will affect the accuracy and variability of this assay.

β-Galactosidase Assay

This method measures the amount of o-nitrophenyl-β-D-galactoside (colorless) that is converted to o-nitrophenol (yellow) by parasites expressing β-galactosidase.[11,31] This assay offers the advantages of being easy, fast, and sensitive. However, in addition to the requirement to remove or kill extracellular parasites, the assay can only be used for parasites expressing β-galactosidase. Although such parasites are available for *T. gondii*[32,33] and *Neospora*,[34] the requirement for β-galactosidase expression may complicate further transgenic manipulation of the parasites, due to the limited number of selectable markers available (summarized in Table 1).

Quantitative PCR Assay

Genomic DNA recovered from invaded parasites can be used as a template for quantitative real-time PCR, which is compared to a dilution series of DNA from a known number of parasites.[74] This assay again requires removal and/or killing of extracellular parasites to ensure that contaminating DNA from these parasites is not included as template for PCR. A strength of this assay, like the ^{3}H-uracil- and β-galactosidase-based assays, is that it allows large numbers of parasites to be analyzed per sample, compared to microscope-based techniques.

Differential Fluorescent Staining

In the most commonly used assay for *T. gondii* invasion, parasites are labeled with different fluorochromes before and after host cell permeabilization, resulting in dual labeled extracellular parasites and singly labeled intracellular parasites (Fig. 2B,F), which are scored by fluorescence microscopy.[11] The use of parasites expressing YFP or GFP further simplifies the assay, by eliminating the need to permeabilize the host cells to visualize intracellular parasites.[50,75] This assay isolates invasion from other aspects of the parasite life cycle and does not require removal of extracellular parasites for accurate counting of invaded parasites. However, the assay suffers from significant field-to-field variability, which is compounded by the limited sample size that can be feasibly obtained, given the extensive hands-on time required for manual counting. An automated microscope has been used to semi-quantitatively assay invasion by differential staining in a high-throughput format.[50] In an attempt to make the differential staining assay more quantitative and reproducible, the assay was recently adapted for scoring by a laser scanning cytometer (LSC). The LSC-based assay significantly increases the number of parasites that can be counted, overcoming much of the field-to-field variability observed when the assay is scored manually. The major drawbacks to the LSC-based assay are that it requires specialized

Table 1. ***β-galactosidase array of intracellular parasites***

Organism	Transfection	Recombination	Selectable Markers	Reporters	Genome Sequenced?
Toxoplasma	Transient[35] Stable[36] Episomes[37]	Homologous[38] Nonhomologous (reviewed in ref. 39)	DHFR[36,40] BLE[41] HXGPRT(+/-)[13,42] CAT[38] UPRT[43] TK[44] Tryp. synthase.[45] Nourseothricin[46]	CAT[47,49] β-Gal[48,49] GFP[51] Luciferase[12,52]	12x www.toxodb.org
Plasmodium	Transient[12,53] Stable[54-56] Long term episomes[57]	Homologous (reviewed in ref. 58) Nonhomologous[56]	DHFR-TS[54] HDHFR[59] BSD[60] NEO[60] CAT[53,61] TK[62] Cyt. Deaminase[62]	Luciferase[63] GFP[64]	Yes www.plasmodb.org
Cryptosporidium	No	NA			Yes www.cryptodb.org
Neospora	Transient[65]	NA	BLE[65]	β-gal[65]	EST database www.cbil.upenn.edu/apidots
Eimeria	Transient[66]	NA		β-Gal[66]	~8x www.genedb.org/gendb/*etenella*
Theileria	Transient[67]	NA		GFP[67]	Yes[68] www.genedb.org/genedb/annulata
Babesia	Transient[69]	NA		Luciferase[69]	underway
Sarcocystis	Transient[70] Stable[72,73]	Homologous[72]	DHFR[72]	YFP[70] β-gal[72]	EST database www.cbil.upenn.edu/apidots

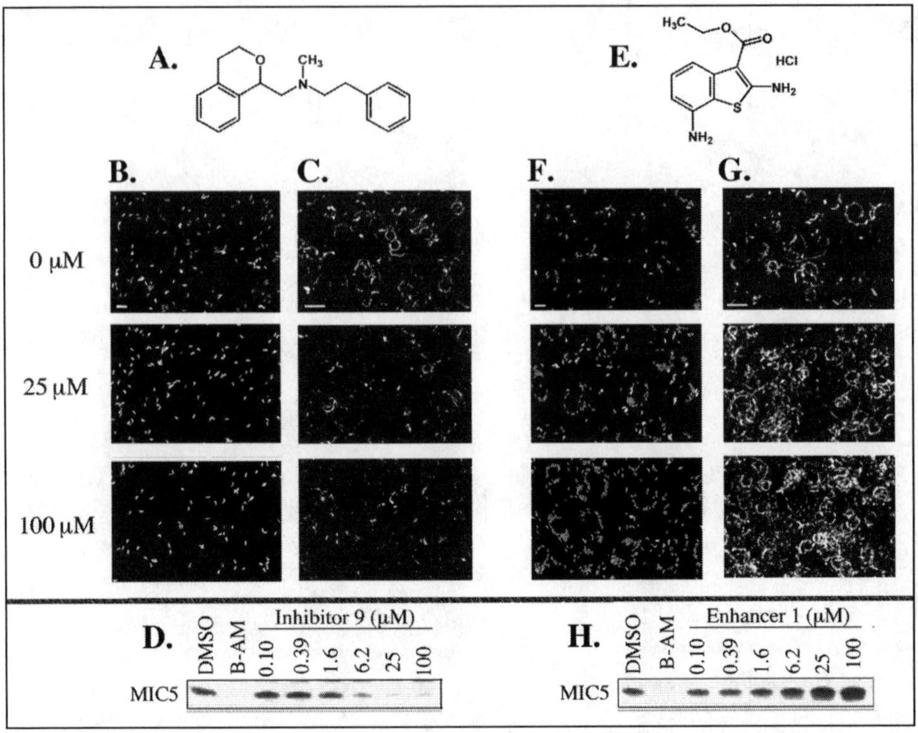

Figure 2. Small molecule inhibitors and enhancers of invasion, motility, and microneme secretion. Representative images from differential fluorescence invasion assays (B,F), trail deposition motility assays (C,G), and microneme secretion assays (D,H) of parasites treated with two small molecules identified by high-throughput screening (A,E; see ref. 50). In the invasion assays (B,F) extracellular parasites appear yellow and invaded parasites appear green. In the motility assays (C,G), the trails deposited by gliding parasites are labeled with an antibody to the abundant surface protein SAG1. In the microneme secretion assays, Western blotting is used to analyze the amount of the microneme protein, MIC5, secreted into the supernatant; BAPTA-AM (B-AM) inhibits microneme secretion and is used as a negative control. The number of invaded parasites (B,F), the number of trails (C,G), and the level of microneme secretion (D,H) decrease or increase in a dose-dependent manner after treatment with inhibitor (A) or enhancer (E), respectively. Scale bars = 20 μm. From: Carey KL et al. Proc Natl Acad Sci USA 101(19):7433-7438; ©2004 with permission from the National Academy of Sciences, USA.[50]

instrumentation and is not well-suited to analyzing many samples at one time.[76] However, it may be possible to adapt these assays for analysis by flow cytometry.[77]

Other Apicomplexans

In the most common invasion assay for *Plasmodium* blood stage parasites, mature schizonts are added to a suspension of erythrocytes and the number of ring stages that develop several hours post infection are visualized in blood smears on a glass slide. Staining of rings is typically accomplished with Giemsa,[78-81] but DAPI[82] and acridine orange[83] have also been used. These assays are complicated by the fact that they measure not only invasion, but several pre and post-invasion events as well, including schizont rupture and morphological differentiation to the ring stage. ^{3}H-hypoxanthine uptake assays (e.g., ref. 84) suffer from a similar problem and are further complicated by the possibility that host erythrocytes and leukocytes may also incorporate ^{3}H-hypoxanthine.[85,86] An enzymatic assay for *Plasmodium* lactate dehydrogenase provides a measure of parasitemia, and can be used as an indirect measure of invasion, with the caveats mentioned above.[87,88] Assays based on parasite staining have recently been adapted for FACS scoring,[89] which significantly reduces operator time but necessitates the quantitative removal of extracellular parasites. A similar FACS assay has been used to measure *Eimeria* invasion.[90] A microscope-based assay that visualizes the incorporation of fluorescent lipid analogs into the PVM during parasite entry can effectively isolate merozoite invasion from other pre and post-invasion events,[91] but has thus far been used only for *P. knowlesi*.

The invasion of *Plasmodium* sporozoites into hepatocytes and other target cells can be assayed by differential fluorescent staining methods similar to those described above for *T. gondii*.[94-96] Some of the other assays described above for *T. gondii* are also directly transferable to other species, e.g., RT-PCR assays for assaying invasion in *Neospora*.[74] The invasion of *Babesia* sporozoites into erythrocytes, which lack a nucleus, can be assayed by nuclear staining.[92] Relatively little is known about *Babesia* invasion, but the recent development of a system that supports efficient in vitro invasion, based upon high voltage release of parasites from previously invaded erythrocytes,[93] may increase our understanding of this process.

Assaying the Individual Steps of Invasion

T. gondii invasion can be readily assayed either in its entirety (using the methods described above), or as a sequence of discrete steps, each of which can be assayed independently. A recently developed protocol for synchronous invasion by *T. gondii*[97] further complements these assays. Parasites are allowed to settle onto host monolayers in a high potassium buffer, which inhibits parasite motility and invasion. The medium is then exchanged for low potassium medium, stimulating motility and invasion. This allows for synchronous and robust invasion over a short period of time. A temperature shift can also be used to synchronize invasion of *T. gondii*.[40]

Assays for the individual steps of *T. gondii* invasion are outlined below, and similar assays for other apicomplexans, when available, are described.

Gliding Motility

T. gondii can be observed gliding over host cells or coverglasses using video-[50,98] or time lapse-microscopy.[10,73] These data can be analyzed either qualitatively[8,50] or quantitatively.[10,98]

Motility can also be assayed by trail deposition; the surface proteins deposited behind *T. gondii* tachyzoites as they glide over host cells and coverglasses can be readily visualized by immunofluorescence microscopy (Fig. 2C,G).[11,48] Alternatively, parasites can be prelabeled with fluorescent antibodies or lipids and trail deposition observed in real time or after fixation.[50] Trail deposition can be quantified by measuring the number and length of trails.[11] Similar trail deposition assays, using species-specific antibodies, have been used for *Cryptosporidium*,[99,100] *Plasmodium*[78] and *Eimeria*[101,102] sporozoites; trails produced by the latter two species have also been observed by electron microscopy.[103,104] A recent study on the motility of *Plasmodium berghei* ookinetes used the dispersal of aggregated parasites as a quantitative measure of ookinete motility.[105]

Conoid Extension

Conoid extension can be experimentally induced by calcium ionophores and inhibited by intracellular calcium chelators.[106] The percentage of *T. gondii* tachyzoites with partially or fully extended conoids can be readily determined by phase microscopy.[50,106]

Microneme Secretion

The secretion of microneme proteins from extracellular parasites occurs constitutively at a basal level and can be upregulated by incubation with calcium ionophores or ethanol.[107] Both constitutive and induced secretion can occur in the absence of host cells; the secreted proteins are recovered from the culture supernatant and analyzed by SDS PAGE and Western blotting (Fig. 2D,H). Similar methods have been used to assay microneme secretion in *Eimeria*,[104] *Cryptosporidium*,[99] *Sarcocystis*,[108] and *Neospora*.[109] The recovery and identification of microneme proteins released into *Plasmodium* culture supernatants was critical in the initial identification and characterization of *Plasmodium* erythrocyte binding proteins (e.g., refs. 110,111).

Direct comparative microneme secretion assays have been developed for *T. gondii*, using 2D-DIGE (2-D Fluorescence Difference Gel Electrophoresis) techniques that differentially label the secreted proteins from two parasite populations with either red or green fluorochromes.[112,113] The samples are mixed in a 1:1 ratio, separated by 2-dimensional gel electrophoresis, and the fluorescent signals analyzed. Proteins that are secreted in equal abundance by both parasite populations appear yellow in a merged image. Spots appear red or green if secretion is up- or down-regulated in one of the populations. The ability to compare two different protein samples on the same gel solves the run-to-run variability problem frequently encountered in 2D gel separations.

Attachment

Both fixed and live host cells have been used for assaying parasite attachment. Parasites cannot invade fixed cells, offering the advantage of having to measure only extracellular parasites. Since fixation may alter the conformation of ligands on the surface of the host cells, it is not known how closely the fixed cell system mimics in vivo attachment and there is some evidence that parasites attach to fixed cells differently than live cells.[76] Live cells offer the advantage of more closely mimicking in vivo conditions. However, since invasion can occur in this case, both attached and internalized parasites must be accounted for when measuring attachment to live cells.

Attachment of *T. gondii* tachyzoites is measured either by counting the number of fluorescently-labeled[114] or Giemsa-stained[115] parasites or by measuring the amount of β–gal[48] or radioactivity[30] associated with the monolayer after infection and washing. Similar assays exist for *Plasmodium*[116] and *Babesia*.[117] Generating quantitative and reproducible attachment data can be difficult due to sampling issues and variations in wash stringency within and between coverslips. A LSC-based attachment assay that counts attached parasites, prelabeled with fluorescent antibodies, overcomes much of this inherent variability.[76] This is accomplished by comparing two differentially labeled populations of parasites on the same coverslip. While highly reproducible and quantitative, the requirement that populations of parasites be compared pairwise means that the assay is not well suited for scoring numerous samples, e.g., high-throughput screening.

In addition to whole parasites, the ability of putative adhesins from *T. gondii*, *Plasmodium*, and *Neospora*, to bind to host cells or host cell components has been tested using purified,[40,109] heterologously expressed,[118] or recombinant radiolabeled proteins.[119,120]

Rhoptry Secretion

T. gondii tachyzoites pretreated with cytochalasinD are unable to move or invade host cells, but they retain the ability to attach to host cells and to secrete the contents of their rhoptries. The secreted rhoptry proteins can be detected as vesicular clusters (termed evacuoles) within the host cell by immunofluorescence microscopy[121] (Fig. 3). The quantity, length, and protein composition of evacuoles can be assessed using different antibodies, as a means of assaying rhoptry secretion.[8,122] Evacuole-like structures are also observed within erythrocytes after *Plasmodium* merozoite invasion is arrested with cytochalasinB, cytochalasinD or staurosporine.[91,124,125]

Approaches to Studying Invasion

Many different experimental approaches are available for the identification and functional analysis of gene products involved in invasion. These include forward genetic screening, reverse genetic analysis, antibody inhibition, small-molecule-based approaches, alteration of the host cell, and genomic/transcriptomic/proteomic studies.

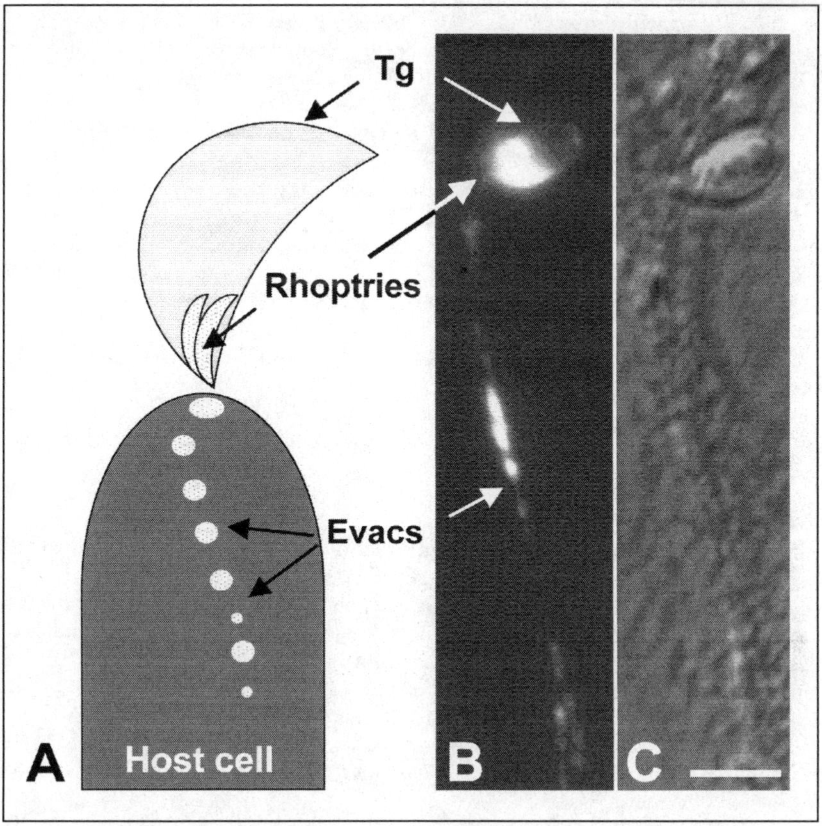

Figure 3. Assaying rhoptry secretion by evacuole formation. *T. gondii* tachyzoites treated with cytochalasinD, which inhibits actin polymerization, attach to host cells but do not invade. In this state, parasites are able to secrete some of the contents of their rhoptries, forming vesicular clusters (evacuoles) within host cells, which can be visualized by indirect immunofluorescence microscopy using antibodies against various rhoptry proteins.[122,123] A) Cartoon illustrating evacuole formation; B) Immunofluorescence image of a parasite with associated evacuoles, visualized here with an antibody against ROP1; C) Corresponding DIC image. Tg = *T. gondii* tachyzoite, Evacs = evacuoles. *Scale* bar = 5 μm. Adapted from reference 8.

Forward Genetics

One of the most powerful means to determine which gene products are involved in a particular process is to introduce mutations into the genome of the organism, screen for mutants which no longer carry out the process, and identify the mutated genes. A number of such forward genetic screens have been undertaken to study different aspects of the *T. gondii* life cycle, utilizing a variety of mutagenic techniques and selection strategies, as described below. In principle, these methods could be applied to the study of invasion. However, in the case of invasion, forward genetic screens must be designed to select for mutations that enhance or decrease invasion, or abolish invasion only under certain conditions, since parasites that are constitutively unable to invade cannot be maintained in culture.

Insertional Mutagenesis

The high frequency of nonhomologous integration in *T. gondii*,[126] coupled with the manageable size of its genome (~65Mbp), have allowed the successful implementation of insertional mutagenesis.[127] Mutagenic saturation, i.e., the theoretical disruption of every gene in the *T. gondii* genome, can be accomplished by mutagenizing approximately $5x10^6$ parasites.[127] Sequence from the dihydrofolate reductase (DHFR) locus appears to increase nonhomologous integration into the *T. gondii* genome (a 10-fold increase compared to non-DHFR sequence[42]) and is therefore often included in insertional mutagenesis plasmids.[128] Interrupting a gene with an insert should alter the function of the disrupted gene, but could also generate toxic truncation product(s) or have unintended effects on adjacent genes, which must be considered when attempting to correlate the interrupted gene with an observed phenotype.

Insertional mutagenesis has been used to study adenosine transport and metabolism,[128] as well as parasite differentiation.[129,130] In principle, this approach could be used to study invasion given a proper selection or screening strategy, i.e., selection or screening for disruptions that enhance or decrease invasion but do not abolish it completely. For example, insertion into the promoter of a critical adhesin might reduce but not completely abolish invasion. Methods have been developed to identify the interrupted gene either by plasmid recovery[129] or by complementing loss of function with a cDNA library that integrates at high frequency and is shuttled by recombination cloning.[131] Complementation with an episomal genomic library has also been reported.[37]

Chemical Mutagenesis

Chemical mutagenesis may be preferable to insertional mutagenesis as a means to genetically modify parasites for invasion screens. Agents such as ethylnitrosourea (ENU)[132-134] and N-methyl-N'-nitro-N-nitrosoguanidine[25] introduce point mutations, which are less likely than insertional mutagenesis to result in complete loss of protein function, and may therefore be more likely to diminish, but not abolish, invasion. One study has produced temperature sensitive mutants that differ in their infectivity.[25] The generation of temperature sensitive mutants should allow for the isolation of parasites with invasion defects, since parasites could be cultured at the permissive temperature and invasion studied at the restrictive temperature.[25] Chemically mutagenized, temperature sensitive *Neospora* have also been screened for invasion and infectivity defects in mouse models.[135]

The major challenge of chemical mutagenesis is mapping and identifying the site of the mutation. The system for cDNA-based complementation cloning mentioned above[131] is one approach to identification of the mutated gene; this approach was recently used to identify a gene involved in cell cycle progression.[136]

In any forward genetic approach, once the interrupted gene has been identified, the specific mutation should be introduced into wild type parasites to confirm and validate the correlation between the mutation and the observed phenotype.[48]

Reverse Genetics

While forward genetic screens are useful for identifying new parasite genes and proteins that function in invasion, reverse genetic approaches have been extensively used to study specific proteins postulated to function in invasion based on other criteria, such as antibody inhibition, subcellular localization, sequence homologies, or the forward genetic approaches mentioned above.

Knockout Studies

Targeted gene disruption by homologous recombination has been widely used to study gene function in *T. gondii*. Increasing the size of the targeting sequence seems to increase the frequency of homologous recombination: one study showed that plasmids containing 1.7, 7.6, or 15.7 kb of targeting sequence resulted in 0, 62, and 82% of the clones harboring homologously integrated plasmid DNA, respectively.[126] In *T. gondii*, where genomic sequence data is available, knockout vectors can be constructed using only the noncoding regions flanking the gene of interest. Utilizing coding sequence to target the knockout construct could result in expression of partial proteins, complicating interpretation of the resulting phenotype. Typically, the coding region of the targeted gene is replaced with a selectable marker (Table 1) flanked by several kb of 5' and 3' noncoding sequence. After selection, individual clones can be screened for gene disruption by Western blotting if antibodies are available.[71,137-139] Alternatively, after selection, pools of clones can be screened by PCR, utilizing one primer from within the genome and one primer from within the knockout construct to identify those parasites with the targeted disruption.[8] In positive clones, gene disruption can be confirmed by Southern blotting, Western blotting and/or immunofluorescence microscopy.[8,71,137-139] Knockout constructs are often introduced as linearized plasmids,[138,139] to promote double crossover-allelic exchange, as opposed to single crossover events, which result in pseudodiploid formation and the possibility of expression of functional protein (Fig. 4). The transfection efficiency of circular and linear plasmids appears similar;[39,41] however, homologous recombination occurs with higher frequency when using circular plasmids, suggesting that pseudodiploid formation is the most prevalent recombination event.[39] Since *DHFR* sequence seems to promote nonhomologous integration in *T. gondii* (see above), inclusion of *DHFR* should be avoided in constructs intended for targeted knockouts.

Using a positive selectable marker within the targeting sequence and a negative selectable marker outside the targeting sequence selects against nonhomologous recombination and pseudodiploid formation,[141] significantly reducing background. A similar double selection strategy is available for *P. falciparum* knockouts.[62]

Attempts to generate a gene knockout could fail either because the gene is essential or because the knockout construct did not properly target the locus of interest. In order to rule out the latter, an exogenous, functional copy of the gene can be randomly integrated into the genome before attempting to knockout the endogenous locus.[142,143] Alternatively, an additional copy (mutated or functional) can be inserted by homologous recombination to create a pseudodiploid (Fig. 4B). The pseudodiploid can subsequently be resolved, under negative selection, by intrachromosomal recombination, resulting in either reconstitution of the wild type locus or formation of the mutated locus (Fig. 4C). Since vector sequence, including markers, will be removed by pseudodiploid resolution, a marker which can be used for either positive or negative selection, such as HXGPRT, is ideal for these types of experiments.[140] The wild type and the mutated loci should be recovered with equal frequencies; recovery of only the wild type locus suggests that the gene is essential.[143] In cases where there is a phenotype associated with a knockout, polar effects can be ruled out by complementing the knockout with an exogenous copy of the gene being studied and observing partial or complete rescue of the affected phenotype.[71,137] The choice of promoter and selectable marker should be carefully considered, since either can affect the level of expression during complementation.

Knockouts in a number of *T. gondii* genes have been generated, some of which result in reduced invasion. For example, MIC2 Associated Protein (M2AP) knockout parasites are significantly defective in invasion, which has been attributed to mistargeting of M2AP's partner protein, MIC2.[71] Given that MIC2 itself cannot be knocked out, presumably because it is essential, this work provided valuable information on the function of both M2AP and MIC2. MIC1 knockout parasites are significantly defective in host cell invasion, but only slightly defective in mouse virulence.[144] In contrast, MIC3 knockout parasites are not defective in invasion and are only slightly defective in mouse virulence. Interestingly, MIC1/MIC3 double knockouts show similar levels of invasion to the MIC1 knockout but significantly reduced virulence compared to the MIC1 or MIC3 single knockouts, indicating a synergistic role for these two proteins in virulence.[144] A number of dense granule proteins have been knocked out,[137,139] none of which affect invasion. However, the GRA2 knockout has reduced virulence, consistent with the hypothesis that dense granule proteins are not necessary for active penetration, but are involved in PV modification and possibly immune modulation. The major glycosylphosphatidylinositol (GPI)-anchored surface proteins, SAG1 and SAG3, have both been successfully knocked out. The SAG1 knockout parasites show no defect in invasion; in fact, the knockout parasites were found to invade more quickly than wild type parasites.[145] In

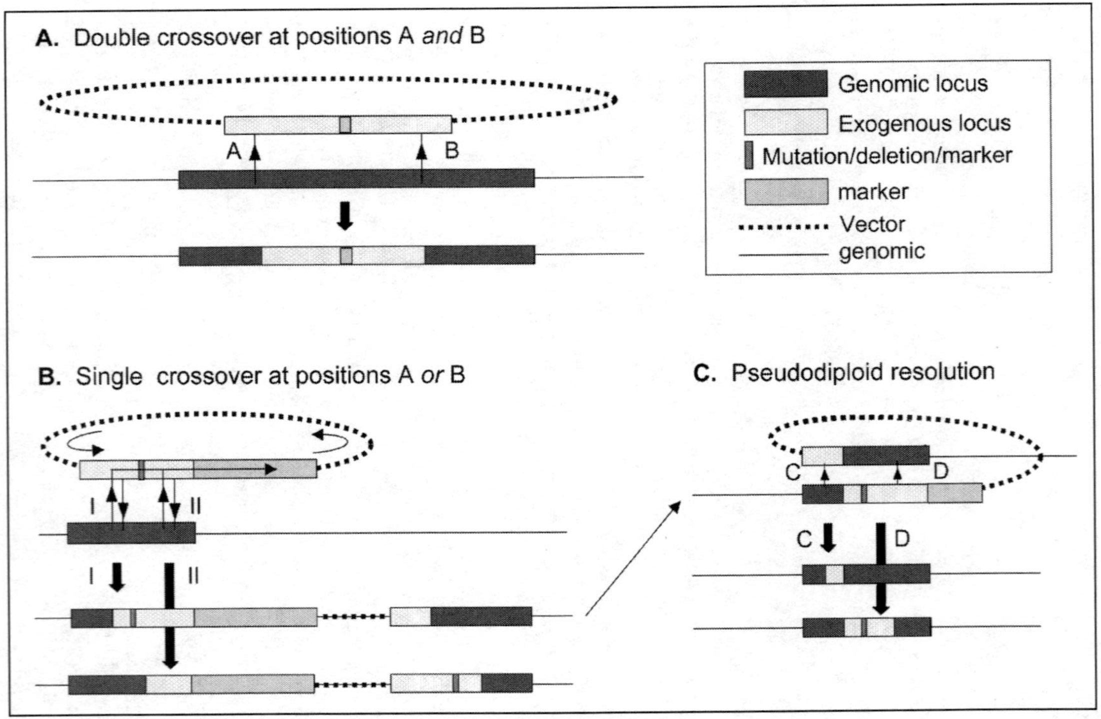

Figure 4. Schematic illustration of recombination and integration. A) Homologous recombination at two sites (double crossover) results in allelic replacement and can be used for targeted gene disruption using circular (shown) or linear plasmids. B) Homologous recombination at one site (single crossover) results in two copies of the gene at the genomic locus, termed a pseudodiploid. For targeted gene disruption the exogenous locus contains a deletion, mutation, or is interrupted by a selectable marker. The type of pseudodiploid formed is determined by whether recombination occurs upstream (Type I) or downstream (Type II) of the mutation/marker. C) Intrachromosomal homologous recombination between the two copies of the gene results in resolution of the pseudodiploid (resolution of a Type I is shown), and can reconstitute the wild type locus (by recombination at site C) or complete a two step knockout (at site D), termed hit and run mutagenesis.[140]

contrast, the SAG3 knockout showed reduced adhesion, invasion, and virulence, supporting the hypothesis that some members of the SAG family of proteins function as adhesins.[138]

When comparing knockout parasites to wild type or complemented parasites in invasion assays, it is important to use parasites in equivalent stages of their life cycle. Parasite invasiveness diminishes progressively after release from the host cell; therefore, care must be taken to ensure that release of the two parasite populations to be compared is synchronous (Mital and Ward, unpublished). Growth assays, comparing the intracellular replication of the parasites under study, can help in the synchronization of different parasite populations.[8]

Knockouts in nonessential genes have been generated and used to study invasion in *Plasmodium* (e.g., refs. 82,146-150 and reviewed in ref. 151). A useful variation on the knockout approach in *Plasmodium* has been to disrupt an essential gene in a life cycle stage that does not require the gene (e.g., the blood stages), and then analyze the invasion phenotype in the stage in which the gene is essential (e.g., mosquito salivary gland sporozoites).[78,152,153] This approach has provided a great deal of functional information about the proteins involved in sporozoite invasion (reviewed in refs. 154-157).

Conditional Knockouts

The recent development of an inducible promoter system for *T. gondii*[10] allows the study of genes essential for invasion, and has been successfully used for four such genes (*MyoA*,[10] *TgAMA1*,[8] *MIC2*,[158] and *ACP*[159]). This system is based upon an anhydrotetracycline- (Atc-) responsive transactivator protein that drives expression of genes downstream of a minimal *SAG1*- or *SAG4*-based promoter containing seven *Tet* operators. Expression occurs in the absence of Atc, and is repressed in the presence of Atc[10] (Fig. 5). Generation of a conditional knockout is accomplished in two steps. First, parasites are transfected with the gene of interest downstream of the regulatable promoter, and clones with regulatable expression of properly localized protein are isolated. These clones are then transfected with a knockout construct to disrupt the endogenous locus of the gene of interest. The resulting parasite's sole source of the essential protein can then be turned off by addition of Atc. Conditional knockout parasites are therefore generated and isolated in the absence of Atc, and phenotypically characterized after incubation with Atc. Since phenotypic characterization is accomplished by comparing conditional knockout parasites in the presence and absence of Atc, all assays begin with the same population of parasites, alleviating the problems associated with comparing two nonequivalent populations of parasites (discussed above), provided that the knockout exhibits no growth defect.[8]

The success of this approach depends critically on the expression levels of the regulatable gene relative to wild type expression; expression in the absence of Atc must be sufficiently high for parasite viability, and expression after Atc treatment must be sufficiently low to reveal a phenotype (Fig. 5A,B). The optimal expression level will vary with the particular protein being studied, and may be modulated by the number of regulatable copies integrated and the choice of promoter driving regulatable expression. Currently, two versions of the regulatable promoter are available: the *SAG1*-based promoter gives lower basal expression, but tighter regulation after Atc addition, compared to the *SAG4*-based construct (personal comm. D. Soldati-Favre). In principle, the endogenous gene could be replaced with a regulatable copy in a single step; however, the two step approach enables one to insert multiple copies of the regulatable gene, which may be necessary to achieve the required expression levels prior to disruption of the endogenous gene, and to assess whether the knockout construct can effectively target the locus of interest. Both approaches require that the endogenous gene can be effectively targeted by homologous recombination. Some residual protein may remain after Atc addition, due either to leaky expression or slow protein turnover, which should be considered when interpreting any observed phenotype, or lack thereof.

The conditional promoter system has recently been adapted for use in *P. falciparum*.[160] The development of a conditional mutagenesis system in *P. berghei*, based upon regulatable site-specific recombination, presents an alternative means of constructing stage-specific knockouts; the utility of this system will increase as more stage-specific promoters are characterized.[161]

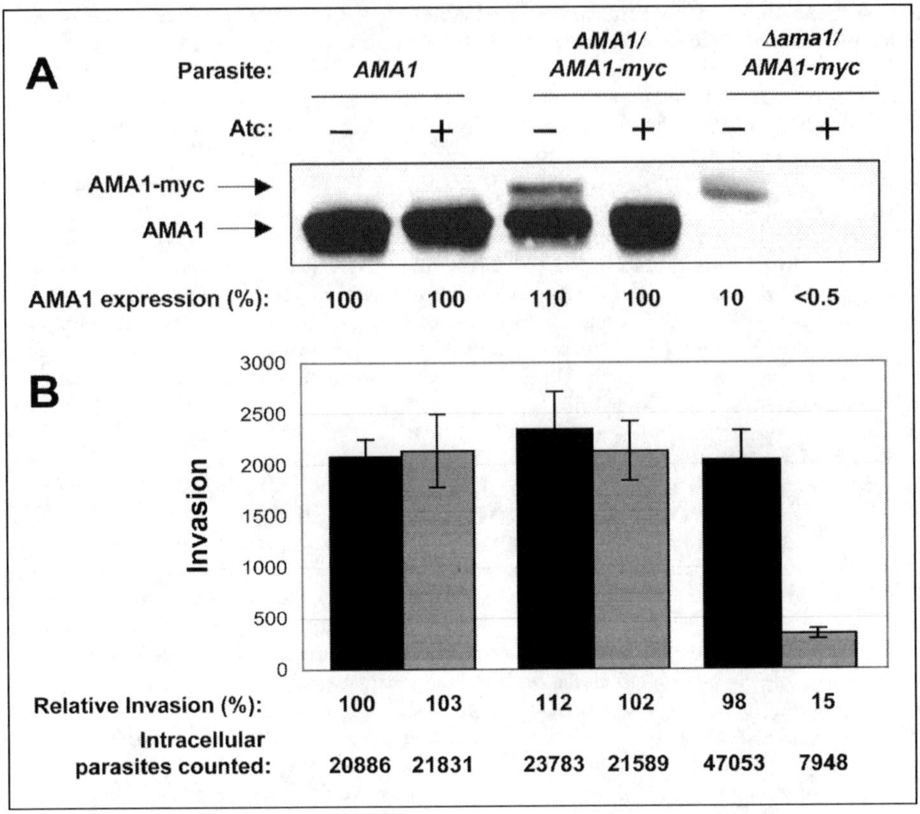

Figure 5. Generation of an AMA1 conditional knockout. A) Wild type *T. gondii* expressing the tetracycline-sensitive transactivator (designated "*AMA1*" parasites) were transfected with a plasmid containing *AMA1* under control of the regulatable promoter. This creates parasites ("*AMA1/AMA1-myc*") that express both endogenous and regulatable *AMA1* in the absence of anhydrotetracycline (Atc), and only endogenous *AMA1* after incubation with Atc. Expression of endogenous and regulatable *AMA1* can be visualized by Western blotting (top panel); the epitope tag on the regulatable copies results in an electrophoretic mobility shift. The endogenous *AMA1* locus was subsequently interrupted by targeted gene disruption, creating parasites ("*Δama1/AMA1-myc*") whose sole source of AMA1 can be regulated by Atc. The relative AMA1 expression levels, as determined by quantitative Western blotting, are shown below each lane. B) Host cell invasion by the *AMA1*, *AMA1/AMA1-myc*, and *Δama1/AMA1-myc* parasites, each grown with or without Atc, was measured using the Laser Scanning Cytometer-based invasion assay. Parasites expressing just 10% of wild type levels of AMA1 (*Δama1/AMA-myc* minus Atc) have wild type levels of invasion. However, parasites expressing undetectable (<0.5%) levels of AMA1 (*Δama1/AMA-myc* plus Atc) are significantly defective in invasion. The LSC-based assay allows for a large sample size (tens of thousands of parasites) to be attained. Adapted from reference 8.

RNA-Based Methods

RNA-based approaches to diminishing, rather than abolishing, expression of a protein are useful for genes that cannot be studied by conditional knockout, either because the required expression levels do not fall within the range of the available conditional promoters, or because the gene cannot be effectively targeted for disruption. In *T. gondii*, RNA-based techniques such as antisense, double stranded RNA interference (RNAi), and delta ribozymes have been successfully used in a limited number of laboratories to modulate gene expression.[162-166] All of these techniques rely on short sequence recognition. The sequence must therefore be carefully chosen to assure knockdown of a specific gene, especially if that gene is part of a gene family or shares significant sequence with other proteins (although this can be an advantage if knockdown of a multigene family is desired.) Optimization of sequence and RNA expression level may be required to achieve a threshold of knockdown that maintains viability of the parasite but results in an invasion phenotype.

As yet, only the antisense RNA technique has been used to study invasion-related genes. Antisense RNA was used to decrease expression of the rhoptry protein ROP2, which was thought to be essential for invasion since previous knockout attempts were unsuccessful.[164] A significant decrease in invasion, among other defects, was seen in ROP2-deficient parasites. Antisense RNAs have also been used to study the *T. gondii* dense granule protein NTPase, which is expressed as two isoforms from different loci[167] and is thought to be essential.[165] Despite these apparent successes, antisense methods have not yet gained widespread use in *T. gondii* due to difficulties with reproducibility. Antisense methods have also been used in *Plasmodium.*[168,169]

Homologues of many of the genes involved in RNAi have been found in *T. gondii,*[170] but not *Plasmodium.*[171] However, RNAi has been reported to work in both *Plasmodium*[172-174] and *T. gondii.*[162] RNAi has been used to show that decreased expression of two independent lactate dehydrogenase genes in *T. gondii* results in reduced differentiation and virulence.[175] Double-stranded RNA has also been used to knockdown a *T. gondii* immunophilin.[176] Since extensive controls are needed to unambiguously attribute downregulation to the RNAi pathway, there remains some uncertainty about whether the effects observed in these studies result from true RNAi.[170]

A recent report on the use of catalytically active delta ribozymes to bind and cleave specific targets[177] indicates that this may also be a useful technique for studying essential genes.

Transgene Expression

Intra- and inter-species expression of wild type or mutated transgenes has been useful for studying apicomplexan invasion-related proteins. Care must be taken when analyzing data from exogenous expression to ensure that expression levels, post-translational modifications, and trafficking are comparable in the exogenous and native proteins. Additionally, codon bias may need to be considered; whereas *T. gondii* genes have been expressed in a variety of systems (e.g., refs. 34,178,179), codon bias in the A/T rich genome of *P. falciparum*[180] can cause difficult, though not insurmountable problems for heterologous expression.[181-184]

Expression of various mutated and truncated *T. gondii* microneme proteins in wild type *T. gondii* has proven to be a useful approach for elucidating the role of these proteins in invasion and for defining their functional domains. For example, exogenous expression of mutated forms of MIC2 in *T. gondii* helped to elucidate MIC2's role as an adhesin and to define the mechanisms governing microneme protein cleavage (reviewed in ref. 185). Flow cytometry can be used to identify mutations that cannot be tolerated by the parasite.[186]

Expression of *T. gondii* proteins in other apicomplexans, or vice-versa, has provided insights into invasion by both the donor and recipient species. For example, functional homology between the cytoplasmic tails of *Toxoplasma* MIC2 and *Plasmodium* TRAP, microneme proteins thought to function in motility and invasion, was demonstrated by the ability of the cytoplasmic tail of *Toxoplasma* MIC2 to rescue the motility and invasion defects caused by deletion of the TRAP cytoplasmic tail in *Plasmodium* sporozoites.[187] Proper expression and localization of a number of *Toxoplasma* proteins in the related apicomplexan *Neospora*[34,65,178] indicate that this may also be a useful system to study functional homology and elucidate protein function. Inter-species expression of proteins will be particularly useful for studying protein function and invasion in organisms that are difficult to culture and/or for which molecular genetic tools are not yet available, such as *Cryptosporidium.*[188] Expression of *Plasmodium* proteins in other species of *Plasmodium* has proven to be a useful way to study the function of secreted and cell

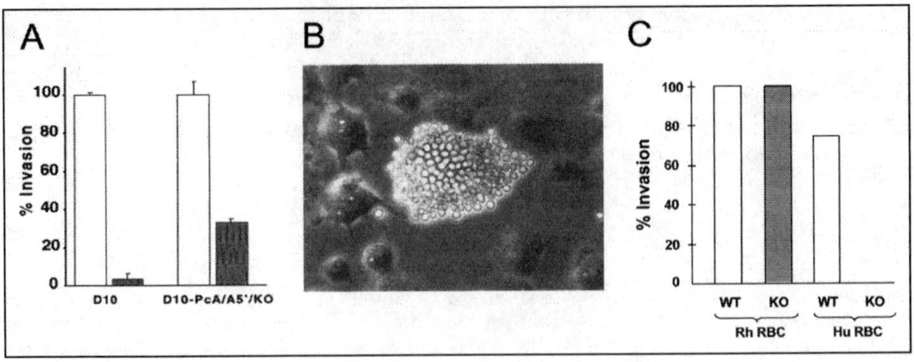

Figure 6. Inter-species approaches to studying invasion. A) Treatment of *P. falciparum* merozoites with an antibody against *P. falciparum* AMA1 inhibits invasion of human erythrocytes (D10: white bar = untreated; grey bar = antibody treated). The expression of *P. chabaudi* AMA1 in these parasites partially restores their ability to invade in the presence of the anti-*P. falciparum* AMA1 antibody (D10-PcA/A5'/KO). Adapted from: Triglia T et al. Mol Microbiol 38(4):706-718; ©2000, with permission from Blackwell Publishing.[189] B) A COS cell transfected with a portion of the Duffy-binding protein from *P. vivax* binds Duffy-positive erythrocytes (center). Untransfected COS cells on the periphery show no measurable erythrocyte binding, and no binding is observed to Duffy-negative erythrocytes. This assay has been used to map the domains of the Duffy-binding protein involved in erythrocyte binding, and to study the erythrocyte-binding capabilities of other merozoite proteins.[118,191-193] Reproduced from: Ward GE et al. The invasion of erythrocytes by malarial merozoites. In: Russell DG, ed. Clinical Infectious Diseases: International Practice and Research, Strategies for Intracellular Survival of Microbes. 1st ed. London: Bailliere Tindall, 155-190; ©1994, with permission from Elsevier Ltd.[194] C) *P. knowlesi* merozoites engineered to contain a deletion of the α gene, which encodes the *P. knowlesi* Duffy-binding protein *(KO)*, show no defect in the invasion of rhesus monkey erythrocytes (Rh RBC), relative to wild type parasites (WT). In contrast, the KO parasites are unable to invade human erythrocytes (Hu RBC), demonstrating that invasion of human erythrocytes by *P. knowlesi* requires Duffy-binding protein. Because the α gene is not essential for the invasion of rhesus erythrocytes, the knockout (lethal if grown in human erythrocytes) could be readily generated in rhesus erythrocytes. Phenotypic characterization of the invasion block with human erythrocytes provided strong evidence that the Duffy-binding protein plays a role in junction formation.[82] Adapted from: Singh AP et al. Mol Microbiol 55(6):1925-1934; ©2005, with permission from Blackwell Publishing.[82]

surface proteins and to identify the parasite proteins responsible for host range specificity (Fig. 6A,C).[189,190] Finally, expression of apicomplexan genes in both yeast[195] and mammalian cells have been useful for determining the role of purported adhesins in host cell attachment (Fig. 6B).[118,191-193,196]

Antibody Inhibition

Another method to determine a particular protein's role during invasion is to alter its function with specific antibodies or peptides. Blocking active sites or inhibiting important conformational changes of invasion-related proteins, such as adhesins, may decrease or abolish invasion. The use of monoclonal antibodies and F_{ab} fragments rather than whole immunoglobulins can decrease nonspecific cross reactivity and cross-linking effects. However, even well executed antibody inhibition experiments need to be interpreted with caution. For example, SAG1, the major surface protein of *T. gondii*, was originally thought to be involved in invasion based on the observation that antibodies and F_{ab} fragments against SAG1 block invasion.[114,197-199] However, subsequent studies showed that a *SAG1* knockout parasite was fully capable of invasion;

in fact, these parasites invade more quickly than wild type parasites, indicating that the antibody-mediated effects were probably due to something other than a direct effect on SAG1.[145] Despite these caveats, antibody inhibition has proven to be a useful tool for studying invasion in many different apicomplexans, including *Neospora*,[200,201] *Eimeria*,[90,202] *Babesia*,[117] *Toxoplasma*,[27] and *Plasmodium*.[203-206] Because antibodies can affect function by different mechanisms, determining the mode of action of the inhibitory antibody can be informative. For example, certain invasion inhibitory antibodies directed against *P. falciparum* surface proteins block invasion indirectly, by interfering with the proteolytic processing of these proteins, illustrating the importance of processing for the function of these proteins during invasion.[203,207,208]

Small-Molecule-Based Approaches

The most common way to use small molecules to study a process such as invasion is to determine whether specific pharmacological agents with known targets in other systems affect the process of interest, thereby implicating the known target in the process. For example, the first evidence implicating actin in the process of invasion, both in *T. gondii*[115,209,210] and *Plasmodium*[125] was the inhibition of invasion by cytochalasinB or D. Similarly, inhibition of microneme secretion by the intracellular calcium chelator BAPTA-AM, together with stimulation of microneme secretion by the calcium ionophore A23187, strongly suggested a role for intracellular calcium in microneme secretion.[31,211]

A more general way to use small molecules, which is not limited to compounds with predefined targets, is "phenotype-based" small molecule screening.[212] In such an approach, libraries of structurally diverse small molecules are screened for those that generate a particular phenotype; these compounds are then used to directly identify the target(s) responsible for generating the phenotype (Fig. 7). The approach is essentially the pharmacological analog of classical forward genetics (compounds that disrupt protein function being analogous to mutations); however, in the small molecule approach, perturbation of protein function is under the investigator's control rather than being permanently encoded in the mutant's genome. When applied to invasion, this circumvents the problem of invasion mutants being nonviable. In a recent screen of 12160 small molecules, 24 novel inhibitors and 6 enhancers of *T. gondii* invasion were identified,[50] most of which were fully reversible. Different compounds had distinctly different effects on microneme secretion, motility, and/or conoid extension.

The most difficult part of any small molecule-based screening project is target identification. A number of biochemical and genetic approaches are available for target identification and have been recently reviewed.[212-215] "Activity-based probes," which can be designed to covalently modify specific subsets of enzymatic targets, can facilitate target identification (reviewed in ref. 113). Although such probes sample only a subset of the proteome, if this subset is of interest (e.g., a particular family of proteases), the selectivity is an advantage, and the ability of the probes to covalently modify their targets can greatly facilitate target identification. In all cases, the identified target must be independently validated as the relevant target in vivo.[212-215]

An alternative way to use small molecules is to screen large collections of structurally diverse small molecules for compounds that affect the activity of an individual recombinant or purified protein ("target-based" screening; Fig. 7), rather than causing a complex, cell-based phenotype. Once a small molecule inhibitor of that protein has been identified, it can be used on whole cells to determine the function of the protein within the context of the cell (the pharmacological analog of reverse genetics). Target-based screening has been used to identify novel inhibitors of *T. gondii* dense granule NTPases[216] and *P. falciparum* dihydroorotate dehydrogenase.[217] One clear benefit of taking a small molecule approach is that it may generate not only new tools for studying invasion, but also new potential lead compounds for drug development.

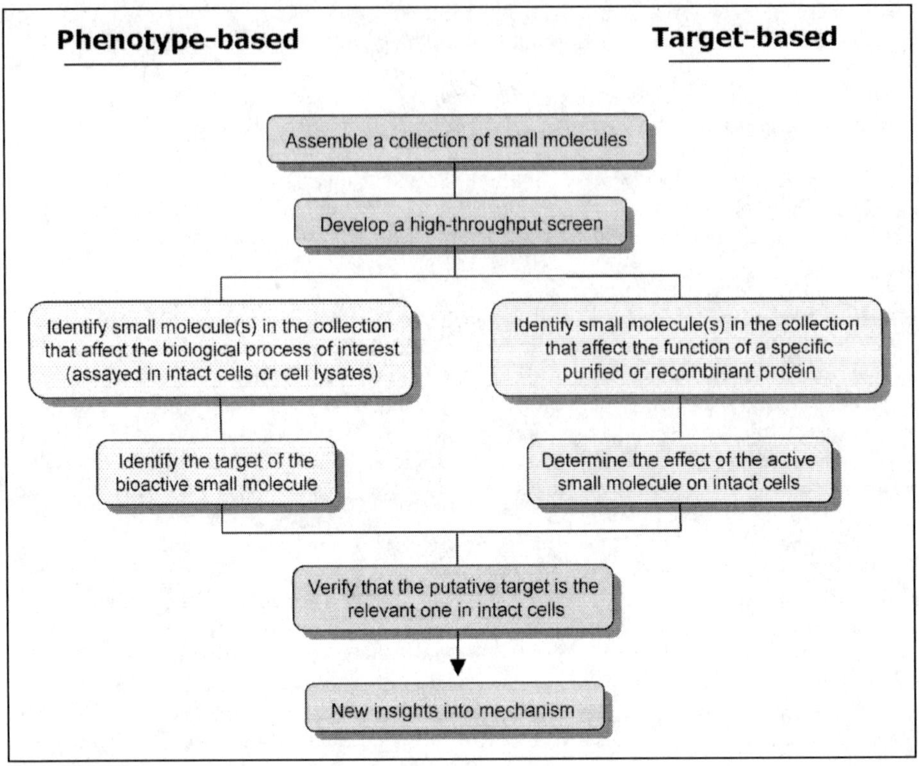

Figure 7. Overview of phenotype- vs. target-based small molecule approaches to studying biological processes. See text for details. Reproduced from: Ward GE et al. Cell Microbiol 4(8):471-482; ©2002, with permission from Blackwell Publishing.[212]

Altering the Target Cell

Analysis of a parasite's ability to invade genetically or enzymatically altered host cells has been a useful approach to studying invasion. For example, a decrease in the ability of *T. gondii* tachyzoites to invade either host cells pretreated with enzymes that cleave glycosaminoglycans (GAGs) or GAG biosynthesis mutants,[218,219] strongly suggests a role for host cell GAGs in *T. gondii* invasion, perhaps as receptors for micronemal adhesins. Cell synchronization studies showed that the ability of *T. gondii* to attach to and invade host cells increases during S-phase of the host cell cycle[220,221] suggesting that a critical host cell receptor may be upregulated in S-phase.[221,222]

Enzyme-treated erythrocytes and mutant erythrocytes lacking specific surface antigens have been tremendously useful experimental tools for studying the receptor-ligand interactions involved in invasion by *Plasmodium* merozoites. For example, the seminal observation that Duffy(-) erythrocytes are refractory to invasion by *P. knowlesi*[223] ultimately led to the identification of the Duffy-Binding Protein family of parasite receptors.[111,193,224] During erythrocyte invasion, the Duffy-binding proteins appear to function in moving junction formation.[82,125] Similarly, enzymatic treatment of erythrocytes, together with a variety of naturally occurring erythrocyte mutants, have helped to elucidate the complex and redundant invasion mechanisms used by *P. falciparum* merozoites.[79,193,225-227] More is currently known about the receptor-ligand interactions that regulate invasion in asexual stage *Plasmodium* parasites than in any other apicomplexan, including *T. gondii* (reviewed in ref. 228; see also Duraisingh, this volume).

Genomic/Transcriptomic/Proteomic Approaches

Genomics

Comprehensive genome sequencing of many apicomplexans (see Table 1) has facilitated almost every aspect of studying invasion, including identification of mutations generated in forward genetic screens, design and implementation of reverse genetic experiments (e.g., knock-out construct design, identification of promoters for exogenous expression, etc.), and identification of unknown proteins by mass spectrometry. The availability of genomic sequence data makes BLAST searching significantly more effective and productive, and allows gene/protein analysis from one apicomplexan to be more broadly interpreted and applied to other apicomplexans (which is particularly important in genetically inaccessible organisms such as *Theileria* and *Cryptosporidium).* It also facilitates target identification in small-molecule-based studies, i.e., if a particular compound has a known target in another organism, the apicomplexan homolog(s) of the known target can be identified by BLAST searching. Genome mining can also be used to search for proteins with proven or suspected functional domains, such as adhesive domains or particular protease cleavage sites.[229]

The discovery of *T. gondii* rhomboid proteases and their potential substrates exemplifies how genomics can stimulate major discoveries and quickly bring new research areas into experimental focus. Early studies of *Drosophila* rhomboids used genome mining to search for non-*Drosophila* proteins that contained consensus rhomboid cleavage recognition motifs, identifying a number of *T. gondii* microneme proteins.[230] Intriguingly, several of these microneme proteins were known to be cleaved intramembranously,[112,231] a defining feature of rhomboid cleavage.[232] Subsequent studies showed that these proteins could indeed be cleaved by *Drosophila* rhomboids.[230] This led to BLAST searches of the *T. gondii* EST and genomic databases (http://ToxoDB.org), querying with sequences of known rhomboids,[233,234] which ultimately resulted in the discovery, cloning, and localization of *T. gondii* rhomboid proteases.[233,234] Rhomboid-like genes have been found in all available apicomplexan genomes,[235] and are currently the focus of numerous studies using reverse genetic techniques to characterize their function during invasion.

Transcriptomics

The up- or down-regulation of gene expression during certain processes (e.g., differentiation or response to specific stimuli) can be used to infer the involvement of those genes in the processes. Expression from a wide spectrum of genes can be examined by determining the change in the amount of transcript (transcriptome) as analyzed by microarray binding or Serial Analysis of Gene Expression (SAGE). These technologies have been used alone[236] and in combination with either insertional[129] or chemical[134] mutagenesis to identify genes involved in *T. gondii* differentiation. A more accurate method of analyzing gene transcription/regulation utilizes a *T. gondii* uracil salvage enzyme (UPRT) to measure only the amount of RNA synthesized during a given interval (via incorporation of thio-substituted uridines), rather than the total amount of transcript present. This technique was initially developed for *T. gondii*,[237] but can be used in any organism or cell in which the salvage enzyme can be expressed.

Because invasion happens so quickly (~20 sec), it is unlikely that changes in transcription during this time frame play a direct role in driving the process. However, changes in the *Plasmodium* transcriptome throughout its life cycle have been used to identify genes that are upregulated in the invasive stages and therefore may be involved in *Plasmodium* invasion.[238-240] Analogous studies of the host cell's response to invasion has identified host cell transcripts upregulated following *T. gondii* invasion.[241] Transcriptome studies have also recently been performed in *Theileria*.[242]

Proteomics

Changes in expression can also be analyzed at the protein level, and proteins that reside in a particular subcellular compartment thought to be involved in invasion can be identified by cell fractionation and proteomic analysis (reviewed in ref. 113). For example, the recent proteomic analysis of purified *T. gondii* rhoptries[243] identified 38 novel proteins, many of which were subsequently confirmed to be rhoptry proteins by fluorescence microscopy. These proteins include a novel subset of rhoptry proteins that localize to the thin, anterior "neck" portion of the rhoptries,[243] and may be discharged differentially from rhoptry "bulb" proteins.[8,13] This study also identified the first *T. gondii* rhoptry proteins with homologs in *Plasmodium*. Proteomic analysis of the excreted/secreted antigens (ESA) of *T. gondii* was recently reported, identifying a number of novel microneme proteins.[244] Proteomic analysis of purified *Eimeria tenella* micronemes has also recently been accomplished.[245] These newly identified rhoptry and microneme proteins can be further studied using the reverse genetic techniques described above.

A combination of proteomic and computational techniques has recently been used to identify novel, GPI-anchored proteins of asexual stage malaria parasites.[246] As GPI-anchored proteins dominate the surface of merozoites, identification and characterization of GPI-anchored proteins will likely be critical to understanding the molecular mechanisms of erythrocyte invasion by *Plasmodium* merozoites.

Combined Approaches

In practice, the most powerful strategy for studying invasion has been to employ a combination of the approaches described above. We will briefly review three of the examples that best illustrate this principle: AMA1, actin/myosin, and PKG.

AMA1

Inhibition of erythrocyte invasion by antibodies and F_{ab} fragments directed against *Plasmodium* AMA1 provided the first clue that this protein might be involved in invasion.[247,248] The *P. falciparum AMA1* gene could be disrupted by homologous recombination, but viable parasites with the disruption could only be isolated when accompanied by transgenic *AMA1* expression, demonstrating that the gene is essential.[189] Remarkably, the introduction of *AMA1* from *P. chabaudi* (a mouse malaria parasite) into *P. falciparum* (a human malaria parasite) generated transgenic parasites that invade mouse erythrocytes more efficiently than wild type *P. falciparum*.[189] Expression of portions of *Plasmodium* AMA1 in COS and CHO cells increased cell binding to erythrocytes, suggesting that AMA1 may act as an adhesin,[118] although neither shed fragments of AMA1 nor properly folded recombinant AMA1 show similar binding. *Plasmodium* invasion was shown to be inhibited by peptides either derived from, or with affinity for, AMA1.[121,249,250] Furthermore, *Plasmodium* merozoites whose invasion was arrested by an anti-AMA1 antibody were shown by electron microscopy to be attached at a distance to the erythrocyte, prior to apical reorientation.[205]

The *T. gondii* homolog of *AMA1* (*TgAMA1*) was identified both by BLAST searches of the *T. gondii* EST database[27] and through a large-scale monoclonal antibody screen.[251] Early failed attempts to knock out *TgAMA1*[27] suggested it was essential, which was confirmed when the gene was knocked out in the presence of exogenous copies of *TgAMA1*.[8] The development of the tetracycline-regulatable promoter in *T. gondii* enabled the generation of a conditional *TgAMA1* knockout. Using these parasites, it was shown that TgAMA1-depleted parasites are significantly impaired in invasion (Fig. 5B). Further analysis showed them to be defective both in forming close interactions with host cell membranes and in rhoptry secretion.[8]

Actin/Myosin

As described above, the inhibition of invasion by cytochalasinB and D was the first indication that actin plays a role in invasion. Invasion studies using cytochalasinD-resistant host cells demonstrated that actin of the parasite, but not host cell actin, was involved.[48] This same study used chemically mutagenized parasites to isolate cytochalasinD-resistant parasites that can invade host cells in the presence of cytochalasinD. A point mutation in the single copy actin gene (*ACT1*) of these mutant parasites correlated with cytochalasinD resistance. Actin's role in invasion was confirmed when exogenous expression of the mutated allele was shown to confer cytochalasinD resistance in wild type parasites.[48] A central role for *T. gondii* myosinA in motility and invasion was subsequently demonstrated by the creation and phenotypic analysis of a *MyoA* conditional knockout parasite.[10]

Protein Kinase G (PKG)

"Compound 1," a trisubstituted pyrrole that was originally identified as an inhibitor of *Eimeria tenella* growth in vitro, was subsequently shown to have activity against other apicomplexan parasites, including *T. gondii*.[252] A tritiated analog was used to purify the major Compound 1-binding protein from extracts of *E. tenella*.[252] This protein was identified as PKG, and PKG was also shown to also be a target for Compound 1 in *T. gondii*.[252] Computer modeling was used to predict the catalytic site residues in *T. gondii* PKG that interact with Compound 1, which were subsequently mutated. An elegant series of molecular genetic experiments, in which wild type *T. gondii PKG* was replaced with a mutant form that was not inhibited by Compound 1, proved that PKG was the relevant target of Compound 1 both in vitro and in animal models of infection.[143] Having validated PKG as the primary molecular target of Compound 1, the compound was used to demonstrate a role for PKG in parasite motility and invasion.[102]

Do We Have All the Tools We Need?

While the approaches available to study invasion are now numerous and powerful, the toolbox is not yet complete. The currently available conditional promoters only work for proteins whose expression levels fall within a certain range; a more complete set of inducible promoters would be very useful. Methods for enhancing the frequency with which genes can be homologously targeted in *T. gondii* would facilitate reverse genetic studies. A method of culturing *T. gondii* in the absence of host cells would also greatly enhance the study of invasion (and many other aspects of the parasite's life cycle) by allowing knockouts or mutations in essential invasion genes to be isolated; the resulting invasive phenotype could then be analyzed in cell culture. Such a cell-free system would also expand the type of screens that could be designed for studying apicomplexan invasion. In this regard, it is encouraging that partial, cell-free development of *Plasmodium* blood stage parasites has been reported.[253]

Toxoplasma as a Model Apicomplexan

The overall process of invasion appears to be relatively well conserved among apicomplexan parasites, even if the particular molecules mediating the process vary. For example, the individual microneme proteins of *Toxoplasma* and *Plasmodium* are different, but many are organized along a similar theme, i.e., modules of adhesive motifs.[254] In fact, portions of some microneme proteins are functionally interchangeable between *Toxoplasma* and *Plasmodium*, despite limited sequence identity.[187] These and other examples[8,50,157,205] clearly demonstrate that findings from one species of apicomplexan parasite can often be extended to others.

One of the main reasons why *T. gondii* has become a model for studies of apicomplexan invasion (reviewed in ref. 255) is its amenability to molecular genetic manipulation (see previous section on "Approaches to Studying Invasion"). Several selectable markers are available, exogenous DNA integrates with high frequency and is readily expressed, and a variety of approaches can be used for gene knockouts, conditional gene expression, and chemical or insertional mutagenesis (Table 1). The genome is completely sequenced and readily accessible (ToxoDB.org). In addition, the *T. gondii* tachyzoite (2 x 7 μm) is considerably bigger than, e.g., the *Plasmodium* merozoite (1 x 1 μm), facilitating subcellular localization of proteins and a variety of other cell biological techniques. *T. gondii* is easily cultured, and the ability of *T. gondii* tachyzoites to invade virtually any nucleated mammalian cell creates powerful opportunities for studying the host cell requirements for invasion.[218,219] As described above, assays have been developed that allow each step of *T. gondii* invasion to be examined in isolation, and a well-developed mouse model is available for in vivo studies. The ability to express non-*T. gondii* genes in *T. gondii*[188,256,257] (see above) also makes it an excellent model for studying genes from less experimentally tractable parasites.

Outstanding Questions

While a great deal of progress has been made in recent years in our understanding of the invasive mechanisms of apicomplexan parasites, many important questions remain. These include:

How Is T. gondii *Able to Invade Such a Wide Variety of Host Cells, When the Host Range of other Apicomplexans (such as Malaria Merozoites) Is so Limited?*

T. gondii's promiscuity might reflect parasite recognition of a ubiquitous host cell ligand, or the presence of redundant receptors on the parasite surface. Alternatively, *T. gondii* might insert its own receptor into the host cell, analogous to translocation of the enteropathogenic *E. coli* TIR protein into the host cell during invasion.[258] One way to begin to address these questions would be to express potential *T. gondii* receptors in other apicomplexans, and determine whether this expands the host range of the recipient species; a similar approach has proven useful in the analysis of AMA1 function in *Plasmodium.*[189]

What Is the Function of the Conoid and Its Repeated Cycles of Extension and Retraction during Parasite Motility and Invasion?

Identification of small molecule inhibitors of conoid extension/retraction[50] and the proteomic analysis of isolated conoids[259] are two potentially powerful approaches to studying conoid function.

How do the Components of the "Moving Junction" Physically Connect the Parasite to the Host Cell During Invasion and Block Host Cell Transmembrane Proteins from being Incorporated into the Developing Parasitophorous Vacuole?

Recent studies identifying components of the *T. gondii* moving junction[13,14] represent an important first step towards understanding the function of the moving junction and the molecular basis of host transmembrane protein exclusion from the PVM.

What Are the Signaling Mechanisms That Underlie Rhoptry Secretion?

The recent analysis of the rhoptry proteome,[243] combined with the generation of a conditional knockout parasite that is defective in rhoptry secretion,[8] will be useful for studying the regulation of rhoptry secretion. The development of a method for assaying rhoptry secretion in the absence of host cells, analogous to the microneme secretion assay, would be extremely helpful, and could be combined with screens designed to identify small molecules or mutants that enhance or inhibit rhoptry secretion.

What Is the Function of Microneme Protein "Shedding" from the Parasite Surface during Invasion?

While several models have been proposed,[55] these models remain to be experimentally tested. One possible approach to this question would be to determine the functional consequences of inhibiting the rhomboid proteases thought to be responsible for microneme protein shedding, using mutagenesis, targeted gene disruption, or small molecules.

Conclusion

Given the powerful combination of experimental tools currently available, it is our hope that we will soon more fully understand the molecular mechanisms of host cell invasion by apicomplexan parasites. Our efforts can then be directed towards developing effective ways of blocking these processes, as a means of preventing and treating the devastating diseases caused by these important human pathogens.

Acknowledgements

We thank Vern Carruthers, Chetan Chitnis, Stacey Gilk, and members of our laboratory for helpful comments on the manuscript. We also gratefully acknowledge the support of the Burroughs-Wellcome Fund and Public Health Service grants AI054961 and AI063276.

References

1. Mercier C, Adjogble KD, Daubener W et al. Dense granules: Are they key organelles to help understand the parasitophorous vacuole of all apicomplexa parasites? Int J Parasitol 2005; 35(8):829-849.
2. Kappe SH, Buscaglia CA, Bergman LW et al. Apicomplexan gliding motility and host cell invasion: Overhauling the motor model. Trends Parasitol 2004; 20(1):13-16.
3. Hakansson S, Morisaki H, Heuser J et al. Time-lapse video microscopy of gliding motility in Toxoplasma gondii reveals a novel, biphasic mechanism of cell locomotion. Mol Biol Cell 1999; 10(11):3539-3547.
4. Hu K, Roos DS, Murray JM. A novel polymer of tubulin forms the conoid of Toxoplasma gondii. J Cell Biol 2002; 156(6):1039-1050.
5. Morrissette NS, Sibley LD. Cytoskeleton of apicomplexan parasites. Microbiol Mol Biol Rev 2002; 66(1):21-38.
6. Carruthers VB, Sibley LD. Sequential protein secretion from three distinct organelles of Toxoplasma gondii accompanies invasion of human fibroblasts. Eur J Cell Biol 1997; 73(2):114-123.
7. Dowse T, Soldati D. Host cell invasion by the apicomplexans: The significance of microneme protein proteolysis. Curr Opin Microbiol 2004; 7(4):388-396.
8. Mital J, Meissner M, Soldati D et al. Conditional expression of Toxoplasma gondii apical membrane antigen-1 (TgAMA1) demonstrates that TgAMA1 plays a critical role in host cell invasion. Mol Biol Cell 2005; 16(9):4341-4349.
9. Suss-Toby E, Zimmerberg J, Ward GE. Toxoplasma invasion: The parasitophorous vacuole is formed from host cell plasma membrane and pinches off via a fission pore. Proc Natl Acad Sci USA 1996; 93(16):8413-8418.
10. Meissner M, Schluter D, Soldati D. Role of Toxoplasma gondii myosin A in powering parasite gliding and host cell invasion. Science 2002; 298(5594):837-840.
11. Dobrowolski JM, Carruthers VB, Sibley LD. Participation of myosin in gliding motility and host cell invasion by Toxoplasma gondii. Mol Microbiol 1997; 26(1):163-173.
12. Sibley LD. Intracellular parasite invasion strategies. Science 2004; 304(5668):248-253.
13. Alexander DL, Mital J, Ward GE et al. Identification of the moving junction complex of Toxoplasma gondii: A Collaboration between distinct secretory organelles. PLoS Pathog 2005; 1(2):e17.
14. Lebrun M, Michelin A, El Hajj H et al. The rhoptry neck protein RON4 relocalizes at the moving junction during Toxoplasma gondii invasion. Cell Microbiol 2005; 7(12):1823-1833.
15. Mordue DG, Hakansson S, Niesman I et al. Toxoplasma gondii resides in a vacuole that avoids fusion with host cell endocytic and exocytic vesicular trafficking pathways. Exp Parasitol 1999; 92(2):87-99.
16. Lingelbach K, Joiner KA. The parasitophorous vacuole membrane surrounding Plasmodium and Toxoplasma: An unusual compartment in infected cells. J Cell Sci 1998; 111(Pt 11):1467-1475.
17. Morisaki JH, Heuser JE, Sibley LD. Invasion of Toxoplasma gondii occurs by active penetration of the host cell. J Cell Sci 1995; 108(Pt 6):2457-2464.
18. Sinai AP, Webster P, Joiner KA. Association of host cell endoplasmic reticulum and mitochondria with the Toxoplasma gondii parasitophorous vacuole membrane: A high affinity interaction. J Cell Sci 1997; 110(Pt 17):2117-2128.
19. Augustine PC. Cell: Sporozoite interactions and invasion by apicomplexan parasites of the genus Eimeria. Int J Parasitol 2001; 31(1):1-8.
20. Buxton D, McAllister MM, Dubey JP. The comparative pathogenesis of neosporosis. Trends Parasitol 2002; 18(12):546-552.
21. Smith HV, Nichols RA, Grimason AM. Cryptosporidium excystation and invasion: Getting to the guts of the matter. Trends Parasitol 2005; 21(3):133-142.
22. Moltmann UG, Mehlhorn H, Schein E et al. Fine structure of Babesia equi Laveran, 1901 within lymphocytes and erythrocytes of horses: An in vivo and in vitro study. J Parasitol 1983; 69(1):111-120.
23. Miller LH, Aikawa M, Johnson JG et al. Interaction between cytochalasin B-treated malarial parasites and erythrocytes. Attachment and junction formation. J Exp Med 1979; 149(1):172-84.
24. Shaw MK. Cell invasion by Theileria sporozoites. Trends Parasitol 2003; 19(1):2-6.
25. Pfefferkorn ER, Pfefferkorn LC. Toxoplasma gondii: Isolation and preliminary characterization of temperature-sensitive mutants. Exp Parasitol 1976; 39(3):365-376.

26. Saffer LD, Long Krug SA, Schwartzman JD. The role of phospholipase in host cell penetration by Toxoplasma gondii. Am J Trop Med Hyg 1989; 40(2):145-149.
27. Hehl AB, Lekutis C, Grigg ME et al. Toxoplasma gondii homologue of Plasmodium apical membrane antigen 1 is involved in invasion of host cells. Infect Immun 2000; 68(12):7078-7086.
28. Schwartzman JD, Pfefferkorn ER. Pyrimidine synthesis by intracellular Toxoplasma gondii. J Parasitol 1981; 67(2):150-158.
29. Camps M, Boothroyd JC. Toxoplasma gondii: Selective killing of extracellular parasites by oxidation using pyrrolidine dithiocarbamate. Exp Parasitol 2001; 98(4):206-214.
30. Soldati D, Kim K, Kampmeier J et al. Complementation of a Toxoplasma gondii ROP1 knock-out mutant using phleomycin selection. Mol Biochem Parasitol 1995; 74(1):87-97.
31. Carruthers VB, Giddings OK, Sibley LD. Secretion of micronemal proteins is associated with Toxoplasma invasion of host cells. Cell Microbiol 1999; 1(3):225-235.
32. Seeber F, Boothroyd JC. Escherichia coli beta-galactosidase as an in vitro and in vivo reporter enzyme and stable transfection marker in the intracellular protozoan parasite Toxoplasma gondii. Gene 1996; 169(1):39-45.
33. Sibley LD, Howe DK, Wan KL et al. Toxoplasma as a model genetic system. In: Smith DF, Parsons M, eds. Molecular Biology of Parasitic Protozoa. New York: Oxford University Press, 1996:55-74.
34. Howe DK, Mercier C, Messina M et al. Expression of Toxoplasma gondii genes in the closely-related apicomplexan parasite Neospora caninum. Mol Biochem Parasitol 1997; 86(1):29-36.
35. Soldati D, Boothroyd JC. Transient transfection and expression in the obligate intracellular parasite Toxoplasma gondii. Science 1993; 260(5106):349-352.
36. Donald RG, Roos DS. Stable molecular transformation of Toxoplasma gondii: A selectable dihydrofolate reductase-thymidylate synthase marker based on drug-resistance mutations in malaria. Proc Natl Acad Sci USA 1993; 90(24):11703-11707.
37. Black MW, Boothroyd JC. Development of a stable episomal shuttle vector for Toxoplasma gondii. J Biol Chem 1998; 273(7):3972-3979.
38. Kim K, Soldati D, Boothroyd JC. Gene replacement in Toxoplasma gondii with chloramphenicol acetyltransferase as selectable marker. Science 1993; 262(5135):911-914.
39. Roos DS, Donald RG, Morrissette NS et al. Molecular tools for genetic dissection of the protozoan parasite Toxoplasma gondii. Methods Cell Biol 1994; 27-63.
40. Garcia-Reguet N, Lebrun M, Fourmaux M et al. The microneme protein MIC3 of Toxoplasma gondii is a secretory adhesin that binds to both the surface of the host cells and the surface of the parasite. Cell Microbiol 2000; 2(4):353-364.
41. Messina M, Niesman I, Mercier C et al. Stable DNA transformation of Toxoplasma gondii using phleomycin selection. Gene 1995; 165(2):213-217.
42. Donald RGK, Carter D, Ullman B et al. Insertional tagging, cloning, and expression of the Toxoplasma gondii hypoxanthine-xanthine-guanine phosphoribosyltransferase gene. Use as a selectable marker for stable transformation. J Biol Chem 1996; 271(24):14010-14019.
43. Donald RG, Roos DS. Insertional mutagenesis and marker rescue in a protozoan parasite: Cloning of the uracil phosphoribosyltransferase locus from Toxoplasma gondii. Proc Natl Acad Sci USA 1995; 92(12):5749-5753.
44. Radke JR, White MW. A cell cycle model for the tachyzoite of Toxoplasma gondii using the Herpes simplex virus thymidine kinase. Mol Biochem Parasitol 1998; 94(2):237-247.
45. Sibley LD, Messina M, Niesman IR. Stable DNA transformation in the obligate intracellular parasite Toxoplasma gondii by complementation of tryptophan auxotrophy. Proc Natl Acad Sci USA 1994; 91(12):5508-5512.
46. Van Tam T, Rooney PJ, Knoll LJ. Nourseothricin acetyltransferease: A positive selectable marker for Toxoplasma gondii. J Parasitol 2006; 92(3):668-670.
47. Mercier C, Lefebvre-Van Hende S, Garber GE et al. Common cis-acting elements critical for the expression of several genes of Toxoplasma gondii. Mol Microbiol 1996; 21(2):421-428.
48. Dobrowolski JM, Sibley LD. Toxoplasma invasion of mammalian cells is powered by the actin cytoskeleton of the parasite. Cell 1996; 84(6):933-939.
49. Black M, Seeber F, Soldati D et al. Restriction enzyme-mediated integration elevates transformation frequency and enables cotransfection of Toxoplasma gondii. Mol Biochem Parasitol 1995; 74(1):55-63.
50. Carey KL, Westwood NJ, Mitchison TJ et al. A small-molecule approach to studying invasive mechanisms of Toxoplasma gondii. Proc Natl Acad Sci USA 2004; 101(19):7433-7438.
51. Striepen B, He CY, Matrajt M et al. Expression, selection, and organellar targeting of the green fluorescent protein in Toxoplasma gondii. Mol Biochem Parasitol 1998; 92(2):325-338.
52. Matrajt M, Nishi M, Fraunholz MJ et al. Amino-terminal control of transgenic protein expression levels in Toxoplasma gondii. Mol Biochem Parasitol 2002; 120(2):285-289.

53. Wu Y, Sifri CD, Lei HH et al. Transfection of Plasmodium falciparum within human red blood cells. Proc Natl Acad Sci USA 1995; 92(4):973-977.
54. van Dijk MR, Waters AP, Janse CJ. Stable transfection of malaria parasite blood stages. Science 1995; 268(5215):1358-1362.
55. Carruthers VB, Blackman MJ. A new release on life: Emerging concepts in proteolysis and parasite invasion. Mol Microbiol 2005; 55(6):1617-1630.
56. Crabb BS, Cowman AF. Characterization of promoters and stable transfection by homologous and nonhomologous recombination in Plasmodium falciparum. Proc Natl Acad Sci USA 1996; 93(14):7289-7294.
57. Kadekoppala M, Cheresh P, Catron D et al. Rapid recombination among transfected plasmids, chimeric episome formation and trans gene expression in Plasmodium falciparum. Mol Biochem Parasitol 2001; 112(2):211-218.
58. Carvalho TG, Menard R. Manipulating the Plasmodium genome. Curr Issues Mol Biol 2005; 7(1):39-55.
59. de Koning-Ward TF, Fidock DA, Thathy V et al. The selectable marker human dihydrofolate reductase enables sequential genetic manipulation of the Plasmodium berghei genome. Mol Biochem Parasitol 2000; 106(2):199-212.
60. Mamoun CB, Gluzman IY, Goyard S et al. A set of independent selectable markers for transfection of the human malaria parasite Plasmodium falciparum. Proc Natl Acad Sci USA 1999; 96(15):8716-8720.
61. Cowman AF, Crabb BS. Invasion of red blood cells by malaria parasites. Cell 2006; 124(4):755-766.
62. Duraisingh MT, Triglia T, Cowman AF. Negative selection of Plasmodium falciparum reveals targeted gene deletion by double crossover recombination. Int J Parasitol 2002; 32(1):81-89.
63. Deitsch K, Driskill C, Wellems T. Transformation of malaria parasites by the spontaneous uptake and expression of DNA from human erythrocytes. Nucleic Acids Res 2001; 29(3):850-853.
64. Waller RF, Reed MB, Cowman AF et al. Protein trafficking to the plastid of Plasmodium falciparum is via the secretory pathway. EMBO J 2000; 19(8):1794-1802.
65. Howe DK, Sibley LD. Development of molecular genetics for Neospora caninum: A complementary system to Toxoplasma gondii. Methods 1997; 13(2):123-133.
66. Kelleher M, Tomley FM. Transient expression of beta-galactosidase in differentiating sporozoites of Eimeria tenella. Mol Biochem Parasitol 1998; 97(1-2):21-31.
67. Adamson R, Lyons K, Sharrard M et al. Transient transfection of Theileria annulata. Mol Biochem Parasitol 2001; 114(1):53-61.
68. Gardner MJ, Bishop R, Shah T et al. Genome sequence of Theileria parva, a bovine pathogen that transforms lymphocytes. Science 2005; 309(5731):134-137.
69. Suarez CE, Palmer GH, LeRoith T et al. Intergenic regions in the rhoptry associated protein-1 (rap-1) locus promote exogenous gene expression in Babesia bovis. Int J Parasitol 2004; 34(10):1177-1184.
70. Vaishnava S, Morrison DP, Gaji RY et al. Plastid segregation and cell division in the apicomplexan parasite Sarcocystis neurona. J Cell Sci 2005; 118(Pt 15):3397-3407.
71. Huynh MH, Rabenau KE, Harper JM et al. Rapid invasion of host cells by Toxoplasma requires secretion of the MIC2-M2AP adhesive protein complex. EMBO J 2003; 22(9):2082-2090.
72. Gaji RY, Zhang D, Breathnach CC et al. Molecular genetic transfection of the coccidian parasite Sarcocystis neurona. Mol Biochem Parasitol 2006; 150(1):1-9.
73. Lovett JL, Sibley LD. Intracellular calcium stores in Toxoplasma gondii govern invasion of host cells. J Cell Sci 2003; 116(Pt 14):3009-3016.
74. Naguleswaran A, Muller N, Hemphill A. Neospora caninum and Toxoplasma gondii: A novel adhesion/invasion assay reveals distinct differences in tachyzoite-host cell interactions. Exp Parasitol 2003; 104(3-4):149-158.
75. Barragan A, Brossier F, Sibley LD. Transepithelial migration of Toxoplasma gondii involves an interaction of intercellular adhesion molecule 1 (ICAM-1) with the parasite adhesin MIC2. Cell Microbiol 2005; 7(4):561-568.
76. Mital J, Schwarz J, Taatjes DJ et al. Laser scanning cytometer-based assays for measuring host cell attachment and invasion by the human pathogen Toxoplasma gondii. Cytometry A 2006; 69(1):13-19.
77. Gay-Andrieu F, Cozon GJ, Ferrandiz J et al. Flow cytometric quantification of Toxoplasma gondii cellular infection and replication. J Parasitol 1999; 85(3):545-549.
78. Sultan AA, Thathy V, Frevert U et al. TRAP is necessary for gliding motility and infectivity of Plasmodium sporozoites. Cell 1997; 90(3):511-522.
79. Reed MB, Caruana SR, Batchelor AH et al. Targeted disruption of an erythrocyte binding antigen in Plasmodium falciparum is associated with a switch toward a sialic acid-independent pathway of invasion. Proc Natl Acad Sci USA 2000; 97(13):7509-7514.

80. Bharara R, Singh S, Pattnaik P et al. Structural analogs of sialic acid interfere with the binding of erythrocyte binding antigen-175 to glycophorin A, an interaction crucial for erythrocyte invasion by Plasmodium falciparum. Mol Biochem Parasitol 2004; 138(1):123-129.
81. Casey JL, Coley AM, Anders RF et al. Antibodies to malaria peptide mimics inhibit Plasmodium falciparum invasion of erythrocytes. Infect Immun 2004; 72(2):1126-1134.
82. Singh AP, Ozwara H, Kocken CH et al. Targeted deletion of Plasmodium knowlesi Duffy binding protein confirms its role in junction formation during invasion. Mol Microbiol 2005; 55(6):1925-1934.
83. Xu L, Chaudhuri A. Plasmodium yoelii: A differential fluorescent technique using Acridine Orange to identify infected erythrocytes and reticulocytes in Duffy knockout mouse. Exp Parasitol 2005; 110(1):80-87.
84. Kocken CH, Withers-Martinez C, Dubbeld MA et al. High-level expression of the malaria blood-stage vaccine candidate Plasmodium falciparum apical membrane antigen 1 and induction of antibodies that inhibit erythrocyte invasion. Infect Immun 2002; 70(8):4471-4476.
85. Kocken CH, Ozwara H, van der Wel A et al. Plasmodium knowlesi provides a rapid in vitro and in vivo transfection system that enables double-crossover gene knockout studies. Infect Immun 2002; 70(2):655-660.
86. Malagon F, Castillo O. On the labeling of malaria parasites In Vivo with 3H-hipoxanthine while developing an infection in the mouse. Acta Protozool 2000; 39:281-288.
87. Makler MT, Ries JM, Williams JA et al. Parasite lactate dehydrogenase as an assay for Plasmodium falciparum drug sensitivity. Am J Trop Med Hyg 1993; 48(6):739-741.
88. Makler MT, Hinrichs DJ. Measurement of the lactate dehydrogenase activity of Plasmodium falciparum as an assessment of parasitemia. Am J Trop Med Hyg 1993; 48(2):205-210.
89. Drew DR, O'Donnell RA, Smith BJ et al. A common cross-species function for the double epidermal growth factor-like modules of the highly divergent Plasmodium surface proteins MSP-1 and MSP-8. J Biol Chem 2004; 279(19):20147-20153.
90. Labbe M, de Venevelles P, Girard-Misguich F et al. Eimeria tenella microneme protein EtMIC3: Identification, localisation and role in host cell infection. Mol Biochem Parasitol 2005; 140(1):43-53.
91. Ward GE, Fujioka H, Aikawa M et al. Staurosporine inhibits invasion of erythrocytes by malarial merozoites. Exp Parasitol 1994; 79(3):480-487.
92. Gaffar FR, Yatsuda AP, Franssen FF et al. Erythrocyte invasion by Babesia bovis merozoites is inhibited by polyclonal antisera directed against peptides derived from a homologue of Plasmodium falciparum apical membrane antigen 1. Infect Immun 2004; 72(5):2947-2955.
93. Franssen FF, Gaffar FR, Yatsuda AP et al. Characterisation of erythrocyte invasion by Babesia bovis merozoites efficiently released from their host cell after high-voltage pulsing. Microbes Infect 2003; 5(5):365-372.
94. Rathore D, Hrstka SC, Sacci Jr JB et al. Molecular mechanism of host specificity in Plasmodium falciparum infection: Role of circumsporozoite protein. J Biol Chem 2003; 278(42):40905-40910.
95. Silvie O, Franetich JF, Charrin S et al. A role for apical membrane antigen 1 during invasion of hepatocytes by Plasmodium falciparum sporozoites. J Biol Chem 2004; 279(10):9490-9496.
96. Brahimi K, Badell E, Sauzet JP et al. Human antibodies against Plasmodium falciparum liver-stage antigen 3 cross-react with Plasmodium yoelii preerythrocytic-stage epitopes and inhibit sporozoite invasion in vitro and in vivo. Infect Immun 2001; 69(6):3845-3852.
97. Kafsack BF, Beckers C, Carruthers VB. Synchronous invasion of host cells by Toxoplasma gondii. Mol Biochem Parasitol 2004; 136(2):309-311.
98. Wetzel DM, Hakansson S, Hu K et al. Actin filament polymerization regulates gliding motility by apicomplexan parasites. Mol Biol Cell 2003; 14(2):396-406.
99. Chen XM, O'Hara SP, Huang BQ et al. Apical organelle discharge by Cryptosporidium parvum is temperature, cytoskeleton, and intracellular calcium dependent and required for host cell invasion. Infect Immun 2004; 72(12):6806-6816.
100. Wetzel DM, Schmidt J, Kuhlenschmidt MS et al. Gliding motility leads to active cellular invasion by Cryptosporidium parvum sporozoites. Infect Immun 2005; 73(9):5379-5387.
101. Entzeroth R, Zgrzebski G, Dubremetz JF. Secretion of trials during gliding motility of Eimeria (Apicomplexa, Coccidia) sporozoites visualized by a monoclonal antibody and immuno-gold-silver enhancement. Parasitol Res 1989; 76(2):174-175.
102. Wiersma HI, Galuska SE, Tomley FM et al. A role for coccidian cGMP-dependent protein kinase in motility and invasion. Int J Parasitol 2004; 34(3):369-380.
103. Stewart MJ, Vanderberg JP. Electron microscopic analysis of circumsporozoite protein trail formation by gliding malaria sporozoites. J Protozool 1992; 39(6):663-671.
104. Bumstead J, Tomley F. Induction of secretion and surface capping of microneme proteins in Eimeria tenella. Mol Biochem Parasitol 2000; 110(2):311-321.

105. Siden-Kiamos I, Ecker A, Nyback S et al. Plasmodium berghei calcium-dependent protein kinase 3 is required for ookinete gliding motility and mosquito midgut invasion. Mol Microbiol 2006; 60(6):1355-1363.
106. Mondragon R, Frixione E. Ca(2+)-dependence of conoid extrusion in Toxoplasma gondii tachyzoites. J Eukaryot Microbiol 1996; 43(2):120-127.
107. Carruthers VB, Moreno SN, Sibley LD. Ethanol and acetaldehyde elevate intracellular [Ca2+] and stimulate microneme discharge in Toxoplasma gondii. Biochem J 1999; 342(Pt 2):379-386.
108. Hoane JS, Carruthers VB, Striepen B et al. Analysis of the Sarcocystis neurona microneme protein SnMIC10: Protein characteristics and expression during intracellular development. Int J Parasitol 2003; 33(7):671-679.
109. Keller N, Riesen M, Naguleswaran A et al. Identification and characterization of a Neospora caninum microneme-associated protein (NcMIC4) that exhibits unique lactose-binding properties. Infect Immun 2004; 72(8):4791-4800.
110. Camus D, Hadley TJ. A Plasmodium falciparum antigen that binds to host erythrocytes and merozoites. Science 1985; 230(4725):553-556.
111. Adams JH, Hudson DE, Torii M et al. The duffy receptor family of Plasmodium knowlesi is located within the micronemes of invasive malaria merozoites. Cell 1990; 63:141-153.
112. Zhou XW, Blackman MJ, Howell SA et al. Proteomic analysis of cleavage events reveals a dynamic two-step mechanism for proteolysis of a key parasite adhesive complex. Mol Cell Proteomics 2004; 3(6):565-576.
113. Phillips CI, Bogyo M. Proteomics meets microbiology: Technical advances in the global mapping of protein expression and function. Cell Microbiol 2005; 7(8):1061-1076.
114. Mineo JR, Kasper LH. Attachment of Toxoplasma gondii to host cells involves major surface protein, SAG-1 (P30). Exp Parasitol 1994; 79(1):11-20.
115. Silva SR, Meirelles SS, De Souza W. Mechanism of entry of Toxoplasma gondii into vertebrate cells. J Submicrosc Cytol 1982; 14(3):471-482.
116. Friedman MJ, Blankenberg T, Sensabaugh G et al. Recognition and invasion of human erythrocytes by malarial parasites: Contribution of sialoglycoproteins to attachment and host specificity. J Cell Biol 1984; 98(5):1672-1677.
117. Mosqueda J, McElwain TF, Stiller D et al. Babesia bovis merozoite surface antigen 1 and rhoptry-associated protein 1 are expressed in sporozoites, and specific antibodies inhibit sporozoite attachment to erythrocytes. Infect Immun 2002; 70(3):1599-1603.
118. Fraser TS, Kappe SH, Narum DL et al. Erythrocyte-binding activity of Plasmodium yoelii apical membrane antigen-1 expressed on the surface of transfected COS-7 cells. Mol Biochem Parasitol 2001; 117(1):49-59.
119. Jacquet A, Coulon L, De Neve J et al. The surface antigen SAG3 mediates the attachment of Toxoplasma gondii to cell-surface proteoglycans. Mol Biochem Parasitol 2001; 116(1):35-44.
120. Urquiza M, Suarez JE, Cardenas C et al. Plasmodium falciparum AMA-1 erythrocyte binding peptides implicate AMA-1 as erythrocyte binding protein. Vaccine 2000; 19(4-5):508-513.
121. Hakansson S, Charron AJ, Sibley LD. Toxoplasma evacuoles: A two-step process of secretion and fusion forms the parasitophorous vacuole. EMBO J 2001; 20(12):3132-3144.
122. Carey KL, Jongco AM, Kim K et al. The Toxoplasma gondii rhoptry protein ROP4 is secreted into the parasitophorous vacuole and becomes phosphorylated in infected cells. Eukaryot Cell 2004; 3(5):1320-1330.
123. Carey KL, Donahue CG, Ward GE. Identification and molecular characterization of GRA8, a novel, proline- rich, dense granule protein of Toxoplasma gondii. Mol Biochem Parasitol 2000; 105(1):25-37.
124. Aikawa M, Miller LH, Rabbege JR et al. Freeze-fracture study on the erythrocyte membrane during malarial parasite invasion. J Cell Biol 1981; 91(1):55-62.
125. Miller LH, Aikawa M, Johnson JG et al. Interaction between cytochalasin B-treated malarial parasites and erythrocytes. Attachment and junction formation. J Exp Med 1979; 149(1):172-184.
126. Donald RG, Roos DS. Homologous recombination and gene replacement at the dihydrofolate reductase-thymidylate synthase locus in Toxoplasma gondii. Mol Biochem Parasitol 1994; 63(2):243-253.
127. Roos DS, Sullivan WJ, Striepen B et al. Tagging genes and trapping promoters in Toxoplasma gondii by insertional mutagenesis. Methods 1997; 13(2):112-122.
128. Sullivan Jr WJ, Chiang CW, Wilson CM et al. Insertional tagging of at least two loci associated with resistance to adenine arabinoside in Toxoplasma gondii, and cloning of the adenosine kinase locus. Mol Biochem Parasitol 1999; 103(1):1-14.
129. Matrajt M, Donald RG, Singh U et al. Identification and characterization of differentiation mutants in the protozoan parasite Toxoplasma gondii. Mol Microbiol 2002; 44(3):735-747.

130. Vanchinathan P, Brewer JL, Harb OS et al. Disruption of a locus encoding a nucleolar zinc finger protein decreases tachyzoite-to-bradyzoite differentiation in Toxoplasma gondii. Infect Immun 2005; 73(10):6680-6688.
131. Striepen B, White MW, Li C et al. Genetic complementation in apicomplexan parasites. Proc Natl Acad Sci USA 2002; 16:16.
132. Black MW, Arrizabalaga G, Boothroyd JC. Ionophore-resistant mutants of Toxoplasma gondii reveal host cell permeabilization as an early event in egress. Mol Cell Biol 2000; 20(24):9399-9408.
133. Radke JR, Guerini MN, White MW. Toxoplasma gondii: Characterization of temperature-sensitive tachyzoite cell cycle mutants. Exp Parasitol 2000; 96(3):168-177.
134. Singh U, Brewer JL, Boothroyd JC. Genetic analysis of tachyzoite to bradyzoite differentiation mutants in Toxoplasma gondii reveals a hierarchy of gene induction. Mol Microbiol 2002; 44(3):721-733.
135. Lindsay DS, Lenz SD, Blagburn BL et al. Characterization of temperature-sensitive strains of Neospora caninum in mice. J Parasitol 1999; 85(1):64-67.
136. White MW, Jerome ME, Vaishnava S et al. Genetic rescue of a Toxoplasma gondii conditional cell cycle mutant. Mol Microbiol 2005; 55(4):1060-1071.
137. Mercier C, Howe DK, Mordue D et al. Targeted disruption of the GRA2 locus in Toxoplasma gondii decreases acute virulence in mice. Infect Immun 1998; 66(9):4176-4182.
138. Dzierszinski F, Mortuaire M, Cesbron-Delauw MF et al. Targeted disruption of the glycosylphosphatidylinositol-anchored surface antigen SAG3 gene in Toxoplasma gondii decreases host cell adhesion and drastically reduces virulence in mice. Mol Microbiol 2000; 37(3):574-582.
139. Mercier C, Rauscher B, Lecordier L et al. Lack of expression of the dense granule protein GRA5 does not affect the development of Toxoplasma tachyzoites. Mol Biochem Parasitol 2001; 116(2):247-251.
140. Donald RG, Roos DS. Gene knock-outs and allelic replacements in Toxoplasma gondii: HXGPRT as a selectable marker for hit-and-run mutagenesis. Mol Biochem Parasitol 1998; 91(2):295-305.
141. Arrizabalaga G, Ruiz F, Moreno S et al. Ionophore-resistant mutant of Toxoplasma gondii reveals involvement of a sodium/hydrogen exchanger in calcium regulation. J Cell Biol 2004; 165(5):653-662.
142. Wichroski MJ, Ward GE. Biosynthesis of glycosylphosphatidylinositol is essential to the survival of the protozoan parasite Toxoplasma gondii. Eukaryot Cell 2003; 2(5):1132-1136.
143. Donald RG, Allocco J, Singh SB et al. Toxoplasma gondii cyclic GMP-dependent kinase: Chemotherapeutic targeting of an essential parasite protein kinase. Eukaryot Cell 2002; 1(3):317-328.
144. Cerede O, Dubremetz JF, Soete M et al. Synergistic role of micronemal proteins in Toxoplasma gondii virulence. J Exp Med 2005; 201(3):453-463.
145. Lekutis C, Ferguson DJ, Grigg ME et al. Surface antigens of Toxoplasma gondii: Variations on a theme. Int J Parasitol 2001; 31(12):1285-1292.
146. Duraisingh MT, Triglia T, Ralph SA et al. Phenotypic variation of Plasmodium falciparum merozoite proteins directs receptor targeting for invasion of human erythrocytes. EMBO J 2003; 22(5):1047-1057.
147. Khater EI, Sinden RE, Dessens JT. A malaria membrane skeletal protein is essential for normal morphogenesis, motility, and infectivity of sporozoites. J Cell Biol 2004; 167(3):425-432.
148. Sijwali PS, Rosenthal PJ. Gene disruption confirms a critical role for the cysteine protease falcipain-2 in hemoglobin hydrolysis by Plasmodium falciparum. Proc Natl Acad Sci USA 2004; 101(13):4384-4389.
149. Sijwali PS, Kato K, Seydel KB et al. Plasmodium falciparum cysteine protease falcipain-1 is not essential in erythrocytic stage malaria parasites. Proc Natl Acad Sci USA 2004; 101(23):8721-8726.
150. Tewari R, Ogun SA, Gunaratne RS et al. Disruption of Plasmodium berghei merozoite surface protein 7 gene modulates parasite growth in vivo. Blood 2005; 105(1):394-396.
151. Cowman AF, Baldi DL, Healer J et al. Functional analysis of proteins involved in Plasmodium falciparum merozoite invasion of red blood cells. FEBS Lett 2000; 476(1-2):84-88.
152. Eksi S, Haile Y, Furuya T et al. Identification of a subtelomeric gene family expressed during the asexual-sexual stage transition in Plasmodium falciparum. Mol Biochem Parasitol 2005; 143(1):90-99.
153. Mueller AK, Camargo N, Kaiser K et al. Plasmodium liver stage developmental arrest by depletion of a protein at the parasite-host interface. Proc Natl Acad Sci USA 2005; 102(8):3022-3027.
154. Yuda M, Ishino T. Liver invasion by malarial parasites—how do malarial parasites break through the host barrier? Cell Microbiol 2004; 6(12):1119-1125.
155. Kappe SH, Buscaglia CA, Nussenzweig V. Plasmodium sporozoite molecular cell biology. Annu Rev Cell Dev Biol 2004; 20:29-59.
156. Mota MM, Rodriguez A. Invasion of mammalian host cells by Plasmodium sporozoites. Bioessays 2002; 24(2):149-156.
157. Menard R. Gliding motility and cell invasion by Apicomplexa: Insights from the Plasmodium sporozoite. Cell Microbiol 2001; 3(2):63-73.

158. Huynh MH, Carruthers VB. Toxoplasma MIC2 is a major determinant of invasion and virulence. PLoS Pathog 2006; 2(8).
159. Mazumdar JE, HW, Masek K et al. Apicoplast fatty acid synthesis is essential for organelle biogenesis and parasite survival in Toxoplasma gondii. Proc Natl Acad Sci USA 2006; 103(35):13192-13197.
160. Meissner M, Krejany E, Gilson PR et al. Tetracycline analogue-regulated transgene expression in Plasmodium falciparum blood stages using Toxoplasma gondii transactivators. Proc Natl Acad Sci USA 2005; 102(8):2980-2985.
161. Carvalho TG, Thiberge S, Sakamoto H et al. Conditional mutagenesis using site-specific recombination in Plasmodium berghei. Proc Natl Acad Sci USA 2004; 101(41):14931-14936.
162. Al-Anouti F, Ananvoranich S. Comparative analysis of antisense RNA, double-stranded RNA, and delta ribozyme-mediated gene regulation in Toxoplasma gondii. Antisense Nucleic Acid Drug Dev 2002; 12(4):275-281.
163. Nakaar V, Ngo EO, Joiner KA. Selection based on the expression of antisense hypoxanthine-xanthine-guanine-phosphoribosyltransferase RNA in Toxoplasma gondii. Mol Biochem Parasitol 2000; 110(1):43-51.
164. Nakaar V, Ngo HM, Aaronson EP et al. Pleiotropic effect due to targeted depletion of secretory rhoptry protein ROP2 in Toxoplasma gondii. J Cell Sci 2003; 116(Pt 11):2311-2320.
165. Nakaar V, Samuel BU, Ngo EO et al. Targeted reduction of nucleoside triphosphate hydrolase by antisense RNA inhibits Toxoplasma gondii proliferation. J Biol Chem 1999; 274(8):5083-5087.
166. Ngo HM, Yang M, Paprotka K et al. AP-1 in Toxoplasma gondii mediates biogenesis of the rhoptry secretory organelle from a post-Golgi compartment. J Biol Chem 2003; 278(7):5343-5352.
167. Bermudes D, Peck KR, Afifi MA et al. Tandemly repeated genes encode nucleoside triphosphate hydrolase isoforms secreted into the parasitophorous vacuole of Toxoplasma gondii. J Biol Chem 1994; 269(46):29252-29260.
168. Gardiner DL, Holt DC, Thomas EA et al. Inhibition of Plasmodium falciparum clag9 gene function by antisense RNA. Mol Biochem Parasitol 2000; 110(1):33-41.
169. Noonpakdee W, Pothikasikorn J, Nimitsantiwong W et al. Inhibition of Plasmodium falciparum proliferation in vitro by antisense oligodeoxynucleotides against malarial topoisomerase II. Biochem Biophys Res Commun 2003; 302(4):659-664.
170. Ullu E, Tschudi C, Chakraborty T. RNA interference in protozoan parasites. Cell Microbiol 2004; 6(6):509-519.
171. Aravind L, Iyer LM, Wellems TE et al. Plasmodium biology: Genomic gleanings. Cell 2003; 115(7):771-785.
172. McRobert L, McConkey GA. RNA interference (RNAi) inhibits growth of Plasmodium falciparum. Mol Biochem Parasitol 2002; 119(2):273-278.
173. Malhotra P, Dasaradhi PV, Kumar A et al. Double-stranded RNA-mediated gene silencing of cysteine proteases (falcipain-1 and -2) of Plasmodium falciparum. Mol Microbiol 2002; 45(5):1245-1254.
174. Mohmmed A, Dasaradhi PV, Bhatnagar RK et al. In vivo gene silencing in Plasmodium berghei—a mouse malaria model. Biochem Biophys Res Commun 2003; 309(3):506-511.
175. Al-Anouti F, Tomavo S, Parmley S et al. The expression of lactate dehydrogenase is important for the cell cycle of Toxoplasma gondii. J Biol Chem 2004.
176. Adams B, Musiyenko A, Kumar R et al. A novel class of dual-family immunophilins. J Biol Chem 2005; 280(26):24308-24314.
177. Sheng J, Al-Anouti F, Ananvoranich S. Engineered delta ribozymes can simultaneously knock down the expression of the genes encoding uracil phosphoribosyltransferase and hypoxanthine-xanthine-guanine phosphoribosyltransferase in Toxoplasma gondii. Int J Parasitol 2004; 34(3):253-263.
178. Beckers CJ, Wakefield T, Joiner KA. The expression of Toxoplasma proteins in Neospora caninum and the identification of a gene encoding a novel rhoptry protein. Mol Biochem Parasitol 1997; 89(2):209-223.
179. Kim K, Bulow R, Kampmeier J et al. Conformationally appropriate expression of the Toxoplasma antigen SAG1 (p30) in CHO cells. Infect Immun 1994; 62(1):203-209.
180. Gardner MJ, Hall N, Fung E et al. Genome sequence of the human malaria parasite Plasmodium falciparum. Nature 2002; 419(6906):498-511.
181. Withers-Martinez C, Carpenter EP, Hackett F et al. PCR-based gene synthesis as an efficient approach for expression of the A+T-rich malaria genome. Protein Eng 1999; 12(12):1113-1120.
182. Flick K, Ahuja S, Chene A et al. Optimized expression of Plasmodium falciparum erythrocyte membrane protein 1 domains in Escherichia coli. Malar J 2004; 3(1):50.
183. Yadava A, Ockenhouse CF. Effect of codon optimization on expression levels of a functionally folded malaria vaccine candidate in prokaryotic and eukaryotic expression systems. Infect Immun 2003; 71(9):4961-4969.

184. Mehlin C, Boni E, Buckner FS et al. Heterologous expression of proteins from Plasmodium falciparum: Results from 1000 genes. Mol Biochem Parasitol 2006; 148(2):144-160.
185. Brossier F, David Sibley L. Toxoplasma gondii: Microneme protein MIC2. Int J Biochem Cell Biol 2005; 37(11):2266-2272.
186. Jewett TJ, Sibley LD. The Toxoplasma proteins MIC2 and M2AP form a hexameric complex necessary for intracellular survival. J Biol Chem 2004; 279(10):9362-9369.
187. Kappe S, Bruderer T, Gantt S et al. Conservation of a gliding motility and cell invasion machinery in Apicomplexan parasites. J Cell Biol 1999; 147(5):937-944.
188. O'Connor RM, Kim K, Khan F et al. Expression of Cpgp40/15 in Toxoplasma gondii: A surrogate system for the study of Cryptosporidium glycoprotein antigens. Infect Immun 2003; 71(10):6027-6034.
189. Triglia T, Healer J, Caruana SR et al. Apical membrane antigen 1 plays a central role in erythrocyte invasion by Plasmodium species. Mol Microbiol 2000; 38(4):706-718.
190. Tewari R, Rathore D, Crisanti A. Motility and infectivity of Plasmodium berghei sporozoites expressing avian Plasmodium gallinaceum circumsporozoite protein. Cell Microbiol 2005; 7(5):699-707.
191. Chitnis CE, Miller LH. Identification of the erythrocyte binding domains of Plasmodium vivax and Plasmodium knowlesi proteins involved in erythrocyte invasion. J Exp Med 1994; 180(2):497-506.
192. Chattopadhyay R, Taneja T, Chakrabarti K et al. Molecular analysis of the cytoadherence phenotype of a Plasmodium falciparum field isolate that binds intercellular adhesion molecule-1. Mol Biochem Parasitol 2004; 133(2):255-265.
193. Sim BK, Chitnis CE, Wasniowska K et al. Receptor and ligand domains for invasion of erythrocytes by Plasmodium falciparum. Science 1994; 264(5167):1941-1944.
194. Ward GE, Chitnis CE, Miller LH. The invasion of erythrocytes by malarial merozoites. In: Russell DG, ed. Clinical Infectious Diseases: International Practice and Research, Strategies for Intracellular Survival of Microbes. 1st ed. London: Bailliere Tindall, 1994:155-190.
195. Saouros S, Edwards-Jones B, Reiss M et al. A novel galectin-like domain from Toxoplasma gondii micronemal protein 1 assists the folding, assembly, and transport of a cell adhesion complex. J Biol Chem 2005; 280(46):38583-38591.
196. Kato K, Mayer DC, Singh S et al. Domain III of Plasmodium falciparum apical membrane antigen 1 binds to the erythrocyte membrane protein Kx. Proc Natl Acad Sci USA 2005; 102(15):5552-5557.
197. Grimwood J, Smith JE. Toxoplasma gondii: The role of a 30-kDa surface protein in host cell invasion. Exp Parasitol 1992; 74(1):106-111.
198. Grimwood J, Smith JE. Toxoplasma gondii: The role of parasite surface and secreted proteins in host cell invasion. Int J Parasitol 1996; 26(2):169-173.
199. Mineo JR, McLeod R, Mack D et al. Antibodies to Toxoplasma gondii major surface protein (SAG-1, P30) inhibit infection of host cells and are produced in murine intestine after peroral infection. J Immunol 1993; 150(9):3951-3964.
200. Uchida Y, Ike K, Kurotaki T et al. Monoclonal antibodies preventing invasion of Neospora caninum tachyzoites into host cells. J Vet Med Sci 2004; 66(11):1355-1358.
201. Augustine PC, Jenkins MC, Dubey JP. Effect of polyclonal antisera developed against dense granule-associated Neospora caninum proteins on cell invasion and development in vitro by N. caninum tachyzoites. Parasitology 1999; 119(Pt 5):441-445.
202. Augustine PC. Reduced invasion of cultured cells pretreated with a monoclonal antibody elicited against refractile body antigens of avian coccidial sporozoites. J Eukaryot Microbiol 1999; 46(3):254-258.
203. Dutta S, Haynes JD, Moch JK et al. Invasion-inhibitory antibodies inhibit proteolytic processing of apical membrane antigen 1 of Plasmodium falciparum merozoites. Proc Natl Acad Sci USA 2003; 100(21):12295-12300.
204. Singh AP, Puri SK, Chitnis CE. Antibodies raised against receptor-binding domain of Plasmodium knowlesi Duffy binding protein inhibit erythrocyte invasion. Mol Biochem Parasitol 2002; 121(1):21-31.
205. Mitchell GH, Thomas AW, Margos G et al. Apical membrane antigen 1, a major malaria vaccine candidate, mediates the close attachment of invasive merozoites to host red blood cells. Infect Immun 2004; 72(1):154-158.
206. Ramasamy R, Ramasamy M, Yasawardena S. Antibodies and Plasmodium falciparum merozoites. Trends Parasitol 2001; 17(4):194-197.
207. Dutta S, Haynes JD, Barbosa A et al. Mode of action of invasion-inhibitory antibodies directed against apical membrane antigen 1 of Plasmodium falciparum. Infect Immun 2005; 73(4):2116-2122.
208. Blackman MJ, Scott-Finnigan TJ, Shai S et al. Antibodies inhibit the protease-mediated processing of a malaria merozoite surface protein. J Exp Med 1994; 180(1):389-393.

209. Ryning FW, Remington JS. Effect of cytochalasin D on Toxoplasma gondii cell entry. Infect Immun 1978; 20(3):739-743.
210. Schwartzman JD, Pfefferkorn ER. Immunofluorescent localization of myosin at the anterior pole of the coccidian, Toxoplasma gondii. J Protozool 1983; 30(4):657-661.
211. Carruthers VB, Sibley LD. Mobilization of intracellular calcium stimulates microneme discharge in Toxoplasma gondii. Mol Microbiol 1999; 31(2):421-428.
212. Ward GE, Carey KL, Westwood NJ. Using small molecules to study big questions in cellular microbiology. Cell Microbiol 2002; 4(8):471-482.
213. Morgan RE, Evans KM, Patterson S et al. Targeting invasion and egress: From tools to drugs? Current Drug Targets 2006, (In Press).
214. Mayer TU. Chemical genetics: Tailoring tools for cell biology. Trends Cell Biol 2003; 13(5):270-277.
215. Stockwell BR. Chemical genetics: Ligand-based discovery of gene function. Nat Rev Genet 2000; 1(2):116-125.
216. Asai T, Takeuchi T, Diffenderfer J et al. Identification of small-molecule inhibitors of nucleoside triphosphate hydrolase in Toxoplasma gondii. Antimicrob Agents Chemother 2002; 46(8):2393-2399.
217. Baldwin J, Michnoff CH, Malmquist NA et al. High-throughput screening for potent and selective inhibitors of Plasmodium falciparum dihydroorotate dehydrogenase. J Biol Chem 2005; 280(23):21847-21853.
218. Carruthers VB, Hakansson S, Giddings OK et al. Toxoplasma gondii uses sulfated proteoglycans for substrate and host cell attachment. Infect Immun 2000; 68(7):4005-4011.
219. Ortega-Barria E, Boothroyd JC. A Toxoplasma lectin-like activity specific for sulfated polysaccharides is involved in host cell infection. J Biol Chem 1999; 274(3):1267-1276.
220. Grimwood J, Mineo JR, Kasper LH. Attachment of Toxoplasma gondii to host cells is host cell cycle dependent. Infect Immun 1996; 64(10):4099-4104.
221. Dvorak JA, Crane MS. Vertebrate cell cycle modulates infection by protozoan parasites. Science 1981; 214(4524):1034-1036.
222. Dutta C, Grimwood J, Kasper LH. Attachment of Toxoplasma gondii to a specific membrane fraction of CHO cells. Infect Immun 2000; 68(12):7198-7201.
223. Miller LH, Mason SJ, Dvorak JA et al. Erythrocyte receptors for (Plasmodium knowlesi) malaria: Duffy blood group determinants. Science 1975; 189(4202):561-563.
224. Su XZ, Heatwole VM, Wertheimer SP et al. The large diverse gene family var encodes proteins involved in cytoadherence and antigenic variation of Plasmodium falciparum-infected erythrocytes. Cell 1995; 82(1):89-100.
225. Miller LH, Haynes JD, McAuliffe FM et al. Evidence for differences in erythrocyte surface receptors for the malarial parasites, Plasmodium falciparum and Plasmodium knowlesi. J Exp Med 1977; 146(1):277-281.
226. Kaneko O, Fidock DA, Schwartz OM et al. Disruption of the C-terminal region of EBA-175 in the Dd2/Nm clone of Plasmodium falciparum does not affect erythrocyte invasion. Mol Biochem Parasitol 2000; 110(1):135-146.
227. Klotz FW, Orlandi PA, Reuter G et al. Binding of Plasmodium falciparum 175-kilodalton erythrocyte binding antigen and invasion of murine erythrocytes requires N-acetylneuraminic acid but not its O-acetylated form. Mol Biochem Parasitol 1992; 51(1):49-54.
228. Gaur D, Mayer DC, Miller LH. Parasite ligand-host receptor interactions during invasion of erythrocytes by Plasmodium merozoites. Int J Parasitol 2004; 34(13-14):1413-1429.
229. Wu Y, Wang X, Liu X et al. Data-mining approaches reveal hidden families of proteases in the genome of malaria parasite. Genome Res 2003; 13(4):601-616.
230. Urban S, Freeman M. Substrate specificity of rhomboid intramembrane proteases is governed by helix-breaking residues in the substrate transmembrane domain. Mol Cell 2003; 11(6):1425-1434.
231. Opitz C, Di Cristina M, Reiss M et al. Intramembrane cleavage of microneme proteins at the surface of the apicomplexan parasite Toxoplasma gondii. EMBO J 2002; 21(7):1577-1585.
232. Urban S, Lee JR, Freeman M. Drosophila rhomboid-1 defines a family of putative intramembrane serine proteases. Cell 2001; 107(2):173-182.
233. Brossier F, Jewett TJ, Sibley LD et al. A spatially localized rhomboid protease cleaves cell surface adhesins essential for invasion by Toxoplasma. Proc Natl Acad Sci USA 2005; 102(11):4146-4151.
234. Dowse TJ, Pascall JC, Brown KD et al. Apicomplexan rhomboids have a potential role in microneme protein cleavage during host cell invasion. Int J Parasitol 2005; 35(7):747-756.
235. Dowse TJ, Soldati D. Rhomboid-like proteins in Apicomplexa: Phylogeny and nomenclature. Trends Parasitol 2005; 21(6):254-258.
236. Cleary MD, Singh U, Blader IJ et al. Toxoplasma gondii asexual development: Identification of developmentally regulated genes and distinct patterns of gene expression. Eukaryotic Cell 2002; 1(3):329-340.

237. Cleary MD, Meiering CD, Jan E et al. Biosynthetic labeling of RNA with uracil phosphoribosyltransferase allows cell-specific microarray analysis of mRNA synthesis and decay. Nat Biotechnol 2005; 23(2):232-237.
238. Kappe SH, Gardner MJ, Brown SM et al. Exploring the transcriptome of the malaria sporozoite stage. Proc Natl Acad Sci USA 2001; 98(17):9895-9900.
239. Le Roch KG, Zhou Y, Blair PL et al. Discovery of gene function by expression profiling of the malaria parasite life cycle. Science 2003; 301(5639):1503-1508.
240. Bozdech Z, Llinas M, Pulliam BL et al. The transcriptome of the intraerythrocytic developmental cycle of Plasmodium falciparum. PLoS Biol 2003; 1(1):E5.
241. Blader IJ, Manger ID, Boothroyd JC. Microarray analysis reveals previously unknown changes in Toxoplasma gondii-infected human cells. J Biol Chem 2001; 276(26):24223-24231.
242. Bishop R, Shah T, Pelle R et al. Analysis of the transcriptome of the protozoan Theileria parva using MPSS reveals that the majority of genes are transcriptionally active in the schizont stage. Nucleic Acids Res 2005; 33(17):5503-5511.
243. Bradley PJ, Ward C, Cheng SJ et al. Proteomic analysis of rhoptry organelles reveals many novel constituents for host-parasite interactions in Toxoplasma gondii. J Biol Chem 2005; 280(40):34245-34258.
244. Zhou XW, Kafsack BF, Cole RN et al. The opportunistic pathogen Toxoplasma gondii deploys a diverse legion of invasion and survival proteins. J Biol Chem 2005; 280(40):34233-34244.
245. Bromley E, Leeds N, Clark J et al. Defining the protein repertoire of microneme secretory organelles in the apicomplexan parasite Eimeria tenella. Proteomics 2003; 3(8):1553-1561.
246. Gilson PR, Nebl T, Vukcevic D et al. Identification and stoichiometry of glycosylphosphatidylinositol-anchored membrane proteins of the human malaria parasite Plasmodium falciparum. Mol Cell Proteomics 2006; 5(7):1286-1299.
247. Thomas AW, Deans JA, Mitchell GH et al. The Fab fragments of monoclonal IgG to a merozoite surface antigen inhibit Plasmodium knowlesi invasion of erythrocytes. Mol Biochem Parasitol 1984; 13(2):187-199.
248. Deans JA. Protective antigens of bloodstage Plasmodium knowlesi parasites. Philos Trans R Soc Lond B Biol Sci 1984; 307(1131):159-169.
249. Li F, Dluzewski A, Coley AM et al. Phage-displayed peptides bind to the malarial protein apical membrane antigen-1 and inhibit the merozoite invasion of host erythrocytes. J Biol Chem 2002; 277(52):50303-50310.
250. Keizer DW, Miles LA, Li F et al. Structures of phage-display peptides that bind to the malarial surface protein, apical membrane antigen 1, and block erythrocyte invasion. Biochemistry 2003; 42(33):9915-9923.
251. Donahue CG, Carruthers VB, Gilk SD et al. The Toxoplasma homolog of Plasmodium apical membrane antigen-1 (AMA-1) is a microneme protein secreted in response to elevated intracellular calcium levels. Mol Biochem Parasitol 2000; 111(1):15-30.
252. Gurnett AM, Liberator PA, Dulski PM et al. Purification and molecular characterization of cGMP-dependent protein kinase from Apicomplexan parasites: A novel chemotherapeutic target. J Biol Chem 2002; 277(18):15913-15922.
253. Trager W, Zung J, Tershakovec M. Initial extracellular development in vitro of erythrocytic stages of malaria parasites (Plasmodium falciparum). Proc Natl Acad Sci USA 1990; 87(15):5618-5622.
254. Tomley FM, Soldati DS. Mix and match modules: Structure and function of microneme proteins in apicomplexan parasites. Trends Parasitol 2001; 17(2):81-88.
255. Kim K, Weiss LM. Toxoplasma gondii: The model apicomplexan. Int J Parasitol 2004; 34(3):423-432.
256. Di Cristina M, Ghouze F, Kocken CH et al. Transformed Toxoplasma gondii tachyzoites expressing the circumsporozoite protein of Plasmodium knowlesi elicit a specific immune response in rhesus monkeys. Infect Immun 1999; 67(4):1677-1682.
257. Charest H, Sedegah M, Yap GS et al. Recombinant attenuated Toxoplasma gondii expressing the Plasmodium yoelii circumsporozoite protein provides highly effective priming for CD8+ T cell-dependent protective immunity against malaria. J Immunol 2000; 165(4):2084-2092.
258. Kenny B, DeVinney R, Stein M et al. Enteropathogenic E. coli (EPEC) transfers its receptor for intimate adherence into mammalian cells. Cell 1997; 91(4):511-520.
259. Hu K, Johnson J, Florens L et al. Cytoskeletal components of an invasion machine-the apical complex of Toxoplasma gondii. PLoS Pathog 2006; 2(2):e13.

CHAPTER 2

Microneme Proteins in Apicomplexans

Vern B. Carruthers* and Fiona M. Tomley

Introduction

The invasive stages (zoites) of most apicomplexan parasites are polarised cells that use their actinomyosin-powered gliding motility or "glideosome" system to move over surfaces, migrate through biological barriers and invade and leave host cells. Central to these processes is the timely engagement and disengagement of specific receptors upon the regulated release of apical invasion proteins from parasite secretory organelles (micronemes, rhoptries). In this short review, we summarise recent progress on identification and functional characterisation of apical invasion proteins mobilised to the parasite surface from the microneme organelles. We have restricted our focus to *Toxoplasma*, *Eimeria*, *Cryptosporidium* and the nonerythrocytic stages of *Plasmodium* because these organisms have been the most intensively studied apicomplexans that invade nucleated cells and because invasion by erythrocytic stages of *Plasmodium* is covered in the next chapter.

Micronemes are the smallest of the apicomplexan secretory organelles that cluster at the apical end of the zoite. The number of micronemes varies enormously between different genera, species and developmental stages with those zoites displaying vigorous and extensive gliding or migration activity generally having the most. Thus, *Theileria* zoites, which are nonmotile, do not migrate and do not display active host cell invasion, have no micronemes;[1] merozoites of *Plasmodium*, which neither glide nor migrate but rapidly and actively invade erythrocytes, have few;[2] sporozoites and merozoites of *Eimeria*, which glide, migrate through intestinal contents and actively invade enterocytes have many;[3] and *Plasmodium* ookinetes, which glide and migrate through the midgut epithelium of the mosquito, but do not classically invade host cells, also have many (and by contrast, do not have rhoptries).[4] This long-standing correlation between micronemes and parasite motility, migration and invasion is well supported by a variety of biochemical and genetic studies which show: (1) that microneme secretion is rapidly up-regulated when parasites make contact with host cells;[5] (2) that some *Plasmodium* microneme proteins are targets of erythrocytic invasion-inhibitory antibodies;[6-10] (3) that parasite invasion is blocked when microneme secretion is chemically inhibited;[11,12] and (4) that genes encoding MICs either alone, or in concert with others, are essential for effective parasite motility, migration and invasion.

MICs have been identified in a variety of approaches (reviewed in ref. 13-14), most recently through the application of proteomics to gradient-purified organelles and excreted-secreted antigens.[15,67] Figure 1 summarises the current repertoires of MICs, including only those genes for which a full sequence and a verified organellar localisation is known. The majority of MICs comprise multiple copies of a limited number of adhesive domain types, which has allowed the identification of a large number of additional putative microneme

*Corresponding Author: Vern B. Carruthers—Department of Microbiology and Immunology, University of Michigan School of Medicine, Ann Arbor, Michigan, USA 48109.
Email: vcarruth@umich.edu

Molecular Mechanisms of Parasite Invasion, edited by Barbara A. Burleigh and Dominique Soldati-Favre. ©2008 Landes Bioscience and Springer Science+Business Media.

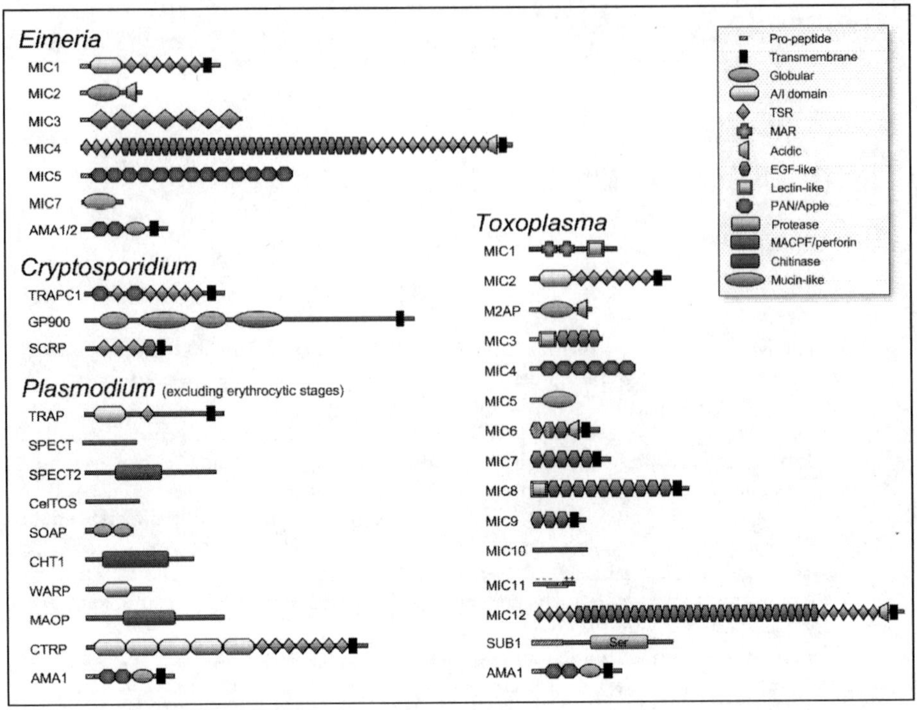

Figure1. Modular MICs. Schematic representations of known microneme proteins from four different apicomplexan genera are depicted (not to scale). Accession numbers, where available: *Eimeria tenella* MIC1, M73495; EtMIC2, Z71755; EtMIC3, AAR87667; EtMIC4, CAC34726; EtMIC5, AJ245536; *Cryptosporidium parvum* TRAP-C1, AAB92609; GP900, AAC98153; CpSCRP AF061328; *Plasmodium falciparum* TRAP, AAC1867; *Plasmodium berghei* SPECT, BAD08209; PbSPECT2, BAD83404; PbCelTOS, BAD97683; PbSOAP, AAL07530; PbCHT1, CAC40151; PbWARP, AAK83296; PbMOAP, AAV28504; PbCTRP, AAF73158; *Toxoplasma gondii* MIC1, CAA96466; TgMIC2, AAB63303; TgM2AP, AAK74070; TgMIC3, CAB56644; TgMIC4 AAD33906; TgMIC5, CAA70921; TgMIC6, AAD28185; TgMIC7, AAK35070; TgMIC8, AAK19757; TgMIC9, AAK19758; TgMIC10, AAG32024; TgMIC11, AAN16379; TgMIC12, AAK58479; TgAMA1, AF010264; TgSUB1, AAK94670. A color version of this figure is available online at www.eurekah.com.

proteins bearing these domains in the parasite databases.[16,17] Based on this it is likely that many more proteins will be shown to occupy the micronemes in future studies.

Ligand Domains and Their Cellular Receptors

Thrombospondin-1 Type 1 Domains (TSR)

Thrombospondin-1 (TSP-1) is a multifunctional, glycoprotein adhesion molecule that mediates a broad range of biological interactions via three distinct repeated domains designated types 1, 2 and 3.[18] The adhesive TSP-1 type 1 domain, TSR, is a small ~60 residue structure found in the extracellular regions of several protein families involved in immunity, cell adhesion and neuronal development, and shown to have binding activity for a number of cellular and matrix molecules (reviewed in ref. 19). The TSR is an ancient eukaryotic module that is found in many nematode and arthropod proteins as well as those from the

Apicomplexa.[20] One or more copies of the TSR are present in several apicomplexan MICs including *Plasmodium* thrombospondin-related adhesive protein (TRAP), circumsporozoite protein (CSP), and circumsporozoite-and-TRAP related protein (CTRP); *Eimeria* EtMIC1, and EtMIC4; *Cryptosporodium* TRAP-C1 and sporozoite cysteine-rich protein (SCRP) and *Toxoplasma* TgMIC2 and TgMIC12. Structures of TSR domains from TSP-1,[21] F-spondin[22] and *Plasmodium* TRAP,[23] representing two different TSR groups in respect to the organisation of their cysteine residues (Groups I and II), have been determined. Despite the different disulphide bonding patterns, these three TSRs share a highly similar elongated structure consisting of an anti-parallel, three-β-stranded fold that is additionally stabilised by stacked tryptophan and arginine residues (Fig. 2). A positively charged groove formed by the arginine stack was proposed to be the site of interaction with ligands and receptors, particularly glycosaminoglycans.[21] Several studies have shown that MICs containing TSRs bind host ligands[24-26] and recently chemical-shift mapping experiments, in which low-molecular weight heparin was titrated into ^{15}N-labelled TRAP-TSR, confirmed a site of interaction in the N-terminal half of the domain on the side of the aligned arginines.[23] Based on the structure, MICs containing TSRs would be predicted to extend out from the parasite surface after secretion from the micronemes and be thus ideally positioned to engage host surface receptors for attachment and invasion (see below).

Toxoplasma TgMIC1 and *Eimeria* EtMIC3 possess domains previously described as TSR-like, since they share some key sequence features with classical TSRs, but are now known to adopt an unrelated novel fold termed MAR (microneme adhesive repeat) (S. Matthews and D. Soldati-Favre, personal communication). The MAR domain is also able to bind host ligands[27,28] (J. Bumstead and F. Tomley, unpublished); TgMIC1 MAR binds specifically to sialic acid (S. Matthews and D. Soldati-Favre, personal communication) whereas the orthologous protein, NcMIC1, from *Neospora caninum* binds glycosaminoglycans.[29]

MIC TSRs have functions other than cell binding, for example the TSRs of TgMIC2 are implicated in its tight association with its partner protein MIC2 associated protein, TgM2AP (J Harper & V. Carruthers, unpublished) and the TSR-like domains of TgMIC1 recruit and interact with TgMIC4 in the TgMIC1-4-6 complex.[28]

Von Willebrand a Domain/ Integrin Inserted (I) Domains

The inserted (I) domain is found in the α- and β- chains of several vertebrate cell-surface integrins and is homologous to the von Willebrand A (WVA) domain, which is present in many extracellular matrix proteins. This ~200 residue A/I domain is ancient, found in proteins derived from eukaryotes, eubacteria and archaebacteria,[30] and adopts a Rossman dinucleotide binding fold consisting of five parallel and one anti-parallel β-strands that collectively are sandwiched by 7 α-helices (Fig. 2). In many A/I domains, a noncontinguous motif of amino acids is exposed on the surface of the structure to form a metal ion-dependent adhesion motif (MIDAS).[31,32] Although the MIDAS is crucial for binding in some cases, such as in type VI collagen dimerisation,[33] it appears irrelevant in others, such as in binding of the third VWA domain of von Willebrand Factor to fibrillar collagen.[34] Several TSR-containing apicomplexan MICs possess one or more A/I domain, and the MIDAS sequence is generally well conserved in these. Experimental studies have shown that the function of these apicomplexan A/I domains may be mediated by both MIDAS-dependent and independent mechanisms. Thus, mutations in the MIDAS of TRAP of *Plasmodium berghei* affect parasite invasion activity,[35] and binding of this domain to hepatocytes and to fetuin is MIDAS dependent;[36] however, binding to glycosaminoglycans is not mediated by MIDAS.[37] Similarly, the A/I domain of *Toxoplasma* TgMIC2 binds heparin in a MIDAS-independent manner.[38] Interestingly, exhaustive searching of the databases of *Cryptosporidium* has failed to identify any proteins containing A/I domains in this member of the phylum (T. Templeton, personal communication).

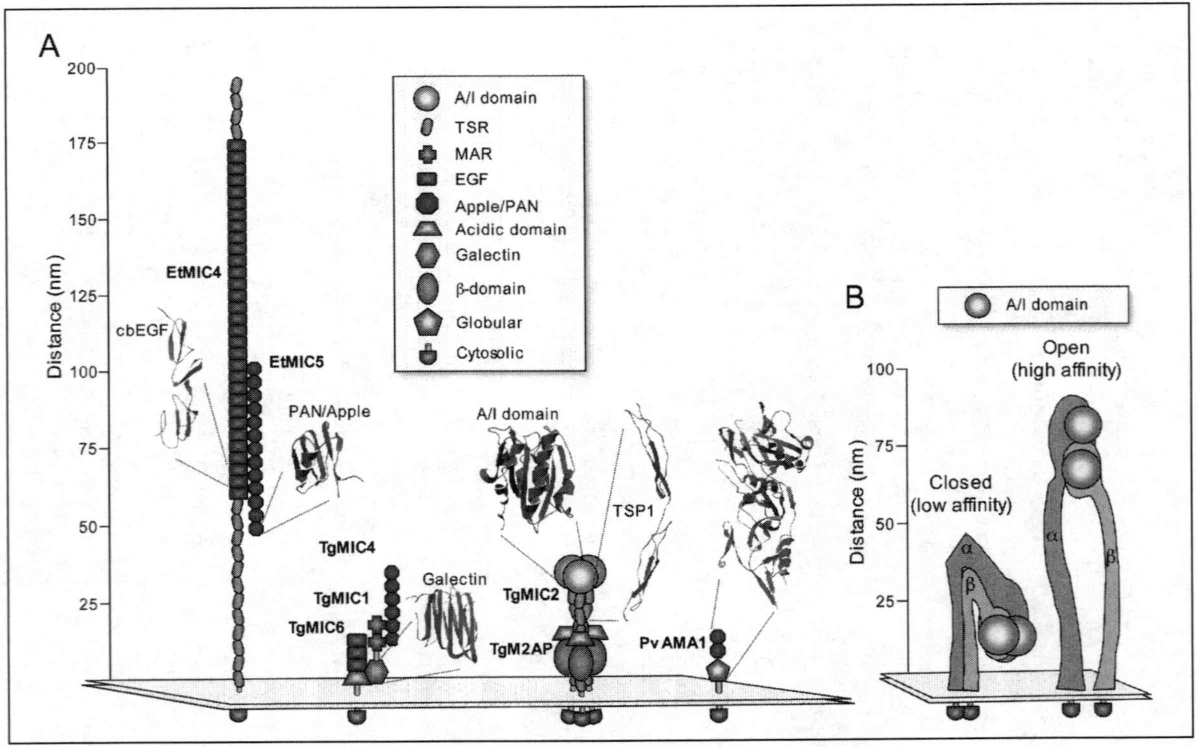

Figure 2. Structural features of micronemal ligands. (A) MICs from *Eimeria* (EtMIC4-MIC5), *Toxoplasma* (TgMIC1-4-6, TgMIC2-M2AP) and *Plasmodium* (PvAMA1) are schematically depicted with a scale (left) illustrating the approximate distance they would project from the parasite membrane, assuming a fully elongated state. Projection estimates are based on the dimensions of the domain types shown as ribbon structures, with α-helices (red), β-sheets (blue), and random coils or turns (grey). Structures are based on the following: cbEGF from fibrillin (Protein Data Bank accession number 1EMO), PAN/Apple from EtMIC5 (1HKY), A/I domain from integrin α_L (1MQA), TSR from thrombospondin (1LSL), ectodomain of PvAMA1 (1W8K). (B) Model based on of an integrin heterodimer in the closed (low affinity) and the open (high affinity) configuration.[80] A color version of this figure is available online at www.eurekah.com.

Apple/PAN Domains

Apple domains, which are a subset of the plasminogen, apple, nematode (PAN) superfamily, have been identified in plasminogen-related proteins such as coagulation factor XI, plasma prekallikrein, hepatocyte growth factor, macrophage stimulation factor and also in several nematode proteins. Apple/PAN domains have three conserved disulphide bridges that are essential for their tertiary structure, but homology in the primary amino acid sequence between domains is generally low, which may contribute to their very different and highly specific ligand binding properties. For example the four Apple/PAN domains from compliment factor XI (FXI) display very different ligand specificities: A1 binds the FXI cofactor H-kininogen and thrombin,[39,40] A2 binds the FXI substrate, FIX,[41] A3 also binds FIX and heparin[42,43] and A4 binds FXIIa.[44] Apple/PAN domains are present in several apicomplexan MIC proteins including EtMIC5, TgMIC4 and TRAP-CI and the solution structure of a single (A9) domain from *Eimeria tenella* confirmed its structural homology to the Apple/PAN superfamily[45] (Fig. 2). Very recently the crystal structures of apical membrane antigens (AMA1) of *Plasmodium* species revealed that the two most N-terminal domains of these MICs are also highly divergent members of the Apple/PAN superfamily[46,47] (Fig. 2). Most functional information on apicomplexan Apple/PAN domains has come from the study of TgMIC4, which contains 6 tandem domains and which exists as a structural heterocomplex with TgMIC1 and TgMIC6.[48,49] The first two Apple domains of TgMIC4 interact directly with the twin MAR domains of TgMIC1 and in the absence of TgMIC1 binding of TgMIC4 to host cells is almost entirely ablated;[28] however, is not known whether this is due to incorrect folding under these conditions or to the inherent lack of cell binding properties of TgMIC4. Interestingly NcMIC4, an orthologue of TgMIC4 in the closely related parasite *Neospora caninum*, is able to bind lactose, a property that is not shared by TgMIC4, which does not bind.[50] The function and binding properties of the newly defined Apple/PAN domains of AMA1 are not well defined, although recent data from *Toxoplasma* indicates that TgAMA1 cooperates with rhoptry neck proteins in the formation and maintenance of the moving junction during host cell invasion.[51,52]

EGF-Like Domains

EGF-like domains are widely distributed in membrane-bound and extracellular eukaryotic proteins and are involved in many different and diverse biological functions including blood coagulation, cell signalling, cell migration and maintenance of extracellular matrix architecture. These domains typically consist of ~50 amino acids with three conserved disulphide bridges and a subclass of EGF-like domains that bind calcium (cbEGFs) has been identified that have a conserved D/N-x-D/N-E/Q-xm-D/N*-xn-Y/F motif, (where m and n are variable and * indicates β-hydroxylation.[53] The first apicomplexan proteins containing EGF-like domains to be identified were GPI-linked proteins from *Plasmodium*[54,55] but more recently a number of apicomplexan MICs with EGF-like domains have been studied including SCRP, EtMIC4, and TgMICs 3, 6, 7, 8, 9, and 12. So far no EGF-like containing MICs have been found in *Plasmodium* species. The majority of the domains described in apicomplexan MICs are regular EGF-like, but the cbEGF motif is present in 22 of the 31 EGF-like domains of EtMIC4[56] and its homologue TgMIC12 (ToxoDB 57.m01872; F. Stavru and D. Soldati-Favre, personal communication). Study of cbEGFs in EtMIC4 has shown that in the presence of calcium these domains adopt a proteinase-resistant, extended structure that would favour the interaction of the N-terminal portion of the molecule with host cell ligands.[57] Interestingly, EtMIC4 forms a stable very high molecular mass heteromeric complex with the soluble Apple/PAN domain containing protein EtMIC5 although the precise sites of interaction between these two MICs are not yet mapped (J. Periz & F. Tomley, unpublished). In *Toxoplasma*, TgMIC3 contains both EGF-like and lectin-like domains and binds to all nucleated cells tested as well as to the tachyzoite surface.[58] The receptor-binding properties of TgMIC3 are attributed to the lectin-like domain, whereas the EGF-like domains are proposed to promote proper folding of the protein in order to expose the binding regions. In addition, they may be involved in heteromeric polymerisation with the transmembrane MIC TgMIC8, which contains 10 EGF-like domains and which functions as an 'escorter' to ensure delivery of TgMIC3 to the micronemes. Similarly, in the TgMIC1-4-6 adhesive complex it is the transmembrane, EGF-like domain containing TgMIC6 that is responsible for targeting to the micronemes but in this case oligomerisation is promoted and stabilised by the interaction of the third EGF of TgMIC6 domain, together with its downstream acidic region, with the galectin domain of TgMIC1.[28]

Lectin Domains

Two types of domains related to lectins have been identified within apicomplexan MICs. Chitin-binding like (CBL) domains are found in a variety of plant lectins including plant defensins that have anti-fungal chitinase activity. CBLs are typically composed of ~30-43 amino acids with four conserved disulphide bridges and several conserved aromatic residues that mediate binding of the domain to N-acetyl glucosamine.[59] CBLs with lectin (or agglutinin) properties are able to bind and cross link GlcNAc-containing polymers and in *Toxoplasma*, TgMIC3 and TgMIC8 each contain a single CBL-domain at their N-termini, followed by several EGF-like domains.[58,60] Binding of the CBL-domain of TgMIC3 to host cell surfaces is dependent upon its dimerisation, which is mediated by the interaction of the C-terminal regions of each monomer,[61] and disruption of the CBL aromatic residues presumed to be important for binding results in lowered parasite virulence.[62] Fusion of the TgMIC3 dimerisation domain to the extracellular domain of TgMIC8 promotes dimerisation and binding of the chimera, indicating that the CBL of TgMIC8 also possesses binding activity when in a dimeric form.[62]

Another lectin-related domain in an apicomplexan MIC was recently identified from the three-dimensional structure of the C-terminal domain of *Toxoplasma* TgMIC1.[28] This domain has a galectin-like fold, which consists of a β-barrel formed by the association of two multi-stranded β-sheets. Galectins are soluble, calcium-independent, carbohydrate-binding animal lectins, however the critical side chains that mediate lectin activity are not conserved in the TgMIC1 galectin domain and no detectable binding to a range of carbohydrate substrates was observed in NMR chemical shift mapping experiments. Instead, the TgMIC1 galectin domain displays a large hydrophobic surface reminiscent of the protein-protein interaction domains seen in bacterial class I chaperones of the type three secretion system and both NMR and biochemical studies indicate that during the biogenesis of the TgMIC1-4-6 adhesive complex, this domain recruits and stabilises TgMIC6 providing a highly specific quality control mechanism for the exit of TgMIC6 from the ER/Golgi and for subsequent trafficking of the adhesive complex to the micronemes.

Adhesive Complexes: Assembly and Organization

Propensity to Form Oligomers

Adhesive proteins often form oligomeric complexes with themselves or other proteins that contribute to adhesion or serve a regulatory function. For example, cadherins are a family of vertebrate adhesive proteins expressed as homodimers that strengthen cell-cell junctions. Integrins are heterodimeric, transmembrane glycoproteins primarily responsible for mediating cell interactions with extracellular matrix (ECM). The propensity to form oligomeric adhesive complexes has been demonstrated in several apicomplexans, although most of the mechanistic studies have been done in *Toxoplasma*. Oligomerization bestows adhesive proteins with several important advantages.

First, oligomerization can promote the proper folding of proteins in a complex, as recently shown for the TgMIC1-4-6 complex.[28] TgMIC1 is a soluble protein that simultaneously associates with TgMIC4 through its two TSR-like MAR domains and with the transmembrane escorter protein TgMIC6 through its C-terminal (CT) galectin-like domain. As mentioned above, NMR spectroscopy revealed that the CT domain is incapable of binding sugars but instead forms an interface with the third EGF-like domain of TgMIC6, which also contains an acidic element (TgMIC6-EGF3acid).[28] When mixed together and monitored by NMR, the TgMIC1 CT domain facilitated the folding and stabilization of the TgMIC6-EGF3acid. The TgMIC1 CT domain also rescued the secretory retention phenotype of TgMIC6 in mic1KO parasites, presumably by navigating through the quality control system that recognizes misfolded proteins. These findings reveal new molecular insights into the interdependence of adhesive proteins for correct folding and movement through the secretory pathway.

Second, assembly into protein complexes allows cooperation in trafficking to the micronemes. MIC complexes typically have one transmembrane (TM) protein. These TM MICs are also referred to as escorters since they accompany and guide the other soluble members to the micronemes based on the targeting signals in their C-terminal tails.[49,60] In TgMIC2, this signal is provided by two tyrosine-based sorting motifs capable of directing a heterologous protein to the micronemes.[63] Genetic disruption of any of the TM MICs results in retention of the other members of the complex along the secretory system or in mistargeting to the default secretory pathway, which in *T. gondii* is secretion via the dense granules. When the level of TgMIC2 expression is experimentally reduced, TgM2AP colocalizes with the dense granules and is secreted into the PV.[64] Similarly, TgMIC6 knockout parasites show a complete misrouting of TgMIC1 and TgMIC4 to the dense granules.[49] Nonetheless, escorters still depend on their cargo for proper trafficking since soluble proteins in the complexes are required for protein folding, as is the function of the galectin-like domain of TgMIC1,[28] or necessary for exiting an endosomal compartment associated with microneme biogenesis, as shown for the TgM2AP propeptide.[28a]

Third, different combinations of partners can expand the receptor repertoire and/or fine-tune the specificity of receptor binding. Humans express eighteen integrin α-subunits and eight β-subunits that form 24 heterodimers for recognition of distinct but overlapping receptors.[65,66] Although there are no firm examples of subunit mixing in the apicomplexa, these parasites often express paralogous families of adhesive proteins with the potential to participate in such a phenomenon. Four closely related putative adhesins were recently identified in a proteomic screen of *Toxoplasma* secretory proteins.[67] These proteins have four Apple/PAN domains but no predicted anchoring sequence, and, by analogy with TgMIC4 and its association with TgMIC1 and TgMIC6, they likely oligomerize with a TM protein, possibly in a manner that would expand their receptor binding capabilities. Three additional genes coding for proteins closely related to TgMIC1 are also present in the *Toxoplasma* genome (D. Soldati-Favre, personal communication).

Fourth, oligomerization allows proteins from distinct compartments to facilitate invasion collaboratively. Two studies[51,52] have recently shown that the microneme protein TgAMA1 oligomerizes with three proteins derived from the rhoptry neck: TgRON2, TgRON4, and TgRON5. Although they are discharged from different organelles during invasion, TgAMA1 and TgRON2/4/5 form an oligomeric complex on the parasite surface within the moving junction, a ring-like constriction that slides over the parasite as it penetrates the host cell. TgAMA1 is a key component of the complex since depletion of this protein causes a failure to form the moving junction and parasite invasion is arrested at the stage of apical attachment.[68] Since TgRON4 is predicted to be an integral membrane protein, this raises the hypothesis that it inserts into the host plasma membrane and acts as an autologous receptor for cell invasion.[51] In this case, oligomerization would allow the parasite to use its own receptor to support invasion of the many cell types susceptible to *Toxoplasma* invasion.

Finally, oligomerization increases valency and avidity, thereby enhancing the formation of a robust binding interface. For example, TgMIC2-M2AP is a heterohexameric complex consisting of a trimer of dimers.[38,69] The corresponding complex in *Eimeria tenella*, EtMIC1-MIC2, presumably also forms a similar hexameric assembly. Such an arrangement could promote tight binding to a complementary oligomeric receptor on the host cell surface, thereby allowing the parasite to grip sufficiently well to power its way into the target cell.

Ligand Organization in Micronemes and on Parasite Surface

It is not known precisely how adhesive ligands are organized within micronemes. However, several features suggest that ligands are packaged in an orderly fashion. First, the contents of *Cryptosporidium* micronemes are arranged in an array of 15 nm cubic crystals framing a pine-cone-like pattern.[70] Although this crystalline appearance is unique to *Cryptosporidium*, micronemes of other apicomplexa are electron dense, implying a high protein concentration. Second, since a number of micronemes are discharged in rapid succession, a strong measure of organization is presumably necessary to achieve efficient deployment. Finally, the internal dimensions of micronemes (~75 nm x 150 nm) might not accommodate some of the larger microneme proteins (e.g., EtMIC4-MIC5) in their fully extended state (see also below) and therefore these proteins are likely packaged in orderly fashion so that they are primed for secretion onto the parasite surface.

It has been proposed that some microneme proteins are involved in organizing the organellar contents. For example, TgMIC10 and TgMIC11 are small, soluble microneme proteins that display a marked charge asymmetry, which may promote electrostatic assembly into higher ordered structures.[71,72] Unlike most other microneme proteins, TgMIC10 and TgMIC11 do not associate with the parasite surface during invasion, consistent with an alternative role independent of adhesion. During transport to the micronemes, TgMIC11 is proteolytically processed to remove an internal propeptide in a manner reminiscent of insulin maturation within nascent secretory granules of pancreatic beta cells. Insulin processing is thought to promote its ordered packaging and retention in maturing secretory granules,[73,74] although this idea is somewhat controversial.[75]

During gliding and invasion the microneme contents are deployed onto the parasite's apical surface where substrate or receptor engagement occurs. Adhesive complexes are not randomly distributed. For example, TRAP is arranged in a cap or ring-like pattern on gliding *Plasmodium* sporozoites.[76] Also, the EtMIC4-MIC5 complex displays a punctate pattern on the surface of invading *Eimeria* sporozoites[77] in a manner similar to TgMIC2-M2AP during *Toxoplasma* tachyzoite invasion.[78] Invading zoites display a particularly high density of ligands at the external boundary of the moving junction. The organization of ligands in this adhesion zone may further promote multivalent, high avidity interactions with host receptors, especially if the receptors have a complementary clustering distribution. Clustering may therefore be an additional level of organization that further promotes the creation of a robust binding interface between the parasite and host cell membranes.

The Surface Ligand Landscape: Does Size (and Conformation) Matter?

Crystal structure analysis of several domain types found in apicomplexan microneme proteins is beginning to reveal both the approximate size and shape of these important ligands. For example, cbEGF domains form an elongated structure that is stabilized by interdomain Ca^{2+} binding and hydrophobic interactions between adjacent domains.[53] Since the majority of EGF domains in the extracellular portion of EtMIC4 are of the cbEGF type, EtMIC4 is predicted to adopt a highly extended conformation that could project nearly 200nm from the parasite surface. However, it is unlikely that this structure is completely rigid there is greater flexibility between noncalcium binding EGF domains.[57] This semi-rigid conformation may allow the molecule to project maximally from the membrane while still retaining some degree of flexibility to "survey" the host cell surface for receptors. The ninth Apple/PAN domain of EtMIC5 adopts a globular α/β structure with the N- and C-termini situated on the same side of the molecule.[45] Although for EtMIC5 the structure of only one domain was solved, *Plasmodium* AMA1 has two PAN/Apple domains that are stacked upon one another,[47] suggesting that EtMIC5 and other multi-PAN/Apple domain containing microneme proteins may also adopt an elongated structure that projects away from the parasite surface. Based on the crystal structures of the A/I domains from various integrins and a pair of TSR domains from thrombospondin, TRAP family members including TgMIC2 are predicted to form a "ball-on-a-stick" type of structure that could extend up to 40 nm from the parasite surface. Six tandem TSR domains that form a highly elongated stalk provide most of the molecule's height. The trimeric arrangement likely imparts a high degree of rigidity and strength in the molecule, which may be important to form a solid connection between extracellular receptors and the parasite's intracellular motility system.

For mammalian cell adhesion, recent studies have also provided new insight into role of conformational shifting in modulating ligand affinity. Molecular electron microscopy of the integrin $\alpha_5\beta_1$ showed that it undergoes a dramatic conformational shift from a "closed" to "open" configuration upon activation by inside-out signaling and/or exposure to certain divalent cations.[79,80] As shown in Figure 2, in the closed, low affinity position the heterodimer is bent over with the paired A/I domans positioned proximal to the cell membrane. However, when Mn^{2+} binds to the MIDAS site the complex "stands up" to project the adhesive A/I domains 2-3 times further away from the cell membrane. Although no direct evidence is available, similarly dramatic conformational changes could occur in micronemal ligands. For example, if EtMIC4 is not exposed to high concentrations of Ca^{2+} during transport and packaging in the micronemes then it would be sufficiently flexible and compact to fit within the microneme lumen. However, upon secretion and exposure to millimolar concentrations of Ca^{2+} in the extracellular milieu, EtMIC4 might unfurl to attain maximum height for long-range interactions with host receptors in the initial apical docking of the parasite. Other large microneme proteins in *Cryptosporidium* (CpGP900) and *Toxoplasma* (TgMIC12) may play a similar role. In this manner the parasite could establish an initial connection between its apical pole and the host surface before using other perhaps higher affinity or more abundant micronemal ligands to strengthen the grip for active penetration.

Role of Micronemal Proteins in Migration across Biological Barriers

For *Plasmodium*, the mosquito midgut and the sinusoidal layer of the liver are two significant biological barriers against infection and cell migration activity is needed for the zoites to breach these barriers.

Ookinetes of *Plasmodium* are highly motile and they migrate through the midgut epithelium of the mosquito causing massive destruction. The microneme proteins CTRP, SOAP (soluble ookinete adhesive protein), MAOP (membrane attack ookinete protein) and CelTOS (cell-traversal protein for ookinetes and sporozoites) have been shown to play crucial roles. CTRP is essential for apical attachment to the midgut epithelial cell,[81] SOAP is involved in mosquito midgut invasion and oocyst development,[82] MAOP which has a MACPF domain is necessary for ookinetes to breach the apical plasma membrane of the epithelial cell[83] and CelTOS is needed for the ookinetes to migrate through the cell cytoplasm to reach the basal lamina where oocyst development occurs.[84]

Sporozoites of *Plasmodium* are able to glide, migrate and invade host cells. Entry of the sporozoite into the hepatocyte is controversial and has been reported to occur following direct parasite migration through cells and by 'classical' invasion, vacuole formation and egress. It has been suggested that sporozoite migration through hepatocytes has an effect on subsequent sporozoite infectivity for new hepatocytes and on permissiveness of surrounding hepatocytes (via release of hepatocyte growth factor, HGF).[85,86] However, gene-targeting experiments on sporozoite microneme proteins contradict this—SPECT (Sporozoite-protein-essential-for-cell-traversal) disrupted sporozoites are deficient in cell migration yet they show normal cell invasion and gliding motility.[87] This indicates that cell migration is not an absolute requirement for cell invasion, although it is clearly important in vivo since disruption of SPECT decreases liver infectivity ~20-fold. This decrease in infectivity was reversed by depletion of Kupfer cells that line the liver sinusoids, leading to the conclusion that the cell migration activity mediated by SPECT is required to cross the liver sinusoidal barrier.[87] Two other sporozoite MICs are implicated in liver invasion. SPECT2 contains a membrane attack complex/perforin domain and disruptants show the same phenotype as SPECT disruptants, thus SPECT2 is presumed also to be necessary for sporozoite traversal of the liver sinusoid.[88] CelTOS is expressed in both ookinetes and sporozoites and again disruption of the gene gives essentially the same phenotype as SPECT and SPECT2 except that the disruptants maintain a low level of cell migration (cell wounding) activity.[84] It is unclear whether the proteins involved in cell migration function by binding specific receptors on the host cell surface or within the cytoplasm, or whether they function in a regulatory or sensory role (for more details see chapter IX, Frevert et al).

While *Toxoplasma* tachyzoites have not been reported to migrate through cells, recent studies suggest that they cross biological barriers by a paracellular route i.e., between host cells, using ICAM1 as a receptor.[89] TgMIC2 was shown to bind ICAM, but only upon proteolytic removal of a short N-terminal extension that preceeds the A/I domain. This proteolytic trimming phenomenon, mediated by a hypothetical surface protease called MPP2, constitutes another means of regulating adhesive activity associated with parasite migration and possibly also attachment.

Summary

Microneme secretion supports several key cellular processes including gliding motility, active cell invasion and migration through cells, biological barriers, and tissues. The modular design of microneme proteins enables these molecules to assist each other in folding and passage through the quality control system, accurately target to the micronemes, oligimerizing with other parasite proteins, and engaging a variety of host receptors for migration and cell invasion. Structural and biochemical analyses of MIC domains is providing new perspectives on how adhesion is regulated and the potentially distinct roles MICs might play in long or short range interactions during parasite attachment and entry. New access to complete genome sequences and ongoing advances in genetic manipulation should provide fertile ground for refining current models and defining exciting new roles for MICs in apicomplexan biology.

References

1. Shaw MK. The same but different: The biology of Theileria sporozoite entry into bovine cells. Int J Parasitol 1997; 27(5):457-474.
2. Langreth SG, Jensen JB, Reese RT et al. Fine structure of human malaria in vitro. J Protozool 1978; 25(4):443-452.
3. Scholtyseck E, Mehlhorn H. Ultrastructural study of characteristic organelles (paired organelles, micronemes, micropores) of sporozoa and related organisms. Z Parasitenkd 1970; 34(2):97-127.
4. Sinden RE, Hartley RH, Winger L. The development of Plasmodium ookinetes in vitro: An ultrastructural study including a description of meiotic division. Parasitology 1985; 91(Pt 2):227-244.
5. Carruthers VB, Sibley LD. Sequential protein secretion from three distinct organelles of Toxoplasma gondii accompanies invasion of human fibroblasts. Eur J Cell Biol 1997; 73(2):114-123.
6. Healer J, Murphy V, Hodder AN et al. Allelic polymorphisms in apical membrane antigen-1 are responsible for evasion of antibody-mediated inhibition in Plasmodium falciparum. Mol Microbiol 2004; 52(1):159-168.
7. Miller LH, Hudson D, Haynes JD. Identification of Plasmodium knowlesi erythrocyte binding proteins. Mol Biochem Parasitol 1988; 31(3):217-222.
8. Narum DL, Haynes JD, Fuhrmann S et al. Antibodies against the Plasmodium falciparum receptor binding domain of EBA-175 block invasion pathways that do not involve sialic acids. Infect Immun 2000; 68(4):1964-1966.
9. Sim BKL, Orlandi PA, Haynes JH et al. Primary structure of the 175K Plasmodium falciparum erythrocyte binding antigen and identification of a peptide which elicits antibodies that inhibit malaria merozoite invasion. J Cell Biol 1990; 111:1877-1884.
10. Singh AP, Puri SK, Chitnis CE. Antibodies raised against receptor-binding domain of Plasmodium knowlesi Duffy binding protein inhibit erythrocyte invasion. Mol Biochem Parasitol 2002; 121(1):21-31.
11. Carruthers VB, Giddings OK, Sibley LD. Secretion of micronemal proteins is associated with Toxoplasma invasion of host cells. Cell Microbiol 1999; 1(3):225-235.
12. Wiersma HI, Galuska SE, Tomley FM et al. A role for coccidian cGMP-dependent protein kinase in motility and invasion. Int J Parasitol 2004; 34(3):369-380.

13. Soldati D, Dubremetz JF, Lebrun M. Microneme proteins: Structural and functional requirements to promote adhesion and invasion by the apicomplexan parasite Toxoplasma gondii. Int J Parasitol 2001; 31(12):1293-1302.
14. Tomley FM, Soldati DS. Mix and match modules: Structure and function of microneme proteins in apicomplexan parasites. Trends Parasitol 2001; 17(2):81-88.
15. Bromley E, Leeds N, Clark J et al. Defining the protein repertoire of microneme secretory organelles in the apicomplexan parasite Eimeria tenella. Proteomics 2003; 3(8):1553-1561.
16. Deng M, Templeton TJ, London NR et al. Cryptosporidium parvum genes containing thrombospondin type 1 domains. Infect Immun 2002; 70(12):6987-6995.
17. Templeton TJ, Lancto CA, Vigdorovich V et al. The Cryptosporidium oocyst wall protein is a member of a multigene family and has a homolog in Toxoplasma. Infect Immun 2004; 72(2):980-987.
18. Lawler J, Hynes RO. The structure of human thrombospondin, an adhesive glycoprotein with multiple calcium-binding sites and homologies with several different proteins. J Cell Biol 1986; 103(5):1635-1648.
19. Adams JC, Tucker RP. The thrombospondin type 1 repeat (TSR) superfamily: Diverse proteins with related roles in neuronal development. Dev Dyn 2000; 218(2):280-299.
20. Tucker RP. The thrombospondin type 1 repeat superfamily. Int J Biochem Cell Biol 2004; 36(6):969-974.
21. Tan KM, Duquette M, Liu JH et al. Crystal structure of the TSP-1 type 1 repeats: A novel layered fold and its biological implication. J Cell Biol 2002; 159(2):373-382.
22. Paakkonen K, Tossavainen H, Permi P et al. Solution structures of the first and fourth TSR domains of F-spondin. Proteins 2006; 64(3):665-72.
23. Tossavainen H, Pihajamaa T, Huttunen TK et al. The layered fold of the TSR domain of P. falciparum TRAP contains a heparin binding site. Protein Sci 2006; 15:1760-1768.
24. Robson KJ, Frevert U, Reckmann I et al. Thrombospondin-related adhesive protein (TRAP) of Plasmodium falciparum: Expression during sporozoite ontogeny and binding to human hepatocytes. EMBO J 1995; 14(16):3883-3894.
25. Spaccapelo R, Naitza S, Robson KJ et al. Thrombospondin-related adhesive protein (TRAP) of Plasmodium berghei and parasite motility. Lancet 1997; 350(9074):335.
26. Sultan AA, Thathy V, Frevert U et al. TRAP is necessary for gliding motility and infectivity of Plasmodium sporozoites. Cell 1997; 90(3):511-522.
27. Fourmaux MN, Achbarou A, Mercereau-Puijalon O et al. The MIC1 microneme protein of Toxoplasma gondii contains a duplicated receptor-like domain and binds to host cell surface. Mol Biochem Parasitol 1996; 83(2):201-210.
28. Saouros S, Edwards-Jones B, Reiss M et al. A novel galectin-like domain from Toxoplasma gondii micronemal protein 1 assists the folding, assembly, and transport of a cell adhesion complex. J Biol Chem 2005; 280(46):38583-38591.
28a. Harper JM, Huynh MH, Coppens I et al. A cleavable propeptide influences Toxoplasma infection by facilitating the trafficking and secretion of the TgMIC2-M2AP invasion complex. Mol Biol Cell 2006; 17(10):4551-4563.
29. Keller N, Naguleswaran A, Cannas A et al. Identification of a Neospora caninum microneme protein (NcMIC1) which interacts with sulfated host cell surface glycosaminoglycans. Infect Immun 2002; 70(6):3187-3198.
30. Tuckwell D. Evolution of von Willebrand factor A (VWA) domains. Biochem Soc Trans 1999; 27:835-840.
31. Humphries M, Liddington R. Molecular basis of integrin-dependent cell adhesion. In: Kleanthous C, ed. Protein-Protein Recognition. Oxford: Oxford University Press, 2000.
32. Whittaker CA, Hynes RO. Distribution and evolution of von Willebrand/integrin a domains: Widely dispersed adhesion and elsewhere. Mol Biol Cell 2002; 13(10):3369-3387.
33. Ball S, Bella J, Kielty C et al. Structural basis of type VI collagen dimer formation. J Biol Chem 2003; 278(17):15326-15332.
34. Romijn RAP, Bouma B, Wuyster W et al. Identification of the collagen-binding site of the von Willebrand factor A3-domain. J Biol Chem 2001; 276(13):9985-9991.
35. Matuschewski K, Nunes AC, Nussenzweig V et al. Plasmodium sporozoite invasion into insect and mammalian cells is directed by the same dual binding system. EMBO J 2002; 21(7):1597-1606.
36. Jethwaney D, Lepore T, Hassan S et al. Fetuin-A, a hepatocyte-specific protein that binds Plasmodium berghei thrombospondin-related adhesive protein: A potential role in infectivity. Infect Immun 2005; 73(9):5883-5891.

37. McCormick CJ, Tuckwell DS, Crisanti A et al. Identification of heparin as a ligand for the A-domain of Plasmodium falciparum thrombospondin-related adhesion protein. Mol Biochem Parasitol 1999; 100(1):111-124.
38. Harper JM, Hoff EF, Carruthers VB. Multimerization of the Toxoplasma gondii MIC2 integrin-like A-domain is required for binding to heparin and human cells. Mol Biochem Parasitol 2004; 134(2):201-212.
39. Baglia FA, Jameson BA, Walsh PN. Localization of the high molecular weight kininogen binding site in the heavy chain of human factor XI to amino acids phenylalanine 56 through serine 86. J Biol Chem 1990; 265(7):4149-4154.
40. Baglia FA, Walsh PN. A binding site for thrombin in the apple 1 domain of factor XI. J Biol Chem 1996; 271(7):3652-3658.
41. Baglia FA, Jameson BA, Walsh PN. Identification and chemical synthesis of a substrate-binding site for factor IX on coagulation factor XIa. J Biol Chem 1991; 266(35):24190-24197.
42. Ho DH, Badellino K, Baglia FA et al. The role of high molecular weight kininogen and prothrombin as cofactors in the binding of factor XI A3 domain to the platelet surface. J Biol Chem 2000.
43. Sun Y, Gailani D. Identification of a factor IX binding site on the third apple domain of activated factor XI. J Biol Chem 1996; 271(46):29023-29028.
44. Baglia FA, Jameson BA, Walsh PN. Identification and characterization of a binding site for factor XIIa in the Apple 4 domain of coagulation factor XI. J Biol Chem 1993; 268(6):3838-3844.
45. Brown PJ, Mulvey D, Potts JR et al. Solution structure of a PAN module from the apicomplexan parasite Eimeria tenella. J Struct Funct Genomics 2003; 4(4):227-234.
46. Bai T, Becker M, Gupta A et al. Structure of AMA1 from Plasmodium falciparum reveals a clustering of polymorphisms that surround a conserved hydrophobic pocket. Proc Natl Acad Sci USA 2005; 102(36):12736-12741.
47. Pizarro JC, Vulliez-Le Normand B, Chesne-Seck ML et al. Crystal structure of the malaria vaccine candidate apical membrane antigen 1. Science 2005; 308(5720):408-411.
48. Brecht S, Carruthers VB, Ferguson DJP et al. The Toxoplasma micronemal protein MIC4 is an adhesin composed of six conserved apple domains. J Biol Chem 2001; 276(6):4119-4127.
49. Reiss M, Viebig N, Brecht S et al. Identification and characterization of an escorter for two secretory adhesins in Toxoplasma gondii. J Cell Biol 2001; 152(3):563-578.
50. Keller N, Riesen M, Naguleswaran A et al. Identification and characterization of a Neospora caninum microneme-associated protein (NcMIC4) that exhibits unique lactose-binding properties. Infect Immun 2004; 72(8):4791-4800.
51. Alexander PL, Mital J, Ward GE et al. Identification of the moving junction complex of Toxoplasma gondii: A collaboration between distinct secretory organelles. PloS Pathogens 2005; 1e17.
52. Lebrun M, Michelin A, El Hajj H et al. The rhoptry neck protein RON4 relocalizes at the moving junction during Toxoplasma gondii invasion. Cell Microbiol 2005; 7(12):1823-1833.
53. Downing AK, Knott V, Werner JM et al. Solution structure of a pair of calcium-binding epidermal growth factor- like domains: Implications for the Marfan syndrome and other genetic disorders. Cell 1996; 85(4):597-605.
54. Blackman MJ, Whittle H, Holder AA. Processing of the Plasmodium falciparum major merozoite surface protein-1: Identification of a 33-kilodalton secondary processing product which is shed prior to erythrocyte invasion. Mol Biochem Parasitol 1991; 49(1):35-44.
55. Kaslow DC, Quakyi IA, Syin C et al. A vaccine candidate from the sexual stage of human malaria that contains EGF-like domains. Nature 1988; 333(6168):74-76.
56. Tomley FM, Billington KJ, Bumstead JM et al. EtMIC4: A microneme protein from Eimeria tenella that contains tandem arrays of epidermal growth factor-like repeats and thrombospondin type-I repeats. Int J Parasitol 2001; 31(12):1303-1310.
57. Periz J, Gill AC, Knott V et al. Calcium binding activity of the epidermal growth factor-like domains of the apicomplexan microneme protein EtMIC4. Mol Biochem Parasitol 2005; 143(2):192-199.
58. Garcia-Reguet N, Lebrun M, Fourmaux MN et al. The microneme protein MIC3 of Toxoplasma gondii is a secretory adhesin that binds to both the surface of the host cells and the surface of the parasite. Cell Microbiol 2000; 2(4):353-364.
59. Wright HT, Sandrasegaram G, Wright CS. Evolution of a family of N-acetylglucosamine binding-proteins containing the disulfide-rich domain of wheat-germ-agglutinin. J Mol Evol 1991; 33(3):283-294.
60. Meissner M, Reiss M, Viebig N et al. A family of transmembrane microneme proteins of Toxoplasma gondii contain EGF-like domains and function as escorters. J Cell Sci 2002; 115(Pt 3):563-574.
61. Cerede O, Dubremetz JF, Bout D et al. The Toxoplasma gondii protein MIC3 requires pro-peptide cleavage and dimerization to function as adhesin. EMBO J 2002; 21(11):2526-2536.

62. Cerede O, Dubremetz JF, Soete M et al. Synergistic role of micronemal proteins in Toxoplasma gondii virulence. J Exp Med 2005; 201(3):453-463.
63. Di Cristina M, Spaccapelo R, Soldati D et al. Two conserved amino acid motifs mediate protein targeting to the micronemes of the apicomplexan parasite Toxoplasma gondii. Mol Cell Biol 2000; 20(19):7332-7341.
64. Huynh MH, Carruthers VB. Toxoplasma MIC2 is a major determinant of invasion and virulence. PLoS Pathogens 2006; 2(8), [epub ahead of print].
65. Arnaout MA, Goodman SL, Xiong JP. Coming to grips with integrin binding to ligands. Curr Opin Cell Biol 2002; 14(5):641-651.
66. Sonnenberg A. Integrins and their ligands. Curr Top Microbiol Immunol 1993; 184:7-35.
67. Zhou XW, Kafsack BFC, Cole RN et al. The opportunistic pathogen Toxoplasma gondii deploys a diverse legion of invasion and survival proteins. J Biol Chem 2005; 280(40):34233-34244.
68. Mital J, Meissner M, Soldati D et al. Conditional expression of Toxoplasma gondii apical membrane antigen-1 (TgAMA1) demonstrates that TgAMA1 plays a critical role in host cell invasion. Mol Biol Cell 2005; 16(9):4341-4349.
69. Jewett TJ, Sibley LD. The Toxoplasma proteins MIC2 and M2AP form a hexameric complex necessary for intracellular survival. J Biol Chem 2004; 279(10):9362-9369.
70. Petry F, Harris JR. Ultrastructure, fractionation and biochemical analysis of Cryptosporidium parvum sporozoites. Int J Parasitol 1999; 29(8):1249-1260.
71. Harper JM, Zhou XW, Pszenny V et al. The novel coccidian micronemal protein MIC11 undergoes proteolytic maturation by sequential cleavage to remove an internal propeptide. Int J Parasitol 2004; 34(9):1047-1058.
72. Hoff EF, Cook SH, Sherman GD et al. Toxoplasma gondii: Molecular cloning and characterization of a novel 18-kDa secretory antigen, TgMIC10. Exp Parasitol 2001; 97(2):77-88.
73. Kuliawat R, Prabakaran D, Arvan P. Proinsulin endoproteolysis confers enhanced targeting of processed insulin to the regulated secretory pathway. Mol Biol Cell 2000; 11(6):1959-1972.
74. Zhang B, Chang A, Kjeldsen TB et al. Intracellular retention of newly synthesized insulin in yeast is caused by endoproteolytic processing in the Golgi complex. J Cell Biol 2001; 153(6):1187-1198.
75. Arvan P, Halban PA. Sorting ourselves out: Seeking consensus on trafficking in the beta-cell. Traffic 2004; 5(1):53-61.
76. Kappe S, Bruderer T, Gantt S et al. Conservation of a gliding motility and cell invasion machinery in Apicomplexan parasites. J Cell Biol 1999; 147(5):937-944.
77. Brown PJ, Billington KJ, Bumstead JM et al. A microneme protein from Eimeria tenella with homology to the Apple domains of coagulation factor XI and plasma prekallikrein. Mol Biochem Parasitol 2000; 107(1):91-102.
78. Carruthers VB, Blackman MJ. A new release on life: Emerging concepts in proteolysis and parasite invasion. Mol Microbiol 2005; 55(6):1617-1630.
79. Luo BH, Springer TA. Integrin structures and conformational signaling. Curr Opin Cell Biol 2006, [ebup ahead of print].
80. Springer TA, Wang JH. The three-dimensional structure of integrins and their ligands, and conformational regulation of cell adhesion. Adv Protein Chem 2004; 68:29-63.
81. Dessens JT, Beetsma AL, Dimopoulos G et al. CTRP is essential for mosquito infection by malaria ookinetes. EMBO J 1999; 18(22):6221-6227.
82. Dessens JT, Siden-Kiamos I, Mendoza J et al. SOAP, a novel malaria ookinete protein involved in mosquito midgut invasion and oocyst development. Mol Microbiol 2003; 49(2):319-329.
83. Kadota K, Ishino T, Matsuyama T et al. Essential role of membrane-attack protein in malarial transmission to mosquito host. Proc Natl Acad Sci USA 2004; 101(46):16310-16315.
84. Kariu T, Ishino T, Yano K et al. CelTOS, a novel malarial protein that mediates transmission to mosquito and vertebrate hosts. Mol Microbiol 2006; 59(5):1369-1379.
85. Mota MM, Hafalla JCR, Rodriguez A. Migration through host cells activates Plasmodium sporozoites for infection. Nat Med 2002; 8(11):1318-1322.
86. Mota MM, Pradel G, Vanderberg JP et al. Migration of Plasmodium sporozoites through cells before infection. Science 2001; 291(5501):141-144.
87. Ishino T, Yano K, Chinzei Y et al. Cell-passage activity is required for the malarial parasite to cross the liver sinusoidal cell layer. PLoS Biol 2004; 2(1):E4.
88. Ishino T, Chinzei Y, Yuda M. A Plasmodium sporozoite protein with a membrane attack complex domain is required for breaching the liver sinusoidal cell layer prior to hepatocyte infection. Cell Microbiol 2005; 7(2):199-208.
89. Barragan A, Brossier F, Sibley LD. Transepithelial migration of Toxoplasma gondii involves an interaction of intercellular adhesion molecule 1 (ICAM-1) with the parasite adhesin MIC2. Cell Microbiol 2005; 7(4):561-568.

CHAPTER 3

Erythrocyte Invasion by *Plasmodium falciparum*: Multiple Ligand-Receptor Interactions and Phenotypic Switching

Manoj T. Duraisingh,* Tiffany DeSimone, Cameron Jennings, Philippe Refour and Chenwei Wu

Introduction

Infection with the protozoan parasite *Plasmodium falciparum* causes the most severe form of human malaria with over two million deaths per year. The clinical symptoms of malaria infection result from the rapid exponential expansion of parasites during the asexual erythrocytic phase of the *P. falciparum* life cycle. Invasion of erythrocytes by merozoites is a tightly controlled process that involves specific receptor-ligand interactions between host and parasite molecules. Virulence of *P. falciparum* parasites has been associated with increased multiplication rates and an ability to invade a greater range of host erythrocytes.[1] Here we focus on our understanding of the molecular mechanisms underlying host cell selection and invasion of the host erythrocyte using parasite adhesive proteins. We will consider the parasite strategy of deploying multiple and variant adhesive ligands for successful invasion. An understanding of the molecular mechanism by which these proteins mediate invasion will facilitate their use in the rational design of vaccine and drug strategies.

P. falciparum Invasion of Human Erythrocytes Is a Specific and Ordered Process

P. falciparum belongs to the phylum Apicomplexa, a group characterized by a highly specialized conoid apical complex containing specialized organelles that play a central role in the invasion of host cells. These include pear-shaped secretory organelles known as rhoptries and smaller tubular organelles known as micronemes that contain proteins required for different aspects of the invasion process. *Plasmodium* spp. are restricted to invading erythrocytic cells in the bloodstream. Specificity is determined by specific ligand-receptor interactions that occur between surface proteins found on invasive forms of the parasite, known as merozoites, and cognate receptors on the host erythrocyte surface.[2] This restricted tropism of *Plasmodium* spp. is in contrast to other members of Apicomplexa, such as *Toxoplasma gondii*, that invade a wide range of cell types from many different species. The sequence of events leading to erythrocyte

*Corresponding Author: Manoj T. Duraisingh—Department of Immunology and Infectious Diseases, Harvard School of Public Health, 665 Huntington Ave. SPH-1, Room 703, Boston, Massachusetts, USA. Email: mduraisi@hsph.harvard.edu

Molecular Mechanisms of Parasite Invasion, edited by Barbara A. Burleigh and Dominique Soldati-Favre.

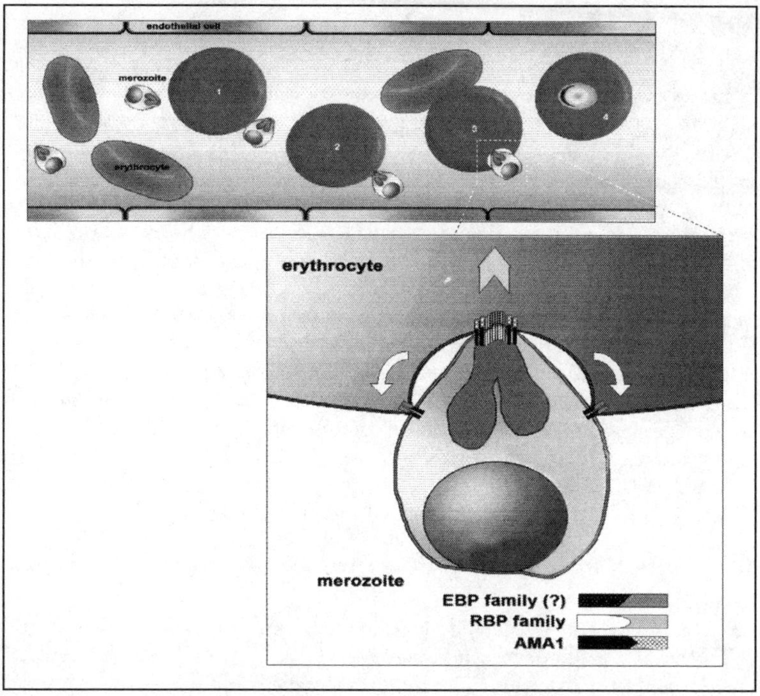

Figure 1. *Plasmodium* invasion of host red blood cells is an active process requiring apical orientation of the parasite and tight junction formation between pathogen and host. In the erythrocytic stage, free-floating merozoites (1) attach to circulating red blood cells, (2) reorient such that the apical complex is in apposition to the host cell membrane, and (3) commit to reinvasion through tight junction formation. The process continues with movement of the tight junction from the anterior to the posterior end of the merozoite and the progressive envelopment of the parasite within the burgeoning parasitophorous vacuole. Invasion concludes with (4) the parasite adopting ring-stage morphology within the newly invaded host red blood cell. Micronemal proteins, such as the erythrocyte binding antigens (EBA) and AMA1 are thought to play a role in tight junction formation, while the RBP family of proteins functions in erythrocyte recognition (INSET).

invasion is postulated to be similar for all *Plasmodium* spp. (Fig. 1) and has been derived from a combination of live video and electron microscopy studies (e.g., refs. 3,4):

STEP 1. Loose attachment and reorientation. Following egress from a host red cell, free merozoites form long distance, reversible, low affinity initial contacts between the parasite and a new red cell. This can occur at any point on the surface of the merozoite.

STEP 2. Recognition and tight junction formation. In an obligatory step, the parasite then reorientates to juxtapose its apex with the erythrocyte surface. A high affinity irreversible contact occurs with the formation of a close membrane-to-membrane contact known as a 'tight junction'. The contents of the micronemes and rhoptries are then discharged and a nascent vacuole forms within the erythrocyte.[4-6] Microneme discharge is regulated by free Ca2+ levels within the cytoplasm of the parasite.[7]

STEP 3. Entry and transformation. The merozoite then begins to enter the red cell with the tight junction bridging the parasite and the erythrocyte moving from the anterior to the posterior pole. This is powered by the parasite's actinomyosin motor, which propels the parasite into the nascent parasitophorous vacuole. The invasion process is complete when the newly formed parasitophorous vacuole closes behind it.

Families of Plasmodial Ligands That Bind to Erythrocytes with High Affinity

Proteins spread over the entire surface of the merozoite are involved in the initial, reversible attachment of the parasite to host erythrocytes (STEP 1). The glycophosphatidylinositol (GPI)-anchored protein merozoite surface protein-1 (Msp-1) is one such protein and there is some evidence for its interaction with Band 3.[54]

Merozoites of the primate malaria *Plasmodium knowlesi* have been shown to attach in an apical orientation to erythrocytes prior to the formation of the tight junction.[4] This implies that recognition of apical orientation precedes microneme and rhoptry discharge. It is therefore likely that specific proteins sense the apical orientation of the merozoite with the erythrocyte surface and signal downstream effector molecules to initiate subsequent steps in the invasion process, such as tight junction formation (STEP 2). The tight junction is formed through irreversible high affinity interactions between multiple parasite adhesive ligands and erythrocyte receptors. Some evidence suggests that proteins such as Apical Membrane Antigen-1 (AMA-1) and the components of the High Molecular Weight Rhoptry (RhopH) complex bind to erythrocytes, but their role in the invasion process remains unclear. Here we will focus on two families of adhesive ligands that bind to the erythrocyte with high affinity: the Duffy-Binding like (DBL) family and the Reticulocyte Binding Protein-like (RBL) family.

Plasmodial Duffy Binding-Like (DBL) Proteins Bind to Host Erythrocytes with High Affinity

The first Plasmodium adhesive ligand identified that binds to erythrocytes with high affinity was the *P. falciparum* erythrocyte binding antigen-175 (EBA-175), which interacts specifically with the glycophorin A erythrocyte receptor via sialic acid moieties.[9,10] EBA-175 is homologous to the *P. knowlesi* and *P.vivax* Duffy binding proteins (PkDBPα and PvDBP), the ligands that bind to the Duffy receptor during invasion.[11,12] EBA-175, PkDBPα and PvDBP are members of a large family of proteins that share structural motifs in Plasmodia that have been termed the Duffy-binding-like protein (DBL) family.[13,14] Paralogs of *EBA-175* have been identified from the *P. falciparum* genome: *EBA-140 (BAEBL), EBA-165, EBA-181 (JESEBL)* and *EBL-1*.[15,16]

EBA molecules are localized to the micronemes and are thought to be involved in tight junction formation.[11,17] They are large type 1 transmembrane proteins (between 140 and 181 kDa) with large ectodomains and short cytoplasmic tails. The cysteine-rich DBL domain found within the ectodomain binds to erythrocyte receptors; recently, the crystal structures of both the *P. knowlesi* PkDBPα and *P. falciparum* EBA-175 DBL receptor-binding domains have been solved. Analysis of the molecular structure reveals that the inner surface of the DBL functions in receptor engagement, while the outer face is implicated in immune evasion; consequently, the latter is comparatively more variant in its amino acid sequence. Critical amino acid residues within the DBL domains of PkDBPα and EBA-175 are associated with binding to alternative host receptors.[18,19]

Studies on Apicomplexan parasites demonstrate that the cytoplasmic tail domains of adhesive micronemal proteins, such as the thrombospondin-related anonymous protein (TRAP) found in both *Plasmodium* sporozoites and *Toxoplasma*, engage the parasite's actinomyosin motor.[20] It is the actinomyosin motor that is required to actively propel the merozoite into the red cell.[21] Since it is known that the EBA ectodomain complexes with receptors found on the erythrocyte surface, it is tempting to speculate that the cytoplasmic tails of these micronemal proteins-by analogy to those found in TRAP-engage the actinomyosin motor either directly or indirectly. Nevertheless, recently a homolog of the TRAP gene has been identified in the *P. falciparum* genome, MTRAP, which has been shown to be expressed in merozoites.[20] The role of this molecule in invasion remains to be determined.

In addition to EBA-175, EBA-140 and EBA-181 have been shown to bind to erythrocytes, but to receptors distinct from glycophorin A.[22-24] *EBA-165* is thought to be a pseudogene.[25] Although protein expression has not been demonstrated in the asexual stage, *EBL-1* has been shown to segregate with increased parasite proliferation in a genetic cross.[13]

A Superfamily of Reticulocyte Binding Protein Homologues Determine Specificity for Red Blood Cells in Plasmodia Species

A second family of high molecular weight adhesive proteins was first identified in *P. vivax* that have been shown to selectively bind immature red blood cells (reticulocytes). As such, these have been termed Reticulocyte Binding Protein-1 and -2 (RBP-1 and RBP-2).[26] They are located at the apical end of merozoites and have been hypothesized to play an important role in the specific recognition of reticulocytes by merozoites.[26] *P. yoelii* proteins homologous to the *P.vivax* RBPs have been identified and are termed Py235; at least one member of this family has been shown to bind mature erythrocytes.[27] Death from a fulminating *P. yoelii* parasitaemia can be prevented by administration of a monoclonal antibody against a Py235 protein.[28] This protection is associated with a limiting of the infection to reticulocytes.

The Reticulocyte Binding Protein homolog (Rh) superfamily of proteins consists of putative type-1 transmembrane proteins (200-380 kDa) that possess a very large ectodomain and a short cytoplasmic tail domain. They are ideally located for sensing the apical orientation of the *P. falciparum* merozoite on the erythrocyte surface through the recognition of specific receptors. It is hypothesized that following engagement of an Rh protein with its cognate receptor, a signal is transmitted that stimulates microneme and rhoptry discharge, leading to the downstream events associated with entry and invasion of the erythrocyte.

There are four *P.falciparum* members of the Rh protein family, PfRh1, PfRh2a, PfRh2b, and PfRh4, and each is encoded by a separate gene. A fifth gene, PfRh3, has also been identified, but it has not yet been shown to encode a protein in the asexual erythrocytic stage.[29]

High affinity erythrocyte binding activity has been observed for the PfRh1 protein,[30,31] and its putative receptor was found to be sialic acid-dependent (neuraminidase-sensitive) and trypsin-resistant. Direct binding of PfRh2a and PfRh2b to erythrocytes has not yet been demonstrated, but this does not exclude a role for these proteins in binding either directly or indirectly to erythrocytes.[32,33] There is some evidence that PfRh2b may form a complex with PfRh1.[30]

PfRh2a and PfRh2b initially localize to the large club-shaped rhoptry organelles during late schizont stages of the parasite and then move to the apical tip of free merozoites forming a crescent-shaped apical cup.[29] A similar localization has been seen for the PvRBP1 and PvRBP2 proteins in *P. vivax* merozoites.[26] A role in invasion for PfRh1, PfRh2a, and PfRh2b has been inferred by the use of specific antibodies.[30,32,34]

Multiple Receptors on the Erythrocyte Surface for Merozoite Invasion

During the process of invasion the parasite encounters numerous proteins on the surface of the erythrocyte, but binds to a limited number of receptors with high affinity. The erythrocyte membrane contains over 200 proteins, and any of these molecules or a combination of these molecules could facilitate erythrocyte invasion by malaria parasites. The erythrocyte protein Band 3 has been postulated as the initial receptor for attachment of merozoites to the erythrocyte surface through interactions with the major surface antigen MSP-1.[8] Following apical orientation, several receptors for high affinity binding have been identified using mutant erythrocytes, enzyme treatments and antibody inhibition experiments. This includes the major glycoproteins on the surface of the erythrocyte, glycophorins A (GlyA) and B (GlyB), as well as the unrelated glycophorin C (GlyC). These are all integral membrane proteins with a single membrane spanning domain, a small cytoplasmic domain that anchors the molecule to the cytoskeleton, and a heavily glycosylated extracellular domain.

Table 1. **Plasmodium falciparum *merozoite adhesive ligands and their cognate receptors***

		Receptor Enzyme Sensitivities			
Parasite Ligand	**RBC Receptor**	**Neuraminidase**	**Trypsin**	**Chymotrypsin**	**Reference**
Duffy-binding-like proteins					
EBA-175	Glycophorin A	S	S	R	8, 41
EBA-140	Glycophorin C	S	S	R	16, 18
EBA-181	Receptor E	S	R	S	37
Reticulocyte-binding-like proteins					
PfRh1	Receptor Y	S	R	R	24, 25
PfRh2a	?				28, 26
PfRh2b	Receptor Z	R	R	S	28, 26
PfRh4	?	R	?	?	42

Several erythrocyte receptors.- X,[35] Y[30] and Z[34]- have been defined by their sensitivity to enzyme treatments such as neuraminidase, chymotrypsin and trypsin.[34] Direct binding of PfRh1 to receptor Y has been demonstrated.[30] Receptor Y is a neuraminidase-sensitive and trypsin/chymotrypsin-resistant protein that is recognized by PfRh1.[30] Receptor Z has been identified as the receptor recognized by PfRh2b by genetic studies and antibody inhibition experiments.[34] Receptor Z is chymotrypsin-sensitive and neuraminidase/trypsin-resistant. Receptor X is a trypsin-sensitive and neuraminidase-resistant molecule whose cognate parasite ligand has not been identified (Table 1).[35]

Interestingly, humans living in malaria-endemic populations contain polymorphisms in many of the putative erythrocyte receptors. These genetic footprints may reflect selection by malaria, as the polymorphims may be inhibitory for erythrocyte invasion. For instance, the Gerbich polymorphism has been selected for in Papua New Guinea and has been associated with a reduction in invasion.[24]

Alternative Invasion Pathways in *P. falciparum* Invasion

Different Parasite Lines use Alternate Erythroycte Receptors for Invasion

The specific recognition of receptors on the erythrocyte is a critical step in the invasion process. *P. vivax* uses the Duffy receptor for invasion into human erythrocytes and this parasite is unable to invade Duffy negative erythrocytes. This phenomenon is thought to account for the lack of *P. vivax* parasite in West Africa, where the prevalence of Duffy negativity approaches 95%.[55]

In contrast, *P. falciparum* invades human erythrocytes using multiple and alternate receptor-ligand interactions that have been defined as invasion pathways.[35,36] A feature of *P. falciparum* parasites is the variability of different lines in their dependence on alternative ligand-receptor interactions. For instance, some strains of *P. falciparum*, such as Dd2, are unable to invade neuraminidase-treated erythrocytes, which are bereft of sialic acid, and are therefore dependent on sialic acid-containing receptors for invasion.[36] However, other *P. falciparum* strains, such as 3D7, are able to invade neuraminidase-treated erythrocytes with high efficiency and are therefore sialic acid-independent, presumably through the use of alternative ligand-receptor interactions. Similarly, different parasite lines are variantly dependent on trypsin-sensitive and/or chymotrypsin-sensitive receptors for invasion. Mutant erythrocytes have also been used to demonstrate dependence on specific receptors, such as glycophorins A and B.[36,37] Field studies have confirmed the use of alternative invasion pathways in fresh isolates without culture adaptation.[38-40]

Variant Expression of Merozoite Ligands

Different laboratory lines vary in their expression of members of the PfRh protein family.[29,34] For example, the sialic acid-independent parasite line, 3D7, strongly expresses PfRh2a and PfRh2b, while very low amounts of PfRh1 are produced. Alternatively, the T994 parasite line strongly expresses PfRh1, but neither PfRh2a nor PfRh2b. In contrast to 3D7, T994 invades erythrocytes via sialic acid-dependent means. Sialic acid dependence has been associated with expression of PfRh1 in a series of laboratory isolates.[31] If each PfRh protein defines a distinct invasion pathway, then variant expression of these proteins will be reflected in the use of alternative invasion pathways among different parasite lines. Differential expression of parasite ligands thus provides a molecular basis for the observed variation in dependency of different parasite lines on certain invasion pathways.

Sequence Polymorphims in Merozoite Ligands

In contrast to the PfRh variant expression, the different EBA paralogs are expressed in all parasite lines with a limited degree of quantitative variation. Nevertheless, sequence polymorphisms have been identified in the adhesive domains of EBA-140 and EBA-181.[15,41,42] These have been shown to result in changes in the binding of these parasite ligands from one erythrocyte receptor to another and hence may contribute to the invasion pathway being utilized.

A Role for the Merozoite Adhesive Proteins in Defining Alternative Invasion Pathways for Invasion into Human Erythrocytes

Antibody Inhibition of Merozoite Adhesive Ligands

Antibody inhibition experiments with specific polyclonal antibodies raised against the different EBA and PfRh antigens have been carried out. These indicate that the parasite ligands are exposed on the surface of the parasite and are functionally important in the invasion process (e. g., refs. 24,30,32,34,43,44).

Genetic Analyses of EBA Paralogs

More recently, a genetic approach has been used to create parasite lines that lack expression of different merozoite adhesive ligands. It appears that there is redundancy in the requirement of each paralog for invasion into normal erythrocytes. Nevertheless, deletion of the parasite ligand EBA-175 results in a switch from the use of a sialic acid-dependent to a sialic acid-independent pathway for invasion, suggesting that mutation of EBA-175 leads to invasion via an alternative ligand-receptor interaction.[45,46] However, it has also been shown that EBA-175 is also functional in the sialic acid-independent parasite line 3D7.[46] In this parasite line, EBA-175, although functional, does not define the predominant pathway utilized for invasion. This corroborates previous work indicating that antibodies specific for EBA-175 are able to inhibit invasion in parasites utilizing both sialic acid-dependent and -independent pathways.[43]

A similar approach has demonstrated that EBA-140 mediates invasion into human erythrocytes through the glycophorin C receptor.[24] Though EBA-140 is an alternative ligand to EBA-175, EBA-140-glycophorin C binding is sialic acid-dependent and therefore cannot account for the sialic acid-independent invasion observed following disruption of EBA-175. When both EBA-140 and EBA-181 are deleted, there is no change in the invasion pathway utilized. Nevertheless, experiments with antibodies specific to EBA-140 indicate that this parasite ligand is functional.[24]

Genetic Analyses of PfRh Paralogs

The PfRh proteins are variantly expressed in different parasite lines.[29,32,34] When *PfRh2b* is deleted from the 3D7 parasite line in which it is expressed, a significant change is observed in the invasion pathway that is utilized. Parasites lacking PfRh2b become more reliant on sialic acid- and trypsin-sensitive receptors, but less reliant on chymotrypsin-sensitve receptors. It can thus be inferred that PfRh2b engages a receptor (known only as Receptor Z) that is both neuraminidase/trypsin-resistant and chymotrypsin-sensitive. Furthermore, loss of PfRh2b expression was associated with an increased ability to invade chymotrypsin-treated erythrocytes.

PfRh1 is expressed at different levels in different parasite lines.[31] A high level of PfRh1 expression has been correlated to some extent with a reliance on sialic acid receptors; increased protein expression was, in turn, found to correlate with *PfRh1* gene copy. *PfRh1* was deleted from the T994 parasite line that possesses a single copy of PfRh1. This was associated with a significant increase in invasion into neuraminidase- or trypsin-treated erythrocytes compared to parental T994.[31] In contrast, 3D7 invasion was comparatively unaffected by lack of PfRh1 expression, as PfRh1 appears to play only a minor role in 3D7 merozoite invasion. It is becoming clear that the genetic background of a parasite, in particular the repertoire of parasite ligands that is expressed, is likely to affect the contribution of any given parasite ligand for invasion.

Phenotypic Switching between Alternative Invasion Pathways

Invasion pathways utilized by each clonal laboratory line is constant over many years of routine culture. However, it has been shown from the use of selective pressures that there exists some degree of plasticity in the ability of certain parasite lines to switch among different invasion pathways. Dolan and colleagues continuously cultured the clonal sialic acid-dependent *P. falciparum* parasite line, Dd2, on sialic acid-depleted erythrocytes. Following an initial drop in parasitemia, a sub-population of Dd2 parasites (Dd2Nm) emerged 10-12 days post-plating capable of growth on the neuraminidase-treated cells. The ability of Dd2Nm to invade enzyme-treated erythrocytes indicates a switch in these parasites from a reliance on a sialic acid-dependent invasion pathway to one of sialic acid-independence.[36] Several lines have been shown to possess this flexibility.[47]

Microarray analysis was performed with the sialic acid-dependent parasite line W2Mef (parental line of Dd2) and with W2mef parasites that had become selected for sialic acid-independence (W2MefNm). This revealed that both EBA-165 and PfRh4 were transcriptionally up-regulated greater than 80-fold in W2mefNm when compared to W2mef and other sialic acid-independent strains, such as 7G8, HB3, and 3D7.[47] Furthermore, disruption of PfRh4 using double-crossover homologous recombination abrogated the ability of W2Mef parasites to invade neuraminidase-treated cells. These data, the increased expression of PfRh4 and the failure of knockout parasites to grow on enzyme-treated cells, imply that PfRh4 plays a central role in switching to a sialic acid-independent invasion phenotype. EBA-165 is a transcribed pseudogene and its increased expression is likely due from its head-to-head orientation with PfRh4 on chromosome 4. These results have been corroborated by the work of Gaur and colleagues on the W2Mef clone, Dd2.[48]

Phenotypic switching is not a phenomenon particular to *P. falciparum*. Preiser and colleagues have discovered that *P. yoelii* Py235 genes, which belong to the RBP superfamily, are variantly expressed. Interestingly, the *P. yoelii* Py235 family is encoded by perhaps up to 50 genes, and individual merozoites originating from a single schizont transcribe distinct members of this gene family.[49] This clonal phenotypic variation has been suggested to provide the murine parasite with a survival strategy in the host: expression of a single, but different Py235 protein in each merozoite allows the parasite to respond efficiently to variations in the host environment. The ability of *P. yoelii* merozoites from a single infected cell to up-regulate distinct Py235 genes is a unique strategy that enables the parasite to combat not only an adapting immune system, but also receptor polymorphisms inherent to the erythrocyte population.[49]

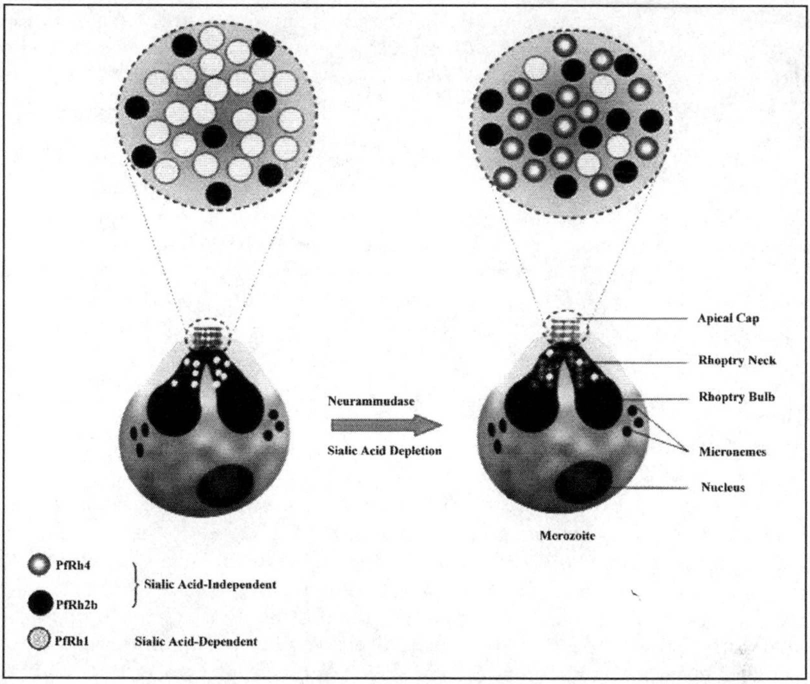

Figure 2. The 'space filling' model dictates merozoite alternate invasion pathways. Whether a merozoite invades a host red blood cell via sialic acid-dependent or -independent means is predicated on the accumulation of parasite ligands at the merozoite apex. PfRh1 is the dominant sialic acid-dependent ligand, while PfRh2b and PfRh4 mediate invasion via sialic acid-independent means. The number and proportion of ligands present at the merozoite apex is a reflection of the relative abundance of these proteins within the rhoptries. In (1), PfRh1 dominates both the store of invasion ligands in the rhoptry neck and at the merozoite apical complex. Following growth on neuraminidase-treated/sialic acid-depleted cells, PfRh4 expression is upregulated. The limited space model proposes that the increase in PfRh4 is concomitant with a decrease in the proportion of PfRh1- both within the rhoptries and at the merozoite apex. In (2), PfRh2b and PfRh4 are therefore relatively more abundant than PfRh1. This phenotypic switch enables the parasite to shift from a sialic acid-dependent to a sialic acid-independent pathway for erythrocyte invasion.

Model for Variant Expression and Phenotypic Switching of Parasite Adhesive Ligands

A model has been proposed to explain our increasing understanding of the variant deployment of multiple adhesive ligands by the parasite. The 'limited space' model assumes that only a finite amount of space exists at the merozoite apex. The final invasion pathway depends on the expression levels of the different parasite ligands. This can explain the changes in invasion pathways associated with the expression of different parasite ligands due either to natural variation or to gene disruption.[34] As space is at a premium in this parasite region, competition exists among the various invasion ligands, DBL and Rh alike, for access to direct erythrocyte receptor engagement and invasion pathway utilization.

Recently, a hierarchy of invasion ligands has been proposed to explain the observation that some genetic knockouts of parasite ligand genes result in the dramatic redeployment of invasion ligands, while others have little effect on the invasion pathway used. Investigating the increased reliance of 3D7 parasites lacking PfRh2b on a chymotrypsin-resistant invasion pathway, Baum and colleagues uncovered, using Affymetrix gene chips, a 60-fold increase in PfRh3 transcription.[50] Frameshift mutations in PfRh3, however, suggest that it is a pseudogene.[29] Neither transcript nor protein levels of the other major invasion ligands (PfRh1, PfRh2a, PfRh2b, EBA-140, EBA-175, and EBA-181) are altered between wild-type and knockout parasites, this suggests that alternative invasion pathway utilization stems from a reorganization of the molecular hierarchy at the merozoite apex.[50] Stated alternatively, absence of PfRh2b at the apical tip allows other parasite ligands, those that are chymotrypsin-resistant, to occupy the merozoite apex and serve a more prominent role in invasion than occurs with the wild-type genotype (i.e., when PfRh2b is present).[34,50] It is not known whether the molecular hierarchy is defined by the absolute levels of expression of the different ligands and/or the relative affinity of each ligand. Additionally, it has been difficult to determine whether the DBLs and Rhs work sequentially or in concert in erythrocyte binding.

Biological Significance: Immune Evasion and/or Receptor Polymorphism

Erythrocyte invasion is a complex process that requires a number of parasite proteins (and their cognate receptors) in an inherently degenerate process. Why does *P. falciparum* utilize multiple ligand-receptors interactions in a variant fashion? This heterogeneity is thought to provide *P. falciparum* parasites with a biological advantage, allowing them to invade erythrocytes with polymorphic receptors and to evade a mounting immune response

In the bloodstream, merozoites are exposed to the host immune system. Immunological selection of surface proteins can lead to antigenic diversity as a means of immune evasion, such as that observed in the major surface antigen MSP-1.[51] Variant expression of PfRh and PfEBA proteins has arisen as a genetic strategy, where parasites expressing one variant ligand survive in the face of an immune response raised against an alternate. This is analogous to the antigenic variation of molecules found on the surface of infected erythrocytes.[52] Alternatively, sequence polymorphisms in and variant expression of these parasite adhesive ligands may reflect coevolution of parasites with novel receptor polymorphisms in human populations. The human erythrocyte is the most polymorphic surface in the human body and the utilization of different parasite ligands may allow the parasite to survive in polymorphic host cells, such as Gerbich erythrocytes.[53] Additionally, receptors may vary in quantity within the lifespan of an individual erythrocyte.

Many questions remain unanswered, some which can be answered using novel post-genomic and proteomic approaches: What are the erythrocyte receptors for the PfRh and EBA molecules? What is the nature of functional protein complexes? How is sensing of apical orientation by multiple parasite ligands integrated prior to rhoptry and microneme discharge? What is the impact of receptor polymorphisms on invasion? To what extent are alternative invasion pathways utilized in different parasite populations? Can different invasion pathways be linked with protective immunity and/or virulence?

Our understanding of the specific roles of the different members of the merozoite adhesive protein families has increased in recent years. During the in vivo invasion process, merozoites are exposed in the bloodstream and antibody recognition of merozoite antigens exposed on the parasite surface or at the tip of the apical complex plays a large part in the development of immunity. An effective vaccine based on these molecules might act to suppress parasitemia and thereby reduce or prevent severe disease. In the long-term, a thorough understanding of the roles of the *Plasmodium* merozoite proteins in defining alternative invasion pathways and in signaling to downstream effectors will prove imperative for the rational design of vaccine and drug strategies used in the prevention and treatment of *falciparum* malaria.

References

1. Chotivanich K, Udomsangpetch R, Simpson JA et al. Parasite multiplication potential and the severity of Falciparum malaria. J Infect Dis 2000; 181(3):1206-1209.
2. Gratzer WB, Dluzewski AR. The red blood cell and malaria parasite invasion. Semin Hematol 1993; 30(3):232-247.
3. Dvorak JA, Miller LH, Whitehouse WC et al. Invasion of erythrocytes by malaria merozoites. Science 1975; 187:748-750.
4. Miller LH, Aikawa M, Johnson JG et al. Interaction between cytochalasin B-treated malarial parasites and erythrocytes. Attachment and junction formation. J Exp Med 1979; 149:172-184.
5. Holder AA, Freeman RR, Uni S et al. Isolation of a Plasmodium falciparum rhoptry protein. Mol Biochem Parasitol 1985; 14:293-303.
6. Howard RF. The lower-molecular-weight protein complex (RI) of the Plasmodium falciparum rhoptries lacks the glycolytic enzyme aldolase. Mol Biochem Parasitol 1990; 42:235-240.
7. Carruthers VB, Sibley LD. Mobilization of intracellular calcium stimulates microneme discharge in Toxoplasma gondii. Mol Microbiol 1999; 31(2):421-428.
8. Goel VK, Li X, Chen H et al. Band 3 is a host receptor binding merozoite surface protein 1 during the Plasmodium falciparum invasion of erythrocytes. Proc Natl Acad Sci USA 2003; 100(9):5164-5169.
9. Camus D, Hadley TJ. A Plasmodium falciparum antigen that binds to host erythrocytes and merozoites. Science 1985; 230:553-556.
10. Sim BKL, Chitnis CE, Wasniowska K et al. Receptor and ligand domains for invasion of erythrocytes by Plasmodium falciparum. Science 1994; 264:1941-1944.
11. Adams JH, Hudson DE, Torii M et al. The duffy receptor family of Plasmodium knowlesi is located within the micronemes of invasive malaria merozoites. Cell 1990; 63(1):141-153.
12. Fang X, Kaslow DC, Adams JH et al. Cloning of the Plasmodium vivax Duffy receptor. Mol Biochem Parasitol 1991; 44:125-132.
13. Peterson DS, Wellems TE. EBL-1, a putative erythrocyte binding protein of Plasmodium falciparum, maps within a favored linkage group in two genetic crosses. Mol Biochem Parasitol 2000; 105(1):105-113.
14. Gardner MJ, Hall N, Fung E et al. Genome sequence of the human malaria parasite Plasmodium falciparum. Nature 2002; 419(6906):498-511.
15. Mayer DC, Kaneko O, Hudson-Taylor DE et al. Characterization of a Plasmodium falciparum erythrocyte-binding protein paralogous to EBA-175. Proc Natl Acad Sci USA 2001; 98(9):5222-5227.
16. Adams JH, Blair PL, Kaneko O et al. An expanding ebl family of Plasmodium falciparum. Trends Parasitol 2001; 17(6):297-299.
17. Singh AP, Ozwara H, Kocken CH et al. Targeted deletion of Plasmodium knowlesi Duffy binding protein confirms its role in junction formation during invasion. Mol Microbiol 2005; 55(6):1925-1934.
18. Singh SK, Hora R, Belrhali H et al. Structural basis for Duffy recognition by the malaria parasite Duffy-binding-like domain. Nature 2006; 439(7077):741-744.
19. Tolia NH, Enemark EJ, Sim BK et al. Structural basis for the EBA-175 erythrocyte invasion pathway of the malaria parasite Plasmodium falciparum. Cell 2005; 122(2):183-193.
20. Baum J, Richard D, Healer J et al. A conserved molecular motor drives cell invasion and gliding motility across malaria life cycle stages and other apicomplexan parasites. J Biol Chem 2006; 281(8):5197-5208.
21. Miller LH, Aikawa M, Johnson JG et al. Interaction between cytochalasin B-treated malarial parasites and erythrocytes. Attachment and junction formation. J Exp Med 1979; 149(1):172-184.
22. Thompson JK, Triglia T, Reed MB et al. A novel ligand from Plasmodium falciparum that binds to a sialic acid-containing receptor on the surface of human erythrocytes. Mol Microbiol 2001; 41(1):47-58.
23. Narum DL, Fuhrmann SR, Luu T et al. A novel Plasmodium falciparum erythrocyte binding protein-2 (EBP2/BAEBL) involved in erythrocyte receptor binding. Mol Biochem Parasitol 2002; 119(2):159-168.
24. Maier AG, Duraisingh MT, Reeder JC et al. Plasmodium falciparum erythrocyte invasion through glycophorin C and selection for Gerbich negativity in human populations. Nat Med 2003; 9(1):87-92.
25. Triglia T, Thompson JK, Cowman AF. An EBA175 homologue which is transcribed but not translated in erythrocytic stages of Plasmodium falciparum. Mol Biochem Parasitol 2001; 116(1):55-63.

26. Galinski MR, Medina CC, Ingravallo P et al. A reticulocyte-binding protein complex of Plasmodium vivax merozoites. Cell 1992; 69:1213-1226.
27. Ogun SA, Holder AA. A high molecular mass Plasmodium yoelii rhoptry protein binds to erythrocytes. Mol Biochem Parasitol 1996; 76:321-324.
28. Holder AA, Freeman RR. Immunization against blood-stage rodent malaria using purified parasite antigens. Nature 1981; 294:361-364.
29. Taylor HM, Triglia T, Thompson J et al. Plasmodium falciparum homologue of the genes for Plasmodium vivax and Plasmodium yoelii adhesive proteins, which is transcribed but not translated. Infect Immun 2001; 69(6):3635-3645.
30. Rayner JC, Vargas-Serrato E, Huber CS et al. A Plasmodium falciparum homologue of Plasmodium vivax reticulocyte binding protein (PvRBP1) defines a trypsin-resistant erythrocyte invasion pathway. J Exp Med 2001; 194(11):1571-1581.
31. Triglia T, Duraisingh MT, Good RT et al. Reticulocyte-binding protein homologue 1 is required for sialic acid-dependent invasion into human erythrocytes by Plasmodium falciparum. Mol Microbiol 2005; 55(1):162-174.
32. Triglia T, Thompson J, Caruana SR et al. Identification of proteins from Plasmodium falciparum that are homologous to reticulocyte binding proteins in Plasmodium vivax. Infect Immun. 2001; 69(2):1084-1092.
33. Rayner JC, Galinski MR, Ingravallo P et al. Two Plasmodium falciparum genes express merozoite proteins that are related to Plasmodium vivax and Plasmodium yoelii adhesive proteins involved in host cell selection and invasion. Proc Natl Acad Sci USA 2000; 97(17):9648-9653.
34. Duraisingh MT, Triglia T, Ralph SA et al. Phenotypic variation of Plasmodium falciparum merozoite proteins directs receptor targeting for invasion of human erythrocytes. EMBO J 2003; 22(5):1047-1057.
35. Dolan SA, Proctor JL, Alling DW et al. Glycophorin B as an EBA-175 independent Plasmodium falciparum receptor of human erythrocytes. Mol Biochem Parasitol 1994; 64:55-63.
36. Dolan SA, Miller LH, Wellems TE. Evidence for a switching mechanism in the invasion of erythrocytes by Plasmodium falciparum. J Clin Invest 1990; 86(2):618-624.
37. Mitchell GH, Hadley TJ, McGinniss MH et al. Invasion of erythrocytes by Plasmodium falciparum malaria parasites: Evidence for receptor heterogeneity and two receptors. Blood 1986; 67(5):1519-1521.
38. Okoyeh JN, Pillai CR, Chitnis CE. Plasmodium falciparum field isolates commonly use erythrocyte invasion pathways that are independent of sialic acid residues of glycophorin A. Infect Immun 1999; 67(11):5784-5791.
39. Baum J, Pinder M, Conway DJ. Erythrocyte invasion phenotypes of Plasmodium falciparum in The Gambia. Infect Immun 2003; 71(4):1856-1863.
40. Lobo CA, de Frazao K, Rodriguez M et al. Invasion profiles of Brazilian field isolates of Plasmodium falciparum: Phenotypic and genotypic analyses. Infect Immun 2004; 72(10):5886-5891.
41. Mayer DC, Jiang L, Achur RN et al. The glycophorin C N-linked glycan is a critical component of the ligand for the Plasmodium falciparum erythrocyte receptor BAEBL. Proc Natl Acad Sci USA 2006; 103(7):2358-2362.
42. Mayer DC, Mu JB, Kaneko O et al. Polymorphism in the Plasmodium falciparum erythrocyte-binding ligand JESEBL/EBA-181 alters its receptor specificity. Proc Natl Acad Sci USA 2004; 101(8):2518-2523.
43. Narum DL, Haynes JD, Fuhrmann S et al. Antibodies against the Plasmodium falciparum receptor binding domain of EBA-175 block invasion pathways that do not involve sialic acids. Infect Immun 2000; 68(4):1964-1966.
44. Pandey KC, Singh S, Pattnaik P et al. Bacterially expressed and refolded receptor binding domain of Plasmodium falciparum EBA-175 elicits invasion inhibitory antibodies. Mol Biochem Parasitol 2002; 123(1):23-33.
45. Reed MB, Caruana SR, Batchelor AH et al. Targeted disruption of an erythrocyte binding antigen in Plasmodium falciparum is associated with a switch toward a sialic acid-independent pathway of invasion. Proc Natl Acad Sci USA 2000; 97(13):7509-7514.
46. Duraisingh MT, Maier AG, Triglia T et al. Erythrocyte-binding antigen 175 mediates invasion in Plasmodium falciparum utilizing sialic acid-dependent and -independent pathways. Proc Natl Acad Sci USA 2003; 100(8):4796-4801.
47. Stubbs J, Simpson KM, Triglia T et al. Molecular mechanism for switching of P. falciparum invasion pathways into human erythrocytes. Science 2005; 309(5739):1384-1387.
48. Gaur D, Furuya T, Mu J et al. Upregulation of expression of the reticulocyte homology gene 4 in the Plasmodium falciparum clone Dd2 is associated with a switch in the erythrocyte invasion pathway. Mol Biochem Parasitol 2006; 145(2):205-215.

49. Preiser PR, Jarra W, Capiod T et al. A rhoptry-protein-associated mechanism of clonal phenotypic variation in rodent malaria. Nature 1999; 398(6728):618-622.
50. Baum J, Maier AG, Good RT et al. Invasion by P. falciparum merozoites suggests a hierarchy of molecular interactions. PLoS Pathog 2005; 1(4):e37.
51. Kimura E, Mattei D, Mana di Santi S et al. Genetic diversity in the major merozoite surface antigen of Plasmodium falciparum: High prevalence of a third polymorphic form detected in strains derived from malaria patients. Gene 1990; 91(1):57-62.
52. Newbold CI. Antigenic variation in Plasmodium falciparum: Mechanisms and consequences. Curr Opin Microbiol 1999; 2:420-425.
53. Patel SS, Mehlotra RK, Kastens W et al. The association of the glycophorin C exon 3 deletion with ovalocytosis and malaria susceptibility in the Wosera, Papua New Guinea. Blood 2001; 98(12):3489-3491.
54. Barnwell J and Galinski M. Invasion of vertebrate cells: Erythrocytes. In: Sherman I, ed. Malaria: Parasite biology, pathogenesis, and protection. Washington: ASM Press, 1998:93.
55. Mourant AE, Kopec AC and Domaniewska-Sobczak K. The distribution of human blood groups and other polymorphisms. 2. London: Oxford university Press, 1976.

Chapter 4

Role of the gp85/Trans-Sialidase Superfamily of Glycoproteins in the Interaction of *Trypanosoma cruzi* with Host Structures

Maria Júlia M. Alves* and Walter Colli

Abstract

Invasion of mammalian cells by *T. cruzi* trypomastigotes is a multi-step and complex process involving several adhesion molecules, signaling events and proteolytic activities. From the blood to the cell target in different tissues the parasite has to interact with different cells and the extracellular matrix (ECM). The review focus on the role of the gp85/trans-sialidase superfamily members in the interaction of the parasite with the host cell, particularly with ECM components, with emphasis on the significant variability among the ligands and receptors involved. Use of the SELEX technique to evolve nuclease-resistant RNA aptamers for receptor identification is briefly discussed.

Introduction

The protozoan *Trypanosoma cruzi* is the etiological agent of Chagas' disease of major medical significance throughout South to Central America. Chagas' disease was discovered by Carlos Chagas almost 100 years ago. In 1909, the Brazilian physician described the parasite (*Trypanosoma cruzi*) in the gut of a bug belonging to the Reduviidae family. Due to the bloodsucking habits of the insect, he hypothesized and demonstrated that the parasite infected different experimental mammals and searched for cases of infection among local inhabitants and their domestic animals. Chagas described the infectivity of the parasite for different mammals, its life cycle in the insect and vertebrate hosts, demonstrated the presence of the parasite in acute cases of the disease and, finally, described the symptoms for the acute and chronic phase of the disease in humans.[1] With some modifications, the complex dixenic life cycle (Fig. 1) of the parasite is basically the same as originally described. Some observations on the parasite's life cycle that have been misinterpreted led Chagas to propose a change of genus to Schyzotrypanum. However, independent experiments conducted by him and others showed that the parasite should indeed belong to the genus Trypanosoma.[2] This is one of the few examples in the literature where the parasite and the insect vector were first described, followed by the description of the disease in humans.

The parasite has a complex life cycle characterized by several developmental forms present in vertebrate and invertebrate hosts.[3] Based mainly on morphological criteria, such as the spindle shape of the parasite, as well as the position of the kinetoplast (mitochondrial DNA, kDNA)

*Corresponding Author: Maria Júlia M. Alves—Departamento de Bioquímica, Instituto de Química, Universidade de São Paulo, Av. Prof. Lineu Prestes 748, 05508-900 São Paulo, Brazil. Email: mjmalves@iq.usp.br

Molecular Mechanisms of Parasite Invasion, edited by Barbara A. Burleigh and Dominique Soldati-Favre.

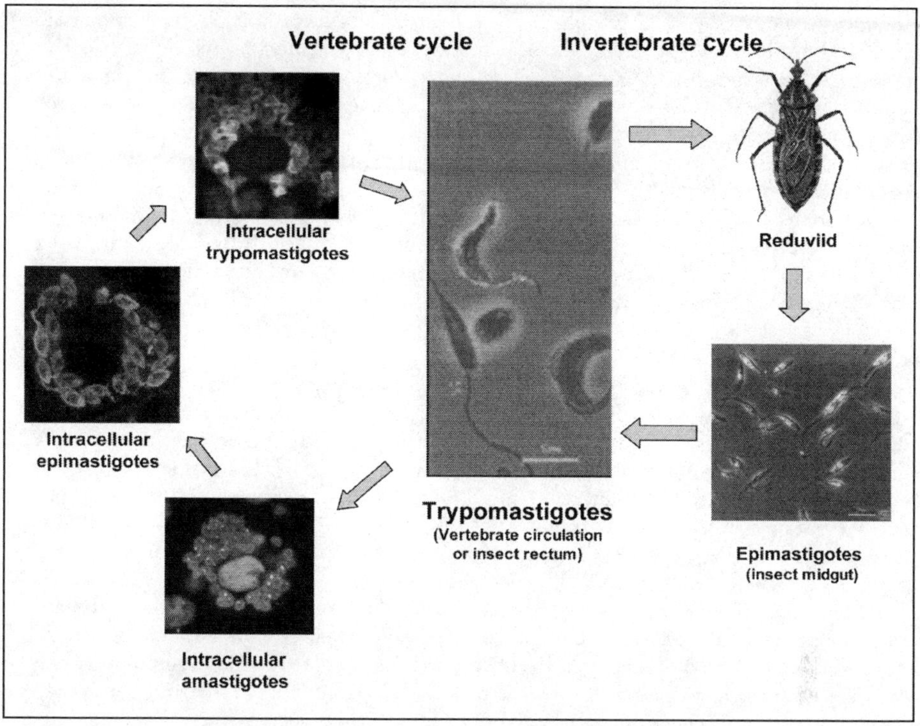

Figure 1. *Trypanosoma cruzi* life cycle. Insect vector releases epimastigotes and metacyclic trypomastigotes with feces and urine infecting mammalian hosts through a skin wound or mucosae. After adhesion to the host cell membrane followed by cell invasion, the parasite escapes from the parasitophorous vacuole and replicates in the cytoplasm as amastigotes. Intracellular differentiation from amastigotes to trypomastigotes occurs through intermediate forms, including a characteristic intracellular epimastigote that precedes the differentiation to trypomastigotes. Trypomastigotes that are released into the bloodstream infect new cells or are taken up by the insect vector during a blood meal. In the digestive tract, trypomastigotes differentiate to epimastigotes that replicate and differentiate to metacyclic trypomastigotes at the end of the digestive tract. Intermediate forms are also found in the digestive tract.

relative to the nucleus and the flagellar emergency region, three forms are classically described: (i) amastigotes (greek, a = without; mastis = whip for flagellum) which are dividing round cells with 2.4—6.5 μm in diameter with a very short flagellum found in the cytoplasm of the vertebrate host cell; (ii) trypomastigotes (trypo = to drill, referring to a property of this cell to attach to glass by one point while making rotatory movements), an infective, flagellated and nondividing form present in vertebrate (blood trypomastigotes) and invertebrate (metacyclic trypomastigotes) hosts; trypomastigotes are spindle-shaped, approximately 18 μm in length (including a 6 μm free flagellum) and 2-3 μm in breadth; (iii) epimastigotes (epi = anterior, from above), an extracellular flagellated, 20-40 μm long and 2-5 μm large, noninfective and dividing stage present typically in the invertebrate host intestine. Trypomastigotes appear in the rectum of the insect as a differentiation product of the epimastigotes. When the insect bites the vertebrate host and initiates blood sucking it concomitantly eliminates in feces and urine the trypomastigotes—in that case called metacyclic trypomastigotes—which are mechanically carried into the wound by the host. Metacyclic means "beyond the cycle" because parasitologists believed that this form was terminal and could not transform into the other forms in laboratory. Trypomastigotes ultimately invade cells and reside for a short period of time inside a parasitophorous vacuole, escaping to the cytoplasm where differentiation to the amastigote form occurs. After a few rounds of division, amastigotes start to differentiate into trypomastigotes in approximately 96 h. This step is mediated by an epimastigote-like form sharing some properties with the epimastigote form, but being considerably shorter (4 μm in length). The existence of this form in the vertebrate part of the cycle has been a matter of debate since 1914, but presently it was well characterized,[4,5] needing proline to complete differentiation.[6] It appears that the parasite has a continuous differentiation cycle following the pattern trypomastigote - amastigote - epimastigote - trypomastigote with one or possibly more intermediate forms. The predominance of a particular form would be dependent mainly on the environmental conditions surrounding the parasite (cf. *T. cruzi* cell cycle movie on www.sbpz.org.br).

Variability

The existence of a large spectrum of clinical manifestations, ranging from asymptomatic to severe cardiac complications, is an important point to be considered when analyzing and interpreting the data available in the literature. Different isolates from *T. cruzi* have demonstrated that the parasite populations are extremely diverse. Attempts to correlate this level of diversity with the pathology observed in different clinical forms and distinct geographic areas have been made. For instance, in Brazil, the asymptomatic form of Chagas' disease predominates (60-70%), followed by the cardiac (20-30%) and digestive forms (8-10%); in Chile, the digestive form predominates, whereas in Argentina digestive symptoms correspond to only 3.5% of the total chagasic patients.[7] The reason for the observed differences is a matter of debate. Probably, genetic variations of the parasite are the major cause. However, physiological conditions and the genetic background of the host[8] have to be considered. Extensive genetic and protein polymorphism profiles were described in the literature when laboratory strains and field isolates were compared, resulting in different suggestions for the distribution of the parasite population. More recently, two principal groups were proposed (*T. cruzi* I and *T. cruzi* II), but the heterogeneity among the groups (more evident in group II) and the presence of an intermediate group awaiting a final classification indicate that this classification/division should be taken as preliminary (reviewed in refs. 9-11). Interestingly, a high prevalence of *T. cruzi* II in chagasic patients is indicating that group II or, at least, a subgroup within, is associated with the human disease.

Infection of Host Cells by Trypomastigotes of *T. cruzi*: More Ways Than One

By definition, *T. cruzi* trypomastigotes are the classical invasive form, although extracellular amastigotes also infect mammalian cells via a less characterized process.[12] To establish a successful infection inside the host cell, the parasite has to (1) interact with the cell surface; (2) escape from the entry vacuole; and (3) undergo differentiation and division in the cytoplasm. Although *T. cruzi* trypomastigotes can invade and proliferate in most of the vertebrate cells tested in vitro, differences in tissue tropism of the various strains in animals,[13-15] and patients,[10,16] as well as the fact that not all *T. cruzi* strains are capable of establishing effective infections in the host[13,17-20] point out to the importance of elucidating—in addition to the cell invasion mechanism itself—the factors responsible for the survival of the parasite inside the host cells. For example, the quantification of live parasites in experimental models, using a transfected clone of Tulahuén strain of *T. cruzi* that stably expresses *E. coli* β-galactosidase, detected a large number of live parasites three weeks post-infection in skeletal muscle, heart, bladder and connective tissues, with smaller numbers in sciatic nerve, liver, spleen, adrenal glands and intestine and rare parasites in kidney and brain. Two weeks later, however, the number of parasites had decreased more that 30-fold and were more abundant in skeletal muscle and bladder for that particular clone.[21] This may be a useful technique to compare tissue tropism/parasite survival in tissues of experimental models.

Different molecules implicated in the attachment and signaling during the invasion of mammalian cells by trypomastigotes from *T. cruzi* have been identified (cf. refs. 22-24). Carbohydrates and proteins are probably involved in multiple interactions between the parasite and the host cell. Parasite recognition of, and attachment to, the host cell surface is followed by induction of a calcium flux in the host, setting off a sequence of events that ends with parasites inside the vacuole. At the molecular level, each parasite-host cell interaction is dependent on the concentration of parasite ligands and host cell receptors that are available to interact at any given moment, as well as to the strength of the association and the nature of the cell target. It was clearly demonstrated that trypomastigotes when invading HeLa cells or cardiomyocytes, but not MDCK cells, induce the formation of actin-rich pseudopodial protrusions around the parasite. This is insensitive to cytochalasin D, a drug that blocks actin polymerization and phagocytosis.[15,25,26] Parasites can invade host cells by two processes: (1) lysosome-dependent, with lysosome recruitment to the cell surface in a calcium-signaling dependent pathway and actin polymerization[5,27,28] and (2) plasma membrane-mediated, with invagination of the plasma membrane in an actin-independent and PI-3 kinase dependent signaling pathway.[3,29-32] The importance of each pathway for the establishment of an effective infection is, as yet, a matter of discussion (cf. refs. 28,29). Most intriguing is the observation that some trypomastigotes can clearly cross host-cells without infecting them.[33] In any case, two main questions should be addressed: (1) is each pathway a consequence of the surface repertoire of adhesion molecules of each individual parasite at that moment? (2) do metacyclic trypomastigotes and blood trypomastigotes share the same mechanisms of invasion?

gp85/Trans-Sialidase Superfamily (gp85/TS): Role of the Tc-85 Family in the Invasion of Nonphagocytic Cells by *T. cruzi*

For a successful infection to occur some steps are necessary including adhesion to the host cell, action of proteolytic enzymes, and intracellular signaling. Discussion here will be restricted to the role of the gp85/trans-sialidase (gp85/TS) glycoprotein superfamily in the parasite interaction with the host cell and the extracellular matrix. Excellent general reviews on related subjects have been published, including the role of surface molecules on the host immune system.[24,28,29,34,35] A general scheme is shown in Figure 2.

The involvement of 85 kDa-glycoproteins in *T. cruzi* infection was suggested more than 20 years ago, using in most of the cases, polyclonal or monoclonal antibodies that inhibited the invasion of tissue cultured cells by trypomastigotes, followed by the identification and molecular mass determination of the molecule recognized by the antibody (cf. ref. 36): Tc85, gp85, gp82, TS, TSA-1, among others, were at that time described in tissue-culture derived trypomastigotes or metacyclic trypomastigotes. Further characterization of the molecules and their classification as part of a novel superfamily baptized as gp85/trans-sialidase family were made possible with the advent of new tools for genomic analysis.

In general, members of the gp85/TS are expressed on the cell surface of metacyclic trypomastigotes, blood trypomastigotes and amastigotes, some more abundant in one of the stages, as Tc-85 (85 kDa, trypomastigote-specific) or gp82 (82 kDa, metacyclic trypomastigotes-specific).

At least 50% of the *T. cruzi* genome consist of large gene families of surface proteins, retrotransposons and subtelomeric repeats.[37] Gp85/trans-sialidase superfamily (gp85/TS, 1430 members), mucin-associated surface proteins (MASPs, 1377 members), mucins (863 members), gp63 surface protease (425 members) are the largest gene families encoding surface molecules, although a sizeable proportion of the sequences are pseudogenes (e.g., half of the gp85/TS superfamily). Gp85/TS genes can be found in subtelomeric repetitive regions and in intrachromosomal arrays, frequently within large clusters of gp85/TS, MASP and mucin genes. In the particular case of gp85/TS superfamily, all genes were assigned under the designation of TS in the genome description instead of the previous—and in our opinion more correct—gp85/TS nomenclature, since they include genes encoding proteins with and without trans-sialidase activity. The superfamily was previously characterized by the presence of two conserved neuraminidase motifs—Asp boxes (SxDxGxTW) and the VTV motif (VTVxNVxLYNR),[36,38-40] distributed in the majority of the chromosome bands separated by pulse-field electrophoresis,[41] but only 371 members in the genome contain the VTV motif. It should be considered that approximately 28 chromosomes and 12,000 genes per haploid genome were estimated in *T. cruzi*, Brener strain, although the exact number is unknown.[42] Genes are polymorphic, displaying different degrees of sequence variation interspersed by more conserved regions, which suggests a strong selective pressure for variability. Similar conclusions are drawn from the analysis of the mucin family.[43] More interestingly, chimeras that contain the C or N-terminal conserved domain of MASP combined with the C or N-terminal domain of mucin or the C-terminal domain of gp85/TS were described.[37] The mechanisms determining this gene shuffling are unknown.

The gp85/TS superfamily was subdivided in a variable number of groups by different authors.[36,38-40] Frasch proposes two main groups:[39] I—GPI-anchored glycoproteins in the 120-200 kDa range with trans-sialidase activity (TS). TS isolated from trypomastigotes contain repeats of 12 amino acids *in tandem* at the C-terminal portion (SAPA) that are absent from TS isolated from epimastigotes; II—Molecules without TS activity. This subgroup includes the Gp85 family (80-90 kDa glycoproteins); a 160 kDa protein (involved in complement regulation) and Tc13 (85 kDa, unknown function). Surface glycoproteins previously described in the literature, following the pioneer Tc85 (trypomastigote-specific, 85kDa),[44] gp82 (82 kDa described in metacyclic trypomastigotes),[45] SA85-1 (85 kDa surface antigen),[46] TSA-1,[47] gp90[48] can be included in the gp85 family. At least part of these molecules are trypomastigote-specific, GPI-anchored and shed to the medium with subsets simultaneously expressed by individual parasites.[46,49]

Trans-Sialidase (TS)

T. cruzi is unable to synthesize sialic acid but possess TS, an enzyme that transfers α-2,3-linked sialic acid from macromolecular donors to β-Gal-containing acceptors, with mucins present on the parasite cell surface being the main acceptors.[50,51] TS, which is more active in trypomastigotes than in metacyclic trypomastigotes, has been implicated in conferring the parasite resistance to complement activation,[52] acting as a B cell mitogen,[53] and participating in *T. cruzi* adhesion to and invasion of the host cell.[54,55] Notwithstanding, the results concerning cell invasion are controversial.[54,56] Inactive TS, with a mutation in the catalytic site, binds sialic acid and galactosyl-residues interacting with host T cells through CD43 and helps in the adhesion step of *T. cruzi* in a carbohydrate-dependent way.[55] No specific receptor in the host cell was described and the role of TS in the invasion waits full clarification. Apparently, TS is not rate limiting for cell entry, but instead it could be required for the formation of the parasitophorous vacuole,[57] possibly controlling the number of parasites effectively reaching the cytoplasm, which indirectly affects *T. cruzi* infection.

T. cruzi Interaction with Host ECM Elements

As described for other parasites, *T. cruzi* makes multiple interactions with ECM to establish an effective infection from the blood to the target cells in different organs inside the vertebrate host. Two main forms of ECM can be considered, the basement membrane and the interstitial matrix (IM). Basement membranes (BM) are thin sheets of highly specialized ECM present at the epithelial/mesenchymal interface of most tissues, surrounding peripheral nerve fibers, muscle and fat cells. These membranes differ from each other in composition and properties, according to the cell localization and the developmental stage of the organism.[58]

Important components of BM as laminins,[41,59-62] collagen IV[63,64] and heparan sulfate proteoglycans[65,66] are implicated in trypomastigote invasion, as well other ECM components, as fibronectin[67-69] and thrombospondin[61] (Fig. 2). RGD is the adhesion motif of fibronectin that binds to a noncharacterized 85kDa glycoprotein from *T. cruzi*.[70] Since, on one hand, RGD (and its variants) is a cell attachment motif of a large number of adhesive proteins as

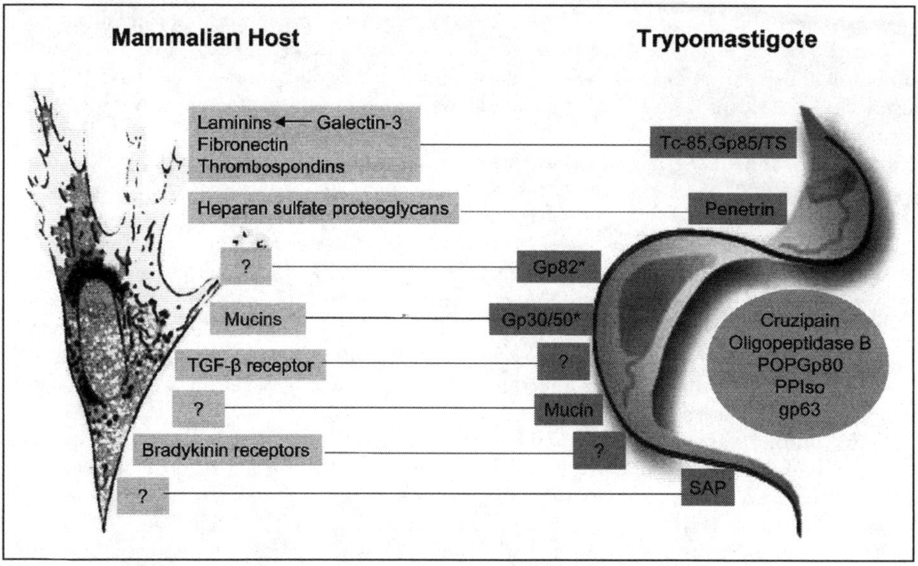

Figure 2. Molecules involved in *T. cruzi* infection of nonprofessional phagocytes. A plethora of adhesion molecules, signaling pathways triggered by parasite components and proteolytic enzymes result in an effective infection of the host. *T. cruzi* binds to ECM through Tc-85 (laminins[41,62]) and other parasite molecules[65,66,70,72] and to the cell surface in multiple carbohydrate[45,82,83] and carbohydrate-independent[78,84,85] interactions. Gp82, a member of the gp85/TS triggers a bi-directional Ca^{2+} signaling and phosphorylation of a 175 kDa protein in the parasite (cf[24]). Stimulation of TGFβ receptors occurs by an unknown molecule released into the medium resulting in activation of TGFβ -responsive genes.[85] *T. cruzi* oligopeptidase B cleaves an inactive precursor resulting in an active Ca^{2+} agonist via IP3 pathway.[34] Cruzipain cleaves high molecular weight kininogen, generating kynins that bind to bradykinin receptors to stimulate IP3-mediated Ca^{2+} release,[84] although a bradykinin independent pathway has been postulated.[86] PPIso (peptidyl-prolyl-cis-trans-isomerase),[87] gp63/metalloproteases,[88] POPGp80 proteinase (collagenase activity)[89] are other proteolytic enzymes involved in parasite infection. *Specific for metacyclic trypomastigotes.

fibronectin, various collagens, vitronectin and thrombospondins and, on the other, approximately one third of the 24 integrins present on the cell surfaces recognize the RGD sequence to a variable extent,[71] it is very attractive to hypothesize that the parasite takes advantage of the ubiquity of the RGD-integrin system to interact with and invade the host cell. Antibodies against the β1-chain of specific integrins inhibit invasion of human macrophages by *T. cruzi*[69] reinforcing the involvement of fibronectin in parasite invasion. Since some members of gp85/TS contain RGD-sequences (cf. genome sequences deposit at DNA Databanks DDBJ/EMBL/Genbank) it should be interesting to see if they play any role in parasite adhesion to the host cell surface. RGD-independent interaction was described in the binding of Tc85-11, one cloned member of the Tc-85 subset, to one of the laminin isoforms, a family of heterotrimeric proteins (αβγ). The binding site of Tc85-11 to laminin-1 (α1β1γ1) was mapped on the amino terminal domain.[62] Although the corresponding binding region of the laminin molecule has not been determined, the binding is probably carbohydrate-independent. Binding of *T. cruzi* to laminin is increased by galectin-3,[72] one isoform of the β-galactoside-binding lectins present in ECM. It is interesting to note that the VTV domain, present in all gp85/TS molecules, binds to cytokeratin 18, a constituent of the intermediate filaments of epithelial cells, enhancing *T. cruzi* invasion by an unknown mechanism.[73]

The infectious agents, including *T. cruzi*, can take advantage of the dynamics and heterogeneity of the ECM, as well as from its molecular complexity represented by a relatively high number of molecular isoforms of the several components and widespread localization in different tissues. These features, associated with the high number of genes coding for gp85/TS members may help to understand the capacity of *T. cruzi* to invade almost any cell and to disseminate through different organs and tissues inside the vertebrate. In this context, it is interesting to note that laminin α-4, IFNγ and lymphocytes are associated with the passage of *T. brucei brucei* through endothelial BM, being responsible for brain infection.[74]

Sequential changes in the main host ECM components (fibronectin, laminin and collagens) were described in *T. cruzi* infections in vivo and in vitro,[64,75,76] as well as an increase in galectin-3[72] and galectin-1,[77] a galectin with immunoregulatory properties. Since trypomastigotes shed into the medium membrane vesicles enriched in Tc-85[49] amongst other surface molecules, it was suggested that the interaction of the vesicles with ECM is a key event for the alteration of ECM elements observed during the infection in vitro.[76] In vivo, ECM changes are being associated with factors secreted by inflammatory cells.[64,75]

Gp82, another member of the Gp85/TS superfamily is the principal molecule involved in the invasion of gastric mucosal epithelium by metacyclic trypomastigotes via binding to gastric mucin,[24] a fact that explains the oral infection. Two peptides were implicated in the binding to the host cell with the activation of a signaling cascade leading to Ca^{2+} mobilization.[45,78] Interestingly, metacyclic trypomastigotes from different strains can invade cells with the involvement of gp82 (more effective) or gp35/50 (mucin-like molecules, less effective) and gp90, a down regulator of invasion. The importance of the down regulation is elusive (cf. ref. 24).

Oligonucleotide Aptamers as Tools for Determination of Adhesion Receptors on the *T. cruzi* Surface

The SELEX method (Systematic Evolution of Ligands by Exponential enrichment) is an oligonucleotide-based combinatorial library approach that has been extensively used to isolate high-affinity ligands (called aptamers) for a wide variety of proteins and small molecules[79,80] (Fig. 3). For this purpose a partial randomized synthetic DNA template is constructed containing a random inner region that is flanked on both sides by constant sequences. The random sequence classically consists of 15 to 75 random positions where all four bases are incorporated with equal probabilities. This pool containing between 10^{12}—10^{15} different sequences can be either directly used for selection or first transcribed to RNA using T7 RNA polymerase. In this case a T7 promoter site needs to be put on the 5′ site of the DNA template. The random DNA / RNA pool is exposed to the protein target. The best fitting molecules in the selection pool are culled and amplified. The procedure will be repeated with increasing stringency until the previous random pool is purified to a few molecules with the desired binding properties. The final pool is cloned into a bacterial vector and individual colonies are sequenced. The previous random regions are aligned and searched for consensus motifs. Consensus motifs often located in stem-loop structures are thought to mediate binding specificity. These aptamers are suitable for applications based on molecular recognition of a target molecule including diagnostics and therapeutics.

RNA aptamers that bind to the receptors of the cell-matrix molecules laminin, thrombospondin, heparan sulfate and fibronectin on *T. cruzi* cell surfaces have been developed.[61,81] Aptamers should only recognize protein epitopes to which they have been selected for. Consensus aptamers displaced by laminin (Kd = 209 ± 99 nM), fibronectin (Kd = 142 ± 94 nM), heparan sulfate (Kd = 40 ± 14 nM) and thrombospondin (Kd = 400 ± 239 nM) were sequenced.[61] A scheme for a selection protocol is sketched in Figure 3. The identification of receptors and ligands on parasite and host cell surfaces may not be only important for elucidating invasion mechanisms used by *T. cruzi* but also for understanding the entry process of other pathogens that depend on adhesion on host-cell membranes.

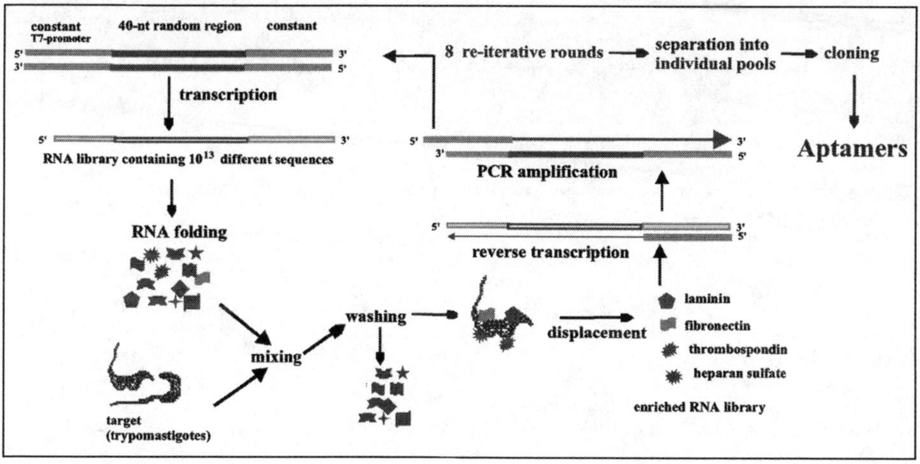

Figure 3. Scheme for selecting consensus aptamers displaceable by ECM components. A partial randomized DNA pool is in vitro transcribed in the presence of 2-OH′ purines and 2-F′ pyrimidines in order to create the initial pool of 10^{13} different nuclease-resistant RNA molecules. The RNA molecules are heated up and cooled slowly down to room temperature to allow for proper folding. The RNA molecules are incubated with live trypomastigotes in a 100:1 RNA: binding sites ratio. Unbound and weakly bound RNA molecules are washed off from the trypomastigotes and specifically bound RNA molecules are displaced using a cocktail of cell-matrix molecules. The eluted RNA molecules are reverse transcribed to cDNA and amplified by PCR. The next generation SELEX pool contains an enriched fraction of RNA molecules that compete with host-cell matrix molecules for their binding sites on *T. cruzi*. The procedure is repeated until no further purification can be achieved. At this stage, RNA molecules are constitutively eluted using individual cell-matrix molecules as displacement agents. Individual aptamer sequences are cloned and searched for consensus motifs. One would expect that aptamers competing with a certain cell-matrix molecule for receptor binding should share a common sequence motif. Determining the minimal RNA sequence that is necessary for receptor binding will help to find the most effective molecule. Reproduced with permission from Ulrich H et al. Braz J Med Biol Res 2001; 34(3):295-300.[81]

Photo-cross linking of radiolabeled RNA ligands to their targets on parasite membranes can be used for ligand-mediated target purification. Using this method, any protein target on the parasite membranes, for which a high affinity RNA ligand has been identified, can be purified. Micro sequencing of the cross-linked RNA:receptor complexes will allow further isolation of cell surface proteins of the gp85 superfamily in *T. cruzi* and possibly also other parasite surface proteins that are involved in the invasion of host cells by these parasites.

Acknowledgements

Part of this work was supported by a grant from FAPESP (project nº 04/03303-5). Authors are grateful for the permission to use the scheme shown in Figure 3 given by the Braz J Med Biol Res.[81]

References

1. Chagas C. Nova tripanosomiaze humana. Mem Inst Oswaldo Cruz 1909; I(II):159.
2. Hoare CA. The Trypanosomes of Mammals. Oxford: Blackwell Scientific Publications, 1972.
3. De Souza W. Basic cell biology of Trypanosoma cruzi. Curr Pharm Des 2002; 8(4):269-285.
4. Almeida-de-Faria M, Freymuller E, Colli W et al. Trypanosoma cruzi: Characterization of an intracellular epimastigote-like form. Exp Parasitol 1999; 92(4):263-274.

5. Tyler KM, Engman DM. The life cycle of Trypanosoma cruzi revisited. Int J Parasitol 2001; 31(5-6):472-481.
6. Tonelli RR, Silber AM, Almeida-de-Faria M et al. L-proline is essential for the intracellular differentiation of Trypanosoma cruzi. Cell Microbiol 2004; 6(8):733-741.
7. WHO. Technical Report Series: WHO Publications, 2002.
8. Williams-Blangero S, VandeBerg JL, Blangero J et al. Genetic epidemiology of Trypanosoma cruzi infection and Chagas' disease. Front Biosci 2003; 8:e337-345.
9. Buscaglia CA, Di Noia JM. Trypanosoma cruzi clonal diversity and the epidemiology of Chagas' disease. Microbes Infect 2003; 5(5):419-427.
10. Macedo AM, Machado CR, Oliveira RP et al. Trypanosoma cruzi: Genetic structure of populations and relevance of genetic variability to the pathogenesis of chagas disease. Mem Inst Oswaldo Cruz 2004; 99(1):1-12.
11. Prata A. Clinical and epidemiological aspects of Chagas disease. Lancet Infect Dis 2001; 1:92-100.
12. Mortara RA, Andreoli WK, Taniwaki NN et al. Mammalian cell invasion and intracellular trafficking by Trypanosoma cruzi infective forms. An Acad Bras Cienc 2005; 77(1):77-94.
13. Lima MT, Lenzi HL, Gattass CR. Negative tissue parasitism in mice injected with a noninfective clone of Trypanosoma cruzi. Parasitol Res 1995; 81(1):6-12.
14. Lenzi HL, Oliveira DN, Lima MT et al. Trypanosoma cruzi: Paninfectivity of CL strain during murine acute infection. Exp Parasitol 1996; 84(1):16-27.
15. Schenkman S, Mortara RA. HeLa cells extend and internalize pseudopodia during active invasion by Trypanosoma cruzi trypomastigotes. J Cell Sci 1992; 101(Pt 4):895-905.
16. Freitas JM, Lages-Silva E, Crema E et al. Real time PCR strategy for the identification of major lineages of Trypanosoma cruzi directly in chronically infected human tissues. Int J Parasitol 2005; 35(4):411-417.
17. Taliaferro H, Pizzi T. Connective tissue reaction in normal and immunized mice to a reticulotropic strains of Trypanosoma cruzi. J Infect Dis 1955; 96:199-226.
18. Bice D, Zeledon R. Comparison of infectivity of strains of Trypanosoma cruzi (Chagas 1909). J Parasitol 1970; 56:663-670.
19. Melo RaB Z. Tissue tropism of different T. cruzi strains. J Parasitol 1978; 64:475-482.
20. Schoemaker J, Hoffman JR, Hoffman DG. Trypanosoma cruzi: Preference for brown adipose tissue in mice by theTulahuen strain. Exp Parasitol 1970; 27:403-407.
21. Buckner FS, Wilson AJ, Van Voorhis WC. Detection of live Trypanosoma cruzi in tissues of infected mice by using histochemical stain for beta-galactosidase. Infect Immun 1999; 67(1):403-409.
22. Burleigh BA, Andrews NW. The mechanisms of Trypanosoma cruzi invasion of mammalian cells. Annu Rev Microbiol 1995; 49:175-200.
23. Burleigh BA, Woolsey AM. Cell signalling and Trypanosoma cruzi invasion. Cell Microbiol 2002; 4(11):701-711.
24. Yoshida N. Molecular basis of mammalian cell invasion by Trypanosoma cruzi. Anais da Academia Brasileira de Ciencias 2005; 77(3):1-25.
25. Barbosa HS, Meirelles MN. Evidence of participation of cytoskeleton of heart muscle cells during the invasion of Trypanosoma cruzi. Cell Struct Funct 1995; 20(4):275-284.
26. Mortara RA. Trypanosoma cruzi: Amastigotes and trypomastigotes interact with different structures on the surface of HeLa cells. Exp Parasitol 1991; 73(1):1-14.
27. Ley V, Robbins ES, Nussenzweig V et al. The exit of Trypanosoma cruzi from the phagosome is inhibited by raising the pH of acidic compartments. J Exp Med 1990; 171(2):401-413.
28. Andrade LO, Andrews NW. The Trypanosoma cruzi-host-cell interplay: Location, invasion, retention. Nat Rev Microbiol 2005; 3(10):819-823.
29. Burleigh BA. Host cell signaling and Trypanosoma cruzi invasion: Do all roads lead to lysosomes? Sci STKE 2005; 2005(293):pe36.
30. Wilkowsky SE, Barbieri MA, Stahl PD et al. Regulation of Trypanosoma cruzi invasion of nonphagocytic cells by the endocytically active GTPases dynamin, Rab5, and Rab7. Biochem Biophys Res Commun 2002; 291(3):516-521.
31. Andrade LO, Andrews NW. Lysosomal fusion is essential for the retention of Trypanosoma cruzi inside host cells. J Exp Med 2004; 200(9):1135-1143.
32. Woolsey AM, Sunwoo L, Petersen CA et al. Novel PI 3-kinase-dependent mechanisms of trypanosome invasion and vacuole maturation. J Cell Sci 2003; 116(Pt 17):3611-3622.
33. Schenkman S, Robbins ES, Nussenzweig V. Attachment of Trypanosoma cruzi to mammalian cells requires parasite energy, and invasion can be independent of the target cell cytoskeleton. Infect Immun 1991; 59(2):645-654.

34. Burleigh BA, Andrews NW. Signaling and host cell invasion by Trypanosoma cruzi. Curr Opin Microbiol 1998; 1(4):461-465.
35. DosReis GA, Freirede-Lima CG, Nunes MP et al. The importance of aberrant T-cell responses in Chagas disease. Trends Parasitol 2005; 21(5):237-243.
36. Colli W. Trans-sialidase: A unique enzyme activity discovered in the protozoan Trypanosoma cruzi. FASEB J 1993; 7(13):1257-1264.
37. El-Sayed NM, Myler PJ, Bartholomeu DC et al. The genome sequence of Trypanosoma cruzi, etiologic agent of Chagas disease. Science 2005; 309(5733):409-415.
38. Schenkman S, Eichinger D, Pereira ME et al. Structural and functional properties of Trypanosoma trans-sialidase. Annu Rev Microbiol 1994; 48:499-523.
39. Frasch AC. Functional diversity in the trans-sialidase and mucin families in Trypanosoma cruzi. Parasitol Today 2000; 16(7):282-286.
40. Cross GA, Takle GB. The surface trans-sialidase family of Trypanosoma cruzi. Annu Rev Microbiol 1993; 47:385-411.
41. Giordano R, Fouts DL, Tewari D et al. Cloning of a surface membrane glycoprotein specific for the infective form of Trypanosoma cruzi having adhesive properties to laminin. J Biol Chem 1999; 274(6):3461-3468.
42. El-Sayed NM, Myler PJ, Blandin G et al. Comparative genomics of trypanosomatid parasitic protozoa. Science 2005; 309(5733):404-409.
43. Campo V, Di Noia JM, Buscaglia CA et al. Differential accumulation of mutations localized in particular domains of the mucin genes expressed in the vertebrate host stage of Trypanosoma cruzi. Mol Biochem Parasitol 2004; 133(1):81-91.
44. Katzin AM, Colli W. Lectin receptors in Trypanosoma cruzi. An N-acetyl-D-glucosamine-containing surface glycoprotein specific for the trypomastigote stage. Biochim Biophys Acta 1983; 727(2):403-411.
45. Ruiz RC, Favoreto Jr S, Dorta ML et al. Infectivity of Trypanosoma cruzi strains is associated with differential expression of surface glycoproteins with differential Ca2+ signalling activity. Biochem J 1998; 330(Pt 1):505-511.
46. Kahn SJ, Nguyen D, Norsen J et al. Trypanosoma cruzi: Monoclonal antibodies to the surface glycoprotein superfamily differentiate subsets of the 85-kDa surface glycoproteins and confirm simultaneous expression of variant 85-kDa surface glycoproteins. Exp Parasitol 1999; 92(1):48-56.
47. Fouts DL, Ruef BJ, Ridley PT et al. Nucleotide sequence and transcription of a trypomastigote surface antigen gene of Trypanosoma cruzi. Mol Biochem Parasitol 1991; 46(2):189-200.
48. Franco FR, Paranhos-Bacalla GS, Yamauchi LM et al. Characterization of a cDNA clone encoding the carboxy-terminal domain of a 90-kilodalton surface antigen of Trypanosoma cruzi metacyclic trypomastigotes. Infect Immun 1993; 61(10):4196-4201.
49. Goncalves MF, Umezawa ES, Katzin AM et al. Trypanosoma cruzi: Shedding of surface antigens as membrane vesicles. Exp Parasitol 1991; 72(1):43-53.
50. Schenkman S, Jiang MS, Hart GW et al. A novel cell surface trans-sialidase of Trypanosoma cruzi generates a stage-specific epitope required for invasion of mammalian cells. Cell 1991; 65(7):1117-1125.
51. Schenkman S, Eichinger D. Trypanosoma cruzi trans-sialidase and cell invasion. Parasitol Today 1993; 9(6):218-222.
52. Tomlinson S, Pontes de Carvalho LC, Vandekerckhove F et al. Role of sialic acid in the resistance of Trypanosoma cruzi trypomastigotes to complement. J Immunol 1994; 153(7):3141-3147.
53. Gao W, Wortis HH, Pereira MA. The Trypanosoma cruzi trans-sialidase is a T cell-independent B cell mitogen and an inducer of nonspecific Ig secretion. Int Immunol 2002; 14(3):299-308.
54. Ming M, Chuenkova M, Ortega-Barria E et al. Mediation of Trypanosoma cruzi invasion by sialic acid on the host cell and trans-sialidase on the trypanosome. Mol Biochem Parasitol 1993; 59(2):243-252.
55. Todeschini AR, Dias WB, Girard MF et al. Enzymatically inactive trans-sialidase from Trypanosoma cruzi binds sialyl and beta-galactopyranosyl residues in a sequential ordered mechanism. J Biol Chem 2004; 279(7):5323-5328.
56. Ciavaglia Mdo C, de Carvalho TU, de Souza W. Interaction of Trypanosoma cruzi with cells with altered glycosylation patterns. Biochem Biophys Res Commun 1993; 193(2):718-721.
57. Lopez M, Huynh C, Andrade LO et al. Role for sialic acid in the formation of tight lysosome-derived vacuoles during Trypanosoma cruzi invasion. Mol Biochem Parasitol 2002; 119(1):141-145.
58. Hohenester E, Engel J. Domain structure and organisation in extracellular matrix proteins. Matrix Biol 2002; 21(2):115-128.

59. Giordano R, Chammas R, Veiga SS et al. Trypanosoma cruzi binds to laminin in a carbohydrate-independent way. Braz J Med Biol Res 1994; 27(9):2315-2318.
60. Giordano R, Chammas R, Veiga SS et al. An acidic component of the heterogeneous Tc-85 protein family from the surface of Trypanosoma cruzi is a laminin binding glycoprotein. Mol Biochem Parasitol 1994; 65(1):85-94.
61. Ulrich H, Magdesian MH, Alves MJ et al. In vitro selection of RNA aptamers that bind to cell adhesion receptors of Trypanosoma cruzi and inhibit cell invasion. J Biol Chem 2002; 277(23):20756-20762.
62. Marroquin-Quelopana M, Oyama Jr S, Aguiar Pertinhez T et al. Modeling the Trypanosoma cruzi Tc85-11 protein and mapping the laminin-binding site. Biochem Biophys Res Commun 2004; 325(2):612-618.
63. Velge P, Ouaissi MA, Cornette J et al. Identification and isolation of Trypanosoma cruzi trypomastigote collagen-binding proteins: Possible role in cell-parasite interaction. Parasitology 1988; 97(Pt 2):255-268.
64. Calvet CM, Meuser M, Almeida D et al. Trypanosoma cruzi-cardiomyocyte interaction: Role of fibronectin in the recognition process and extracellular matrix expression in vitro and in vivo. Exp Parasitol 2004; 107(1-2):20-30.
65. Herrera EM, Ming M, Ortega-Barria E et al. Mediation of Trypanosoma cruzi invasion by heparan sulfate receptors on host cells and penetrin counter-receptors on the trypanosomes. Mol Biochem Parasitol 1994; 65(1):73-83.
66. Calvet CM, Toma L, De Souza FR et al. Heparan sulfate proteoglycans mediate the invasion of cardiomyocytes by Trypanosoma cruzi. J Eukaryot Microbiol 2003; 50(2):97-103.
67. Ouaissi MA, Cornette J, Capron A. Trypanosoma cruzi: Modulation of parasite-cell interaction by plasma fibronectin. Eur J Immunol 1985; 15(11):1096-1101.
68. Ouaissi MA, Afchain D, Capron A et al. Fibronectin receptors on Trypanosoma cruzi trypomastigotes and their biological function. Nature 1984; 308(5957):380-382.
69. Fernandez MA, Munoz-Fernandez MA, Fresno M. Involvement of beta 1 integrins in the binding and entry of Trypanosoma cruzi into human macrophages. Eur J Immunol 1993; 23(2):552-557.
70. Ouaissi MA. Role of the RGD sequence in parasite adhesion to host cells. Parasitol Today 1988; 4(6):169-173.
71. Ruoslahti E. The RGD story: A personal account. Matrix Biol 2003; 22(6):459-465.
72. Vray B, Camby I, Vercruysse V et al. Up-regulation of galectin-3 and its ligands by Trypanosoma cruzi infection with modulation of adhesion and migration of murine dendritic cells. Glycobiology 2004; 14(7):647-657.
73. Magdesian MH, Giordano R, Ulrich H et al. Infection by Trypanosoma cruzi. Identification of a parasite ligand and its host cell receptor. J Biol Chem 2001; 276(22):19382-19389.
74. Masocha W, Robertson B, Rottenberg ME et al. Cerebral vessel laminins and IFN-gamma define Trypanosoma brucei brucei penetration of the blood-brain barrier. J Clin Invest 2004; 114(5):689-694.
75. Andrade SG, Grimaud JA, Stocker-Guerret S. Sequential changes of the connective matrix components of the myocardium (fibronectin and laminin) and evolution of cardiac fibrosis in mice infected with Trypanosoma cruzi. Am J Trop Med Hyg 1989; 40(3):252-260.
76. Pinho RT, Vannier-Santos MA, Alves CR et al. Effect of Trypanosoma cruzi released antigens binding to noninfected cells on anti-parasite antibody recognition and expression of extracellular matrix components. Acta Trop 2002; 83(2):103-115.
77. Zuniga E, Gruppi A, Hirabayashi J et al. Regulated expression and effect of galectin-1 on Trypanosoma cruzi-infected macrophages: Modulation of microbicidal activity and survival. Infect Immun 2001; 69(11):6804-6812.
78. Ramirez MI, Ruiz Rde C, Araya JE et al. Involvement of the stage-specific 82-kilodalton adhesion molecule of Trypanosoma cruzi metacyclic trypomastigotes in host cell invasion. Infect Immun 1993; 61(9):3636-3641.
79. Ellington AD, Szostak JW. In vitro selection of RNA molecules that bind specific ligands. Nature 1990; 346(6287):818-822.
80. Tuerk C, Gold L. Systematic evolution of ligands by exponential enrichment: RNA ligands to bacteriophage T4 DNA polymerase. Science 1990; 249(4968):505-510.
81. Ulrich H, Alves MJ, Colli W. RNA and DNA aptamers as potential tools to prevent cell adhesion in disease. Braz J Med Biol Res 2001; 34(3):295-300.
82. Turner CW, Lima MF, Villalta F. Trypanosoma cruzi uses a 45-kDa mucin for adhesion to mammalian cells. Biochem Biophys Res Commun 2002; 290(1):29-34.

83. Marcipar IS, Welchen E, Roodveldt C et al. Purification of the 67-kDa lectin-like glycoprotein of Trypanosoma cruzi, LLGP-67, and its evaluation as a relevant antigen for the diagnosis of human infection. FEMS Microbiol Lett 2003; 220(1):149-154.
84. Scharfstein J, Schmitz V, Morandi V et al. Host cell invasion by Trypanosoma cruzi is potentiated by activation of bradykinin B(2) receptors. J Exp Med 2000; 192(9):1289-1300.
85. Ming M, Ewen ME, Pereira ME. Trypanosome invasion of mammalian cells requires activation of the TGF beta signaling pathway. Cell 1995; 82(2):287-296.
86. Aparicio IM, Scharfstein J, Lima AP. A new cruzipain-mediated pathway of human cell invasion by Trypanosoma cruzi requires trypomastigote membranes. Infect Immun 2004; 72(10):5892-5902.
87. Moro A, Ruiz-Cabello F, Fernandez-Cano A et al. Secretion by Trypanosoma cruzi of a peptidyl-prolyl cis-trans isomerase involved in cell infection. EMBO J 1995; 14(11):2483-2490.
88. Cuevas IC, Cazzulo JJ, Sanchez DO. gp63 homologues in Trypanosoma cruzi: Surface antigens with metalloprotease activity and a possible role in host cell infection. Infect Immun 2003; 71(10):5739-5749.
89. Santana JM, Grellier P, Schrevel J et al. A Trypanosoma cruzi-secreted 80 kDa proteinase with specificity for human collagen types I and IV. Biochem J 1997; 325(Pt 1):129-137.

CHAPTER 5

Calcium Regulation and Signaling in Apicomplexan Parasites

Kisaburo Nagamune, Silvia N. Moreno, Eduardo N. Chini and L. David Sibley*

Abstract

Apicomplexan parasites rely on calcium-mediated signaling for a variety of vital functions including protein secretion, motility, cell invasion, and differentiation. These functions are controlled by a variety of specialized systems for uptake and release of calcium, which acts as a second messenger, and on the functions of calcium-dependent proteins. Defining these systems in parasites has been complicated by their evolutionary distance from model organisms and practical concerns in working with small, and somewhat fastidious cells. Comparative genomic analyses of *Toxoplasma gondii, Plasmodium* spp. and *Cryptosporidium* spp. reveal several interesting adaptations for calcium-related processes in parasites. Apicomplexans contain several P-type Ca^{2+} ATPases including an ER-type reuptake mechanism (SERCA), which is the proposed target of artemisinin. All three organisms also contain several genes related to Golgi PMR-like calcium transporters, and a Ca^{2+}/H^{+} exchanger, while plasma membrane-type (PMCA) Ca^{2+} ATPases and voltage-dependent calcium channels are exclusively found in *T. gondii.* Pharmacological evidence supports the presence of IP_3 and ryanodine channels for calcium-mediated release. Collectively these systems regulate calcium homeostasis and release calcium to act as a signal. Downstream responses are controlled by a family of EF-hand containing calcium binding proteins including calmodulin, and an array of centrin and caltractin-like genes. Most surprising, apicomplexans contain a diversity of calcium-dependent protein kinases (CDPK), which are commonly found in plants. *Toxoplasma* contains more than 20 CDPK or CDPK-like proteases, while *Plasmodium* and *Cryptosporidium* have fewer than half this number. Several of these CDPKs have been shown to play vital roles in protein secretion, invasion, and differentiation, indicating that disruption of calcium-regulated pathways may provide a novel means for selective inhibition of parasites.

Defining Calcium Regulation in an Early Branching Eukaryote

Apicomplexan parasites are most similar to ciliates and dinoflagellates and only distantly related to plants, fungi, and animals typically used as model organisms.[1] Apicomplexans contain a remnant plastid derived from a secondary endosymbiont, called the apicoplast.[2] A number of plant-like metabolic systems are found in apicomplexans either due to retention of

*Corresponding Author: L. David Sibley—Department of Molecular Microbiology, Washington University School of Medicine, 660 S. Euclid Ave., St. Louis, MO 63110, USA.
Email: sibley@wustl.edu

Molecular Mechanisms of Parasite Invasion, edited by Barbara A. Burleigh and Dominique Soldati-Favre.

plastid-related functions or due to the acquisition of nuclear genes from the endosymbiont. For example, plant like enzymes have been identified in glycolysis[3] and some structural elements such as microtubules are more closely related to plants than animals.[4] As a consequence, comparison to model organisms is not always informative about the biology of apicomplexans either because their early origin predates the specialization of systems found in yeast, plants, and animals, or because it is not always certain where the closest affinities lie, even in systems that are conserved. Fortunately, there are a large number of whole genomes completed for the Apicomplexa, making it possible to delineate many conserved pathways through sequence-based phylogenetic comparisons.[5] Additionally, experimental tools for direct manipulation have advanced significantly, making it possible to explore specific systems directly in parasites.

Calcium is an important second messenger for signaling cascades that regulate protein secretion, motility, gene expression, and cellular development in eukaryotic cells. While calcium is an abundant cation in the environment, it is maintained at a very low resting level in the cytosol of eukaryotic cells. Rapid release of stored calcium is thus a potent signal for controlling a variety of downstream effectors. Calcium-binding motifs in several key cytosolic proteins, (e.g., the EF-hand found in calmodulin and other related proteins), allows this signal to be propagated through different pathways. Because calcium is such a potent signal, it must be rapidly dampened or it becomes highly toxic for cells. Consequently, cells have developed elaborate ways to control release in response to environmental stimuli, and then to rapidly sequester calcium to shut off the signal.

Defining the molecular pathways for the regulation of intracellular calcium in parasitic protozoa is of importance for three primary reasons. First calcium controls several vital cellular processes including secretion of adhesive proteins and motility. Second, the homeostatic mechanisms for controlling calcium may provide drugable targets as shown in the case of artemisinin, which acts on the calcium reuptake pump in the ER (i.e., SERCA).[6] Finally, as an early branching eukaryotic group, apicomplexans occupy a useful position phylogenetically that provides insights into the early origin of complex signaling pathways. This review summarizes what is known about calcium homeostasis, including clues derived form the recently completed genomes, and outlines pathways for calcium signaling in *Toxoplasma*, *Plasmodium*, and *Cryptosporidium*.

Calcium Homeostasis

Regulation of Intracellular Ca^{2+} Concentration

In eukaryotic cells, the intracellular Ca^{2+} concentration ($[Ca^{2+}]_i$) is maintained at 10,000-fold lower levels than the extracellular environment.[7,8] Maintaining this steep calcium gradient requires the concerted operation of several mechanisms present in the plasma membrane and intracellular organelles. Calcium enters through the plasma membrane, which in eukaryotic cells contains a number of calcium channels, some of which are under the control of receptors (receptor-operated Ca^{2+} channels), the electrical potential across the plasma membrane (voltage-gated Ca^{2+} channels), and the content of intracellular Ca^{2+} stores (store-operated Ca^{2+} channels), in addition to nonselective channels.[9]

In the case of the *T. gondii* tachyzoites, the $[Ca^{2+}]_i$ is about 100 nM, as detected in fura-2-loaded cells in the presence of 1 mM extracellular Ca^{2+}.[10] Similar $[Ca^{2+}]_i$ levels have been reported for extracellular *P. chabaudi* and *P. falciparum* parasites using fura-2 loading and ratiometric imaging.[11] Studies of intracellular *P. falciparum* parasites using Fura-red for ratiometric imaging indicate much higher levels of $[Ca^{2+}]_i$ (370-480 nM), which is likely an average of the cytosol and extensive ER compartment within these cells.[12] It is likely the cytosolic concentration of calcium in parasites is within the concentration range observed in other eukaryotic cells (i.e., 90-100 nM).[13] Red blood cells also maintain a low level of $[Ca^{2+}]_i$; however, following infection with malaria, the cytosolic concentration of calcium appears to rise, due to enhanced permeability and reduced export via the membrane Ca^{2+} ATPases (reviewed in ref. 11).

Storage Organelles

Intracellular organelles contain transporters, channels, and pumps that contribute to homeostasis by releasing calcium or taking it up according to the cellular requirements. The most important compartments are the mitochondria, acidocalcisome, and endoplasmic reticulum. While mitochondria are important to control large fluctuations of cytosolic calcium and the acidocalcisomes are important for storage, the endoplasmic reticulum serves as the primary source of mobilizeable calcium for signaling.

Mitochondria

Mitochondria possess a high capacity to sequester calcium. The inner mitochondrial membrane possesses a uniport carrier for Ca^{2+}, which allows the electrogenic entry of the cation driven by the electrochemical gradient generated by respiration or ATP hydrolysis. Calcium efflux takes place by a different pathway, which appears to catalyze the electrically neutral exchange of internal calcium by external sodium or protons. Previous studies have provided biochemical evidence for mitochondrial Ca^{2+} uptake in malaria parasites[14,15] and preliminary evidence suggests the presence of a uniport mechanism in *T. gondii* (Vercesi and Moreno, unpublished observations).

Acidocalcisomes

A large store of Ca^{2+} in *T. gondii* and *Plasmodium spp.* is found in the acidocalcisomes,[10,16,17] which are acidic calcium-storage organelles found in a diverse range of microorganisms from bacteria to man.[18] Acidocalcisomes are characterized by their acidic nature, high density (both in weight and by electron microscopy), and high content of pyrophosphate, polyphosphate, calcium, magnesium, and other elements.[18] Acidocalcisomes in *T. gondii* have been shown to possess a plasma membrane type ATPase (PMCA), involved in Ca^{2+} influx, with similarity to vacuolar Ca^{2+}-ATPases of other unicellular eukaryotes.[16,19] Acidocalcisomes also contain two proton pumps, a vacuolar H^{+}-ATPase (V-H^{+}-ATPase) and a vacuolar H^{+}-pyrophosphatase (V-H^{+}-PPase), involved in their acidification.[16,20,21] Although the Ca^{2+} content of acidocalcisomes is very high (probably in the molar range), most of it is bound to polyphosphate and can only be released upon alkalinization[22] or after polyphosphate hydrolysis.[16] No second messengers have been demonstrated to be involved in Ca^{2+} release from acidocalcisomes of *T. gondii*, and it seems likely this store is not directly involved in calcium-mediated signaling.

In addition to the biochemical[17] and morphological[23] evidence for the presence of acidocalcisomes in malaria parasites it has been proposed that the acidic food vacuole of *Plasmodium falciparum* is also a dynamic internal Ca^{2+} store, which possess a thapsigargin (and cyclopiazonic acid)-sensitive Ca^{2+} pump.[24] However, these studies were based on the changes in fluorescence of vacuole-localized Fluo-4 after addition of inhibitors or alkalinizing agents. More recent studies using more quantitative methods questioned the validity of the results using Fluo-4 and failed to confirm that the food vacuole acts as a major internal Ca^{2+} store containing a thapsigargin-sensitive Ca^{2+} pump.[12] Importantly, these studies did reveal the presence of electron-dense organelles rich in phosphorus (i.e., the acidocalcisomes) in *P. falciparum*.[12]

Endoplasmic Reticulum

The endoplasmic reticulum also contains a large store of Ca^{2+} with local concentration reaching millimolar levels, and this pool is generally considered to be the primary source of mobilizable calcium for signaling. The ER possesses two pathways for calcium efflux that are stimulated by IP_3 and ryanodine or cyclic ADPribose (described further below). The influx is catalyzed by the very well known sarco-endoplasmic reticulum Ca^{2+} -ATPase (SERCA), which actively translocates two Ca^{2+} ions for the hydrolysis of one ATP molecule.

Calcium Transporters and Channels

The active export of calcium across the plasma membrane of eukaryotic cells is accomplished by the action of Na^+/Ca^{2+} or Ca^{2+}/H^+ exchangers or P-type Ca^{2+}-ATPases, which are classified as plasma membrane transporters (PMCA), secretory compartment transporters (i.e., PMR1-like Golgi Ca^{2+} ATPase), and ER or SER transporters.[25]

A PMCA-type Ca^{2+}-ATPase (TgA1) has been characterized in *T. gondii*, and this protein is located in the plasma membrane and acidocalcisomes.[16] TgA1 is able to complement yeast deficient in the vacuolar Ca^{2+}-ATPase gene *PMC1*, providing genetic evidence for its function.[26] Mutants deficient in *tga1* were shown to have decreased infectivity in vitro and in vivo due to their deficient invasion of host cells.[26] Biochemical analysis revealed that the polyphosphate content of *tga1* mutants was drastically reduced, and that the basal Ca^{2+} levels were increased and unstable. These defects lead to impaired microneme secretion in response to agonists, thus demonstrating the importance of homeostasis in this calcium-activated pathway. In addition, a second PMCA-like gene is found in the *T. gondii* genome.[27] The deduced amino acid sequence (1200 residues) of this protein shows 45% identity with TgA1.[16] Surprisingly, no homologues of PMCAs are recognizable in the genomes of *Plasmodium* or *Cryptosporidium*, indicating they do not rely on this pump for calcium homeostasis.[27]

Apicomplexans contain two genes with similarity to the yeast PMR1 transporter, which is a Golgi-type Ca^{2+} ATPase.[27] One of these has previously been named ATPase4 in *P. falciparum* where it has been localized to the plasma membrane in asexual stages of malaria,[28] suggesting it may function more like a PMCAs. Characterization of in vitro activity in *Xenopus* revealed that it is inhibited by vanadate and cyclopiazonic acid but not thapsigargin or ouabain.[29] Apicomplexans contain well-conserved homologues of SERCA-type Ca^{2+} ATPases.[27] Apicomplexans also contain a gene previously named PfATPase2, which is most similar to phospholipids transporting ATPases and may not play a direct role in calcium homeostasis.[27] Finally, all three apicomplexans encode a transporter with similarity to Ca^{2+}/H^+ exchangers;[27] homologues are found in the vacuole in plants, and also present in bacteria and fungi.[25] The location of this exchanger in apicomplexans is uncertain, but it may be involved in calcium homeostasis in internal organelles.

Voltage-gated Ca^{2+} channels (VGCC) have been detected in free-living protozoa, including ciliates.[30] VGCCs regulate influx of calcium from the extracellular medium to the cell cytosol, usually in response to changes in membrane potential. No such channels have been characterized in apicomplexans, although several orthologues are present in the genome of *T. gondii*[27] One of these is similar to the previously described two-pore channel 1 (TPC1), the *Arabidopsis thaliana* Ca^{2+}-dependent Ca^{2+}-release channel.[27,31] The predicted protein has similarities to the channels present in the plant vacuoles and it is possible that this channel might be present in the acidocalcisomes or in other intracellular membranes. These calcium channels are apparently absent in *Plasmodium* spp. and *Cryptosporidium* spp.

Ca^{2+} release from the endoplasmic reticulum of eukaryotic cells is mediated by ryanodine receptors (RyR) and inositol 1,4,5-trisphosphate (IP_3) channels.[32,33] RyR are activated by a rise in $[Ca^{2+}]_i$ (Ca^{2+}-induced Ca^{2+} release, CICR).[34] In addition, there are RyR-like channels activated by cyclic ADP-ribose (cADPR), sphingosine, and a distinct Ca^{2+}-release pathway activated by nicotinic acid adenine dinucleotide phosphate (NAADP).[35] NAADP also mobilizes calcium from lysosomal like stores in sea urchin eggs,[36] and mammalian cells, although this pathway has not been explored in parasites. Pharmacological evidences indicate the presence of a IP_3/ryanodine-sensitive stores in *T. gondii*.[37] However, searching the genomes of apicomplexans failed to reveal genes for IP_3 or ryanodine channels as defined in metazoans. This apparent absence could be due to lack of homology with the channels of animal cells, as occurs in plants that also respond to these second messengers.[25]

SERCA: Structure and Function

SERCA is a large membrane protein of approximately 110-130 kDa with 10 transmembrane regions. SERCA is widely conserved from protozoa to mammals, although direct homologues are absent in yeast and fungi, which instead have a plasma membrane type Ca^{2+} ATPase and a secretory Ca^{2+} ATPase (summarized in ref. 38). Vertebrates contain three SERCA genes each of which exhibits multiple splice variants. SERCA1 is found in muscle, SERCA2 is expressed in muscle and a variety of cell types and is evolutionarily oldest, while SERCA3 is widely expressed but has a lower affinity for calcium.[38] Several human disease states are associated with dysfunction in SERCAs including Brody's disease (defect in ATP2A1) and Darier's disease (defect in ATP2A2). Plants also contain SERCA homologues with conserved transmembrane, phosphorylation and ATP binding domains.[25]

SERCA is one of the most well-characterized P-type ATPases, which are defined by the existence of phosphorylated intermediate.[38,39] Structurally, SERCA consists of 10 transmembrane regions (M1-M10) and three cytoplasmic domains (A domain, actuator; N domain, nucleotide binding; P domain, phosphorylation).[40] The N domain binds to ATP, and after the ATP hydrolysis, residue Asp351 in the P-domain becomes phosphorylated. Extensive X-Ray crystallographic studies have defined amino acids residues that form the ATP-binding and Ca^{2+}-binding pockets.[41,42] SERCA transports two Ca^{2+} molecules from the cytoplasm to the lumen of the ER during hydrolysis of one ATP molecule. The reaction mechanism involves transformation between two conformational states, known as E1 and E2, which bind calcium on the cytoplasmic side and release it into the lumen.[38,39]

SERCA as a Target for Artemisinin

The SERCA orthologue of *P. falciparum* (PfATPase6) encodes a 1228 amino acid protein that shares 41% identity and 65% similarity with human SERCA.[43] All of the key amino acids residues that are predicted to form the Ca^{2+}-binding pocket and ATP-binding site are conserved in PfATPase6[43] and the TgSERCA homologue.[27] Thapsigargin is a sesquiterpene lactone, derived from the plant *Thapsia garganica* that was first reported as a tumor promoter.[44] Thapsigargin inhibits mammalian SERCA at subnanomolar concentrations by locking the molecule in the E2 form, which is unable to bind calcium.[45,46] Biochemical evidence that calcium homeostasis depends on the action of the SERCA pump has been provided by previous studies showing that *T. gondii*[10,17,47] and *P. falciparum*[48] are sensitive to thapsigargin. Consistent with this, 10 of 14 amino acids that are thought to interact with thapsigargin are conserved in PfATPase6[43] and TgSERCA.[27]

Other plant alkaloids also offer potential as selective antiparasitic compounds due to inhibition of SERCA. Artemisinin is an antimalarial agent isolated from sweet wormwood (*Artemisia annua*)[49,50] and artemisinin and derivatives are used to treat multidrug-resistant malaria. Artemisinin has a unique peroxide bridge that is essential for anti-malarial activity. The peroxide is activated to a radical by the presence of Fe^{2+}, and this activation step is essential for its potent anti-malarial activity.[49,50] One potential mechanism for its antimalarial activity is the demonstration that artemisinin inhibits malarial SERCA (PfATPase6) expressed in *Xenopus laevis* oocytes.[6] More recently, Uhlemann et al[51] reported that a single amino acid residue could determine the sensitivity of PfATPase6 to artemisinin in *X. laevis* oocytes. An artemisinin-binding cleft was predicted by computer modeling based on the thapsigargin-binding pocket, which was previously demonstrated by structural studies of the mammalian SERCA.[51] Transmembrane domains M3, M5, and M7 form the putative artemisinin binding-site. Within this region, Leu263 in malarial SERCA was found to be essential for determining the sensitivity to artemisinin. When this residue was mutated to Glu, which is the corresponding residue to mammalian SERCA, the K_i value to artemisinin increased more than 300-fold. Furthermore, the K_i values of artemisinin for SERCA harboring L263A and L263S mutations (corresponding to *P. vivax* and *P. berghei* orthologues, respectively) were in proportion to the sensitivity of these *Plasmodium* species to drug treatment.[51] The sequence of SERCA in *T. gondii* contains

Glu at the residue corresponding to Leu263, consistent with the ~30-fold lower sensitivity of *T. gondii*.[52,53] to artemisinin vs. *P. falciparum*.[54] While these results suggest that SERCA is a possible target for artemisinin, further experiments are required to confirm that the inhibition of SERCA is responsible for the major antimalarial effects of atremisinin.

Calcium-Regulated Signaling Pathways in Apicomplexans

Motility, cell invasion, and egress from infected cells have all been linked to intracellular calcium in *T. gondii*. The relatively robust nature of *T. gondii* tachyzoites has made it possible to probe these events in real-time using time-lapse video recording with dyes such as fluo-4, and to provide quantitative measurements of calcium levels using fura-2. Moreover, efficient systems are available to quantitatively monitor gliding motility, cell invasion, and egress of the parasite. The genome of *T. gondii* has recently been completed and excellent systems for genetic manipulation are available to test the role of specific genes by knock-out[55] or knock-down studies.[56] In this regard, *T. gondii* provides a model system for studying the role of intracellular calcium in apicomplexans.

Microneme Secretion Pathway

During interaction of the host cell, microneme proteins are secreted at the apical tip of the parasite where they engage host cell receptors and mediate polarized attachment to the host cell.[57] Discharge of microneme proteins is triggered by treatment with calcium ionophores and conversely blocked by chelation of intracellular calcium with BAPTA-AM, thus demonstrating that a rise in intracellular calcium is both necessary and sufficient for secretion.[57] Since this initial demonstration of the requirement for calcium in secretion in *T. gondii*, similar studies have revealed that this is likely a conserved process in *C. parvum*[58] and *P. berghei*.[59] Treatment with short chain alcohols and acetaldehyde are also potent triggers of microneme release in *T. gondii*, a property that may result from activation of phospholipase C.[47] Activation of microneme secretion also requires protein kinases, as initially shown by studies that the general S/T kinase inhibitor staurosporine blocks microneme secretion, host cell attachment, and invasion in both *T. gondii*[57] and *P. knowlesi*.[60] Subsequently, two classes of protein kinases have been identified that act downstream of the calcium signal and which are essential for microneme secretion: plant-like calmodulin-like protein kinases (CDPKs)[61] and cyclic GMP-dependent protein kinase (cGMP-PK)[62] (described further below).

Induction of microneme secretion involves both IP_3-like and ryanodine-like response channels for elevating intracellular calcium (Fig. 1). Treatment with ethanol increased IP_3 and $[Ca^{2+}]_i$ and this pathway is sensitive to inhibitors of IP_3 channels.[37] Recent evidence indicates that IP_3 is produced in *T. gondii* by a phosphoinositide-specific phospholipase C, delta type.[63] Studies with *T. gondii* indicate that apicomplexans also responded to agonists of cADPR-gated channels such as ryanodine and caffeine[37] (Fig. 1). Consistent with this model, *T. gondii* contains cADPR cyclase and hydrolase activities, the two enzymes that control cADPR levels.[64] Pharmacological evidence suggests that both IP_3 and ryanodine response pathways contribute to calcium-mediated secretion (Fig. 1).

Calcium Oscillations and Motility

Calcium-dependent secretion of microneme proteins is also required for motility of *T. gondii*, presumably due to the ability of these adhesive proteins to bind to the substratum, cross the membrane, and link to the cytoskeleton.[65] Imaging of gliding parasites with the calcium-sensitive dye fluo-4 revealed a surprising result that intracellular calcium levels undergo periodic fluctuations in *T. gondii*, rising and falling in repeated cycles.[66] Calcium oscillations are associated with gliding and treatment with several agonists of calcium release revealed that these oscillations control motility.[67] Treatment with caffeine causes a prolonged elevation in intracellular calcium and blocks motility. In contrast, treatment with calmidazolium, which decreases the amplitude but increases the frequency of calcium spikes, increases microneme secretion and

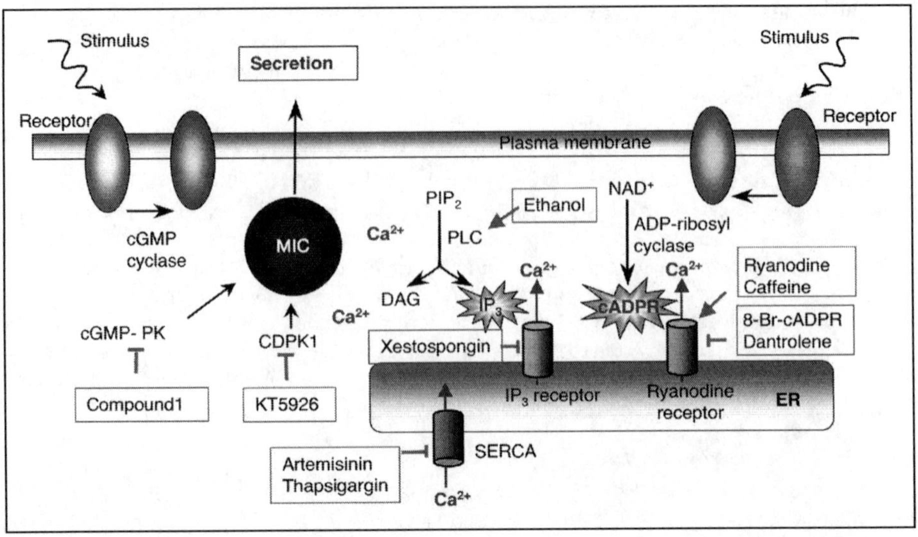

Figure 1. Calcium-mediated secretions in *T. gondii.* Elevated calcium controls microneme secretion in *T. gondii.*[47,57] Two pathways for calcium-mediated release have been described in *T. gondii*: generation of cyclic ADP ribose (cADPR) by ADP-ribose cyclase[64] and generation of IP_3 [37] Although specific release channels have not been identified, pharmacological evidence supports the presence of channels that respond to caffeine and ryanodine.[37] Ethanol is a potent trigger of secretion and is thought to act by stimulating phospholipase C (PLC) to generate IP_3. Ryanodine response channels are blocked specifically by 8-Bromo-cADP (8-Br-cADP) and dantrolene,[64] while IP_3 channels are blocked by xestospongin.[37] A calcium reuptake mechanism is provided by SERCA. Artemisinin and thapsigargin are thought to act by inhibiting this reuptake mechanism ([6] and unpublished data). Downstream of calcium release, two different kinases play an important role in secretion: calmodulin-like domain kinase 1 (CDPK1)[61] and cGMP- protein kinase (cGMP-PK).[62] The role of these kinases has been partially defined by the selective inhibition of cGMP-PK by compound 1[62] and CDPK1 by KT5926.[61] The natural agonists and cell surface receptors involved in sensing have not been defined.

prolongs motility.[67] While increased calcium levels within the parasite cytosol are important for motility and invasion of host cells by *T. gondii*, extracellular calcium appears to play little role.[66,68] Following completion of intracellular replication, the parasite exits the host cell through an active process of egress, which can be stimulated by calcium ionophores.[69] A decrease in extracellular K levels (which occurs in host cells due to loss of membrane integrity) has been linked to increases in intracellular calcium in the parasite and activation of egress.[70] However, this transition is not linked to changes in membrane potential, but rather to the activation of PLC and subsequent release of intracellular calcium.[70]

Ca^{2+}-Binding Proteins

Release of intracellular calcium or influx from outside the cell can activate a number of downstream signaling pathways that are mediated by calcium-binding proteins. Calmodulin (CaM) was initially characterized in *T. gondii* as a small (16 kDa) acidic calcium-binding protein with four calcium-binding sites (EF-hands) and with a high level of identity (92.5%) with human CaM.[71] By immunofluorescence analysis using monoclonal antibodies reactive against CaM from different species, *T. gondii* CaM was found in the apical end of released tachyzoites and also beneath the membrane in intracellular parasites.[71] Immunogold electron microscopy using monoclonal antibodies against mammalian CaM confirmed the localization of CaM in the anterior region of tachyzoites.[72] Such heterologous antibodies may in fact recognize multiple CaM-like proteins in *T. gondii*. Whole genome analysis indicates there is a single conventional CaM, three centrins (CETN), and an additional 9 CETN-like genes in *T. gondii*.[27] Several of these more divergent CETN-like genes have been referred to as CAM1 and CAM2 and have been localized to the conoid.[73] Calmodulin has also been cloned in *P. falciparum*[74] and inhibitors of CaM block invasion of red blood cells, consistent with its apical location.[75]

Apicomplexans also contain a variety of centrin and caltractin-like genes, defined by the presence of EF-hands.[27] Centrins are a key component of centrioles at the core of the centrosome, which serves as a microtubular organizing center.[76] Centrioles also play an important role in cytokinesis, and centrin has been shown to localize to the dividing apicoplast in *Toxoplasma*.[77] It is unclear why apicomplexans have multiple centrins, but divergent centrins have been implicated in controlling voltage-gated calcium channels in *Paramecium*,[78] homologous DNA recombination and excision repair in *Arabidopisis*,[79] and axonemal functions in *Tetrahymena*.[80] *Toxoplasma* also contains a homologue of calnexin, a calcium-binding lectin in the ER that functions as a chaperone. None of the apicomplexans examined here have obvious homologues of other calcium sequestering proteins including calcineurin, endoplasmin, or troponin C.[27]

CDPKs Control Secretion and Development

Apicomplexans resemble plants in that they contain a large number of calcium-dependent protein kinases (CDPK) while they do not contain calmodulin-dependent kinases, which are abundant in animal cells. CDPKs contain an N-terminal S/T kinase domain, a linker region and up to four partially conserved EF-hands that are similar to calmodulin. CDPKs are abundant in plants, for example *Arabidopsis* contains ~30 CDPK genes, which have been implicated in diverse functions including stress response, transport, and cytoskeletal functions.[81] Comparative genome analysis reveals a large number of CDPKs are found in the *T. gondii* genome, along with several members of a related family of CDPK-related kinases that lack conserved EF-hand domains.[27] Remarkably, *Toxoplasma* contains 22 members of this family of CDPK and CRK kinases, suggesting a wide range of cellular functions. *Plasmodium* (contains 12) and *Cryptosporidium* (contains 7) by comparison have fewer members of these families of kinases, but still show this bias to plant-like CDPKs.[27]

Among the diverse number of CDPKs in apicomplexans, only a few have been studied in any detail. CDPK1 has been described in *T. gondii* as a calcium-dependent kinase that phosphorylates several potential substrates required for microneme secretion and/or motility.[61] CDPK1 was shown to be the most probable target of KT5926, a S/T kinase inhibitor that blocks parasite motility and microneme secretion.[61,82] The substrates of CDPK1 have not yet been identified but may involve machinery that is required for vesicle transport or fusion to the membrane. Importantly, the requirement for kinase activity of CDPK cannot be bypassed by elevating intracellular calcium, indicating that protein phosphorylation is a downstream event that follows increases in intracellular calcium (Fig. 1). Additionally, cGMP-PK has been shown to play an essential role in microneme secretion, motility, and invasion.[62] cGMP-PK does not appear to be involved in generation of calcium release but likely acts downstream of this signal or at an independent step (Fig. 1).

In *Plasmodium*, CDPK1 and 2 are expressed in asexual stages while CDPK 3 and 4 are expressed in gametocytes and ookinetes. Recent data indicate that CDPK4 is essential for progression of the cell cycle during differentiation of microgametocytes of *P. berghei*,[83] while CDPK3 is necessary for ookinete migration and midgut invasion.[84,85] Intracellular calcium is an important signal for gametocyte development, and microgametocytes respond to xanthurenic acid by raising intracellular calcium.[83,86] Induction of gametocyte exflagellation by pH shift has been shown to induce IP_3, suggesting a route for release of Ca^{2+} from internal stores,[87] likely from the ER based on analogy to the studies mentioned above for *T. gondii*.

Future Challenges

Apicomplexan parasites exhibit several calcium-dependent activities that are vital to their survival including protein secretion, motility, and differentiation. Understanding the molecular basis for these processes will be greatly aided by the recent completion of parasite genomes, which allow predictions about possible functions of cation transporters, channels, and calcium-dependent effector proteins. Apicomplexans sit at an early branch point in the eukaryotic tree and yet they contain both IP_3-like and ryanodine-like calcium response channels that are characteristic of higher mammals. Hence studies of calcium storage and release mechanism in these parasites may inform us about the origins of complex signaling networks in eukaryotes. One of the most unusual features of calcium metabolism in apicomplexans is the presence of a diverse array of plant-like calcium-dependent protein kinases. Calcium signaling pathways offer multiple targets for selective inhibition either due to the presence of unusual members of these calcium response proteins in parasites or due to key molecular and structural differences in conserved machinery.

Acknowledgements

Supported by grants from the NIH (AI067051, AI34306 to L.D.S.), (AI68467 to S.N.M.) the Mayo Foundation of Anesthesia Research (E.N.C.), the Uehara Memorial Foundation (K.N.). We are grateful to Gary Ward, Kami Kim, Vern Carruthers, Oliver Billker, and Con Beckers for helpful comments and discussions.

References

1. Baldauf SL. The deep roots of eukaryotes. Science 2003; 300:1703-1706.
2. Foth BJ, McFadden GI. The apicoplast: A plastid in Plasmodium falciparum and other apicomplexan parasites. Int Rev Cytol 2003; 224:57-110.
3. Dzierszinski F, Popescu O, Toursel C et al. The protozoan parasite Toxoplasma gondii expresses two functional plant- like glycolytic enzymes. Implications for evolutionary origin of apicomplexans. J Biol Chem 1999; 274:24888-24895.
4. Morrissette NS, Mitra A, Sept D et al. Dinitroanalines bind alpha-tubulin to disrupt microtubules. Molec Bio Cell 2004; 15:1960-1968.
5. Huang J, Mullapudi N, Sicheritz-Ponten T et al. A first glimpse into the pattern and scale of gene transfer in the Apicomplexa. Intl J Parasitol 2004; 34:265-274.
6. Eckstein-Ludwig U, Webb RJ, van Goethem IDA et al. Artemisinins target the SERCA of Plasmodium falciparum. Nature 2003; 424:957-961.
7. Berridge MJ, Lipp P, Bootman MD. Signal transduction. The calcium entry pas de deux. Science 2000; 287:1604-1605.
8. Berridge MJ, Lipp P, Bootman MD. The versatility and universality of calcium signaling. Nat Rev Mol Cell Biol 2000; 1:11-21.
9. Tsien RW. Calcium channels, stores, and oscillations. Annu Rev Cell Biol 1990; 6:715-760.
10. Moreno SNJ, Zhong L. Acidocalcisomes in Toxoplasma gondii tachyzoites. Biochem J 1996; 313:655-659.
11. Garcia CRS. Calcium homeostasis and signaling in the blood-stage malaria parasite. Parasitol Today 1999; 15:11-17.
12. Rohrback P, Freidrich O, Hentschel J et al. Quantitative calcium measurements in subcellular compartments of Plasmodium falciparum infected erythrocytes. J Biol Chem 2005; 280:27960-27969.
13. Grynkiewicz G, Poenie M, Tsien RY. A new generation of Ca^{2+} indicators with greatly improved fluorescence properties. J Biol Chem 1985; 260:3440-3450.
14. Gazarini ML, Garcia CR. The malaria parasite mitochondrion senses cytosolic Ca^{2+} fluctuations. Biochem Biophys Res Comm 2004; 321:138-144.
15. Uyemura SA, Luo S, Moreno SN et al. Oxidative phosphorylation, Ca^{2+} transport, and fatty acid-induce uncoupling in malaria parasites mitochondria. J Biol Chem 2000; 275:9709-9715.
16. Luo S, Vieira M, Graves J et al. A plasma membrane-type Ca^{2+}-ATPase colocalizes with a vacuolar H^+-pyrophosphatase to acidocalcisomes of Toxoplasma gondii. EMBO J 2001; 20:55-64.
17. Marchesini N, Luo S, Rodrigues CO et al. Acidocalcisomes and vacuolar H^+-pyrophosphatase in malaria parasites. Biochem J 2000; 347(Pt 1):243-253.

18. Docampo R, Souza W, Miranda K et al. Acidocalcisomes—conserved from bacteria to man. Nat Rev Microbiol 2005; 3:251-261.
19. Bouchot A, Jaillet JD, Bonhomme A et al. Detection and localization of a Ca^{2+}-ATPase activity in Toxoplasma gondii. Cell Struct Funct 2001; 26:49-60.
20. Moreno SNJ, Zhong L, Lu HG et al. Vacuolar-type H^+-ATPase regulates cytoplasmic pH in Toxoplasma gondii tachyzoites. Biochem J 1998; 330:853-860.
21. Rodrigues CO, Scott DA, Bailey BN et al. Vacuolar proton pyrophosphatase activity and pyrophosphate $[PP_i]$ in Toxoplasma gondii as possible chemotherapeutic targets. Biochem J 2000; 349:737-745.
22. Rodrigues CO, Ruiz FA, Rohloff P et al. Characterization of isolated acidocalcisomes from Toxoplasma gondii tachyzoites reveals a novel pool of hydrolyzable polyphosphate. J Biol Chem 2002; 277:48650-48656.
23. Ruiz FA, Luo S, Moreno SN et al. Polyphosphate content and fine structure of acidocalcisomes of Plasmodium falciparum. Microsc Microanal 2004; 10:563-567.
24. Biagini GA, Bray PG, Spiller DG et al. The digestive food vacuole of the malaria parasite is a dynamic intracellular Ca^{2+} store. J Biol Chem 2003; 278:27910-27915.
25. Nagata T, Iizumi S, Satoh K et al. Comparative analysis of plant and animal calcium signal transduction element using plant full-length cDNA data. Molec Biol Evol 2004; 21:1855-1870.
26. Luo S, Ruiz FA, Moreno SN. The acidocalcisome Ca^{2+} ATPase (TgA1) of Toxoplasma gondii is required for polyphosphate storage, intracellular calcium homeostasis and virulence. Molec Micro 2005; 55:1034-1045.
27. Nagamune K, Sibley LD. Comparative genomic analysis of calcium ATPases and calcium-regulated proteins in the Apicomplexa. Molec Biol Evol 2006; 23:1613-1627.
28. Dyer M, Jackson M, McWhinney C et al. Analysis of a cation-transporting ATPase of Plasmodium falciparum. Mol Biochem Parasitol 1996; 78:1-12.
29. Krishna S, Woodrow C, Webb R et al. Expression and functional characterization of a Plasmodium falciparum Ca^{2+}-ATPase (PfATP4) belonging to a subclass unique to apicomplexan organisms. J Biol Chem 2001; 276:10782-10787.
30. Haynes WJ, Vaillant B, Preston RR et al. The cloning by complementation of the pawn-A gene in Paramecium. Genetics 1998; 149:947-957.
31. Furuichi T, Cunningham KW, Muto S. A putative two pore channel AtTCP1 mediate Ca^{2+} flux in Arabidopsis leaf cells. Plant Cell Physiol 2001; 42:900-905.
32. Berridge MJ. Inositol triphosphate and calcium signalling. Nature 1993; 361:315-325.
33. Guse AH. Cyclic ADP-ribose: A novel Ca^{2+}-mobilising second messenger. Cell Signal 1999; 11:309-316.
34. Berridge MJ. Capacitive calcium entry. Biochem J 1995; 312:1-11.
35. Chini EN, De Toledo FGS. Nicotinic acid adenine dinucleotide phosphate: A new intracellular second messenger? Amer J Physiol 2002; 282:C1191-1198.
36. Churchill GC, Okada Y, Thomas JM et al. NAADP mobilizes Ca^{2+} from reserve granules, lysosome-related stores, in sea urchin eggs. Cell 2002; 111:703-708.
37. Lovett JL, Marchesini N, Moreno SN et al. Toxoplasma gondii microneme secretion involves intracellular Ca^{2+} release from IP_3 / ryanodine sensitive stores. J Biol Chem 2002; 277:25870-25876.
38. Wuytack F, Raeymaekers L, Missiaen L. Molecular physiology of the SERCA and SPCA pumps. Cell Calcium 2002; 35:279-305.
39. Toyoshima C, Inesi G. Structural basis of ion pumping by Ca^{2+}-ATPase of the sarcoplasmic reticulum. Annu Rev Biochem 2004; 73:269-292.
40. Toyoshima C, Nakasako M, Nomura H et al. Crystal structure of the calcium pump of sarcoplasmic reticulum at 2.6 Å resolution. Nature 2000; 405:647-655.
41. Toyoshima C, Mizutani T. Crystal structure of the calcium pump with a bound ATP analogue. Nature 2004; 430:529-535.
42. Toyoshima C, Nomura H. Structural changes in the calcium pump accompanying the dissociation of calcium. Nature 2002; 418:605-611.
43. Kimura M, Yamaguchi Y, Takada S et al. Cloning of a Ca^{2+}-ATPase gene of Plasmodium falciparum and comparison with vertebrate Ca^{2+}-ATPases. J Cell Sci 1993; 104:1129-1136.
44. Thastrup O, Cullen PJ, Drobak BK et al. Thapsigargin, a tumor promotor, discharges intracellular Ca^{2+} stores by specific inhibition of the endoplasmic reticulum Ca^{2+}-ATPase. Proc Natl Acad Sci USA 1990; 87:2766-2470.
45. Sagara Y, Inesi G. Inhibition of the sarcoplasmic reticulum Ca^{2+} transport ATPase by thapsigargin at subnanomolar concentrations. J Biol Chem 1991; 266:13503-13506.
46. Sagara Y, Wade JB, Inesi G. A conformational mechanism for formation of a dead-end complex by the sarcoplasmic reticulum ATPase with thapsigargin. J Biol Chem 1992; 267:1286-1292.

47. Carruthers VB, Moreno SNJ, Sibley LD. Ethanol and acetaldehyde elevate intracellular [Ca^{2+}] calcium and stimulate microneme discharge in Toxoplasma gondii. Biochem J 1999; 342:379-386.
48. Varotti FP, Beraldo FH, Gazarini ML et al. Plasmodium falciparum malaria parasites display a THG-sensitive Ca^{2+} pool. Cell Calcium 2003; 33:137-144.
49. O'Neill PM. Medicinal chemistry: A worthy adversary for malaria. Nature 2004; 430:838-839.
50. Haynes RK, Krishna S. Artemisinins: Activities and actions. Microb Infect 2004; 6:1339-1346.
51. Uhlemann AC, Cameron A, Eckstein-Ludwig U et al. A single amino acid residue can determine the sensitivity of SERCAs to artemisinins. Nat Struct Molec Biol 2005; 12:628-629.
52. Berens RL, Krug EC, Nash PB et al. Selection and characterization of Toxoplasma gondii mutants resistant to artemisinin. J Infect Dis 1998; 177:1128-1131.
53. Sarciron ME, Saccharin C, Petavy AF et al. Effects of artesunate, dihydroartemisinin, and an artesunate-dihydroartemisinin combination against Toxoplasma gondii. Am J Trop Med Hyg 2000; 62:73-76.
54. Ngo T, Duraisingh M, Reed MB et al. Analysis of PFCRT, PFMDR1, DHFR, and DHPS mutations and drug sensitivities in Plasmodium falciparum isolates from patietns in Vietnam before and after treatment with artemisinin. Am J Trop Med Hyg 2003; 68:350-356.
55. Roos DS, Donald RGK, Morrissette NS et al. Molecular tools for genetic dissection of the protozoan parasite Toxoplasma gondii. Methods Cell Biol 1994; 45:28-61.
56. Meisner M, Brecht S, Bujard H et al. Modulation of myosin A expression by a newly established tetracycline repressor based inducible system in Toxoplasma gondii. Nuc Acids Res 2001; 29:E115.
57. Carruthers VB, Giddings OK, Sibley LD. Secretion of micronemal proteins is associated with Toxoplasma invasion of host cells. Cell Microbiol 1999; 1:225-236.
58. Chen XM, O'Hara SP, Huang BQ et al. Apical organelle discharge by Cryptosporidium parvum is temperature, cytoskeleton, and intracellular calcium dependent and required for host cell invasion. Infect Immun 2004; 72:6806-6816.
59. Gantt S, Persson C, Rose K et al. Antibodies against thrombospondin-related anonymous protein do not inhibit Plasmodium sporozoite infectivity in vivo. Infect Immun 2000; 68:3667-3673.
60. Ward GE, Fujioka H, Aikawa M et al. Staurosporine inhibits invasion of erythrocytes by malarial merozoites. Exper Parasitol 1994; 79:480-487.
61. Kieschnick H, Wakefield T, Narducci CA et al. Toxoplasma gondii attachment to host cells is regulated by a calmodulin- like domain protein kinase. J Biol Chem 2001; 276:12369-12377.
62. Wiersma HI, Galuska SE, Tomley FM et al. A role for coccidian cGMP-dependent protein kinase in motility and invasion. Intl J Parasit 2004; 34:369-380.
63. Fang J, Marchesini N, Moreno SNJ. A Toxoplasma gondii phosphoinositde phospholipase C (TgPI-PLC) with high affinity for phosphatidylinositol. Biochem J 2006; 394:417-425.
64. Chini EN, Nagamune K, Wetzel DM et al. Evidence that the cADPR signaling pathway controls calcium-mediated secretion in Toxoplasma gondii. Biochem J 2005; 389:269-277.
65. Sibley LD. Invasion strategies of intracellular parasites. Science 2004; 304:248-253.
66. Lovett JL, Sibley LD. Intracellular calcium stores in Toxoplasma gondii govern invasion of host cells. J Cell Sci 2003; 116:3009-3016.
67. Wetzel DM, Chen LA, Ruiz FA et al. Calcium-mediated protein secretion potentiates motility by Toxoplasma gondii. J Cell Sci 2004; 117:5739-5748.
68. Vieira MCF, Moreno SNJ. Mobilization of intracellular calcium upon attachment of Toxoplasma gondii tachyzoites to human fibroblasts is required for invasion. Mol Biochem Parasitol 2000; 106:157-162.
69. Endo T, Sethi KK, Piekarski G. Toxoplasma gondii: Calcium ionophore A23187-mediated exit of trophozoites from infected murine macrophages. Exp Parasitol 1982; 53:179-188.
70. Moudy R, Manning TJ, Beckers CJ. The loss of cytoplasmic potassium upon host cell breakdown triggers egress of Toxoplasma gondii. J Biol Chem 2001; 276:41492-41501.
71. Seeber F, Beuerle B, Schmidt HH. Cloning and functional expression of the calmodulin gene from Toxoplasma gondii. Mol Biochem Parasitol 1999; 99:295-299.
72. Song HO, Ahn MH, Ryu JS et al. Influence of calcium ion on host cell invasion and intracellular replication by Toxoplasma gondii. Korean J Parasitol 2004; 42:185-193.
73. Hu K, Johnson J, Florens L et al. Cytoskeletal components of an invasion machine—The apical complex of Toxoplasma gondii. PLos Path 2006; 2:121-138.
74. Robson KJH, Jennings MW. The structure of calmodulin gene in Plasmodium falciparum. Molec Biochem Parasitol 1991; 46:19-34.
75. Matsomoto Y, GP, Scheibel LW et al. Role for calmodulin in Plasmodium falciparum: Implications for erythrocyte invasion by the merozoites. Eur J Cell Biol 1987; 45:36-43.
76. Salisbury JL. Centrin, centrosomes, and mitotic spindle poles. Curr Opin Cell Biol 1995; 7:39-45.

77. Striepen B, Crawford MJ, Shaw MK et al. The plastid of Toxoplasma gondii is divided by association with the centrosomes. J Cell Biol 2000; 151:1423-1434.
78. Gonda K, Yoshida A, Oami K et al. Centrin is essential for the activity of the ciliary reversal-coupled voltage-gated Ca^{2+} channels. Biochem Biophys Res Commun 2004; 323:891-897.
79. Molinier J, Ramos C, Fritsch O et al. CENTRIN2 modulates homologous recombination and nuclear excision repair in Arabidopsis. Plant Cell 2004; 16:1633-1643.
80. Guerra C, Wada Y, Leick V et al. Cloning, localization, and axonemal function of Tetrahymena centrin. Mol Biol Cell 2003; 14:251-261.
81. Cheng SH, Willmann MR, Chen HC et al. Calcium signaling through protein kinases. The Arabidopsis calcium-dependent protein kinase gene family. Plant Physiol 2002; 129:469-485.
82. Dobrowolski JM, Carruthers VB, Sibley LD. Participation of myosin in gliding motility and host cell invasion by Toxoplasma gondii. Mol Microbiol 1997; 26:163-173.
83. Billker O, Tewari R, Franke-Fayard B et al. Calcium and a calcium-dependent protein kinase regulate gamete formation and mosquito transmission in a malaria parasite. Cell 2004; 117:503-514.
84. Siden-Kiamos I, Ecker A, Nyback S et al. Plasmodium berghei calcium-dependent protein kinase 3 is required for ookinete gliding and mosquito midgut invasion. Molec Micro 2006; 60:1355-1363.
85. Ishino T, Orito Y, Chinzei Y et al. A calcium-dependent protein kinase regulates Plasmodium ookinete access to the midgut epithelial cell. Molec Micro 2006; 59:1175-1184.
86. Billker O, Lindo V, Panico M et al. Identification of xanthurenic acid as a putative inducer of malaria development in the mosquito. Nature (Lond.) 1998; 392:289-292.
87. Martin SK, Jett M, Schneider I. Correlation of phosphoinositide hydrolysis with exflagellation in the malaria microgametocyte. J Parasitol 1994; 80:371-378.

CHAPTER 6

Trypanosoma cruzi: Parasite and Host Cell Signaling during the Invasion Process

Nobuko Yoshida* and Mauro Cortez

Abstract

Mammalian cell invasion by *Trypanosoma cruzi* is a complex process in which various parasite and host cell components interact, triggering the activation of signaling cascades and Ca^{2+} mobilization in both cells. Using metacyclic trypomastigotes (MT) generated in vitro and tissue culture-derived trypomastigotes (TCT), as counterparts of insect-borne and bloodstream parasites, respectively, the mechanisms of host cell invasion by *T. cruzi* have been partially elucidated. Distinct sets of molecules are engaged by MT and TCT to enter target cells. MT make use of surface glycoproteins with dual Ca^{2+} signaling activity, in a manner dependent of *T. cruzi* isolate. In highly infective MT, the binding of gp82 to its receptor triggers a signaling cascade involving protein tyrosine kinase, phospholipase C and production of inositol 1,4,5-triphosphate, whereas in poorly invasive MT, the mucin-like gp35/50 induces the activation of a signaling route in which adenylate cyclase, generation of cAMP and Ca^{2+} mobilization from acidocalcisomes are implicated. The host cell signaling pathways activated by MT remain to be determined. Differently from MT, the TCT surface molecules that bind to host cells as a prelude to invasion, such as the glycoproteins of gp85 family, appear to be devoid of signaling properties, but they may induce TCT enzymes, such as oligopeptidase B and cruzipain, to generate Ca^{2+} signaling factors of parasite or host cell origin. Host cell responses mediated by TGF-β receptor or integrin family member may also be triggered by TCT. A more complete and detailed picture of *T. cruzi* invasion needs further investigations.

Introduction

Host cell invasion is critical for the establishment of *Trypanosoma cruzi* infection in mammals. The hallmark of this process, which requires the engagement of various parasite and host cell components, in a concerted series of events, is the activation of signaling cascades and the increase in cytosolic Ca^{2+} concentration in both cells.[1-3]

In natural infections, metacyclic trypomastigotes (MT) from insect vectors are the parasite forms responsible for the first contact with host cells. Through the ocular mucosa, lesions in the skin or by oral route, MT reach the target cells and are internalized in a membrane-bounded vacuole. Following the escape to the cytoplasm, differentiation into amastigotes and replication, the parasites transform into trypomastigotes that are released in the circulation upon host cell rupture. These circulating parasites, which can be transmitted congenitally or by blood

*Corresponding Author: Nobuko Yoshida—Departamento de Microbiologia, Imunologia e Parasitologia, Escola Paulista de Medicina, Universidade Federal de São Paulo, R. Botucatu, 862, 04023-062 São Paulo, S.P., Brasil. Email: nyoshida@ecb.epm.br

Molecular Mechanisms of Parasite Invasion, edited by Barbara A. Burleigh and Dominique Soldati-Favre.

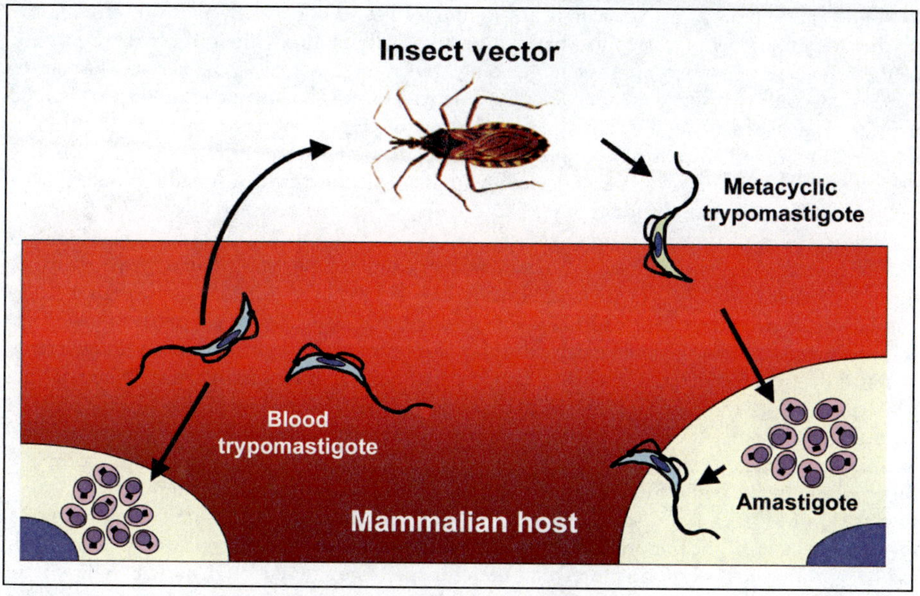

Figure 1. *T. cruzi* life cycle. Upon entering the mammalian host through the ocular mucosa, lesions in the skin or by oral route, insect-derived metacyclic trypomastigotes invade cells, differentiate into amastigotes, which replicate by binary fission and subsequently transform into trypomastigotes that are released in the circulation. These circulating parasites, which can be transmitted congenitally or by blood transfusion, disseminate to diverse organs and tissues, where they go through additional rounds of cell invasion and intracellular multiplication. The cycle goes on when the circulating trypomastigotes are ingested by the triatomine vectors during their blood meal.

transfusion, disseminate to diverse organs and tissues, where they go through additional rounds of cell invasion and intracellular multiplication (Fig. 1).

Most studies aimed at elucidating the mechanisms of mammalian cell invasion by *T. cruzi* have used MT generated in liquid media and tissue culture-derived trypomastigotes (TCT) as counterparts of insect-borne and bloodstream parasites, respectively. The available data indicate that during host cell entry MT and TCT engage distinct sets of molecules to induce the activation of signal transduction pathways leading to Ca^{2+} mobilization. In addition, different *T. cruzi* isolates may trigger different signaling routes to invade cells. Here we summarize the data from in vitro experiments of mammalian cell invasion by MT and TCT, focusing on the molecular events underlying the process.

Host Cell Invasion by Metacyclic Trypomastigotes Expressing Distinct Ca^{2+} Signaling Molecules

The process of MT internalization, which is contact-dependent, initiates with the binding of parasite surface molecules to as yet undefined host cell receptors, and triggering of bi-directional signaling cascades leading to Ca^{2+} mobilization.[4] Three MT molecules with Ca^{2+} signal-inducing activity have been identified so far: the glycoproteins gp82, gp30 and gp35/50. The role of these molecules in host cell invasion was deduced from experiments in which MT internalization was inhibited either by the native glycoprotein or by specific monoclonal antibody.[5-8] Depending on *T. cruzi* isolate, gp82, gp30 or gp35/50 is engaged and this determines the signaling routes that are activated in the parasite and the host cell.

Gp82 is a MT-specific glycoprotein containing N-linked oligosaccharides,[5] inserted in the plasma membrane by glycosylphosphatidylinositol (GPI) anchor.[9] The carbohydrate portion of the molecule is not required for target cell interaction. Located in the central domain of gp82 is the conformation-dependent cell binding site, formed by juxtaposition of two charged peptide sequences separated by a hydrophobic stretch.[10] Gp82 triggers a transient Ca^{2+} mobilization in mammalian cells susceptible to *T. cruzi* infection, such as HeLa and Vero cells, but not in *T. cruzi*-resistant K562 cells. It remains to be determined which signaling route is activated in target cells upon binding of gp82.

Binding of gp82 to its receptor triggers in MT of *T. cruzi* isolate CL, which depend on this surface molecule to enter host cells, a non transient Ca^{2+} response inhibitable by the native or the recombinant protein.[11] From studies using well known inhibitors of mammalian cell signaling cascades for treatment of MT, it was inferred that downstream to gp82, a protein tyrosine kinase (PTK) is activated, with phosphorylation of p175, a protein that is undetectable in non infective epimastigote forms.[12] Another component of this signaling cascade is phospholipase C (PLC), which generates inositol 1,4,5-triphosphate ($InsP_3$) from phosphatidylinositol-4,5-biphosphate, as deduced from inhibition of MT infectivity by PLC inhibitor U73122, as well as by drugs that affect Ca^{2+} release from IP_3-sensitive compartments.[11] Figure 2A depicts the possible signaling pathway leading to Ca^{2+} mobilization that is activated in MT of *T. cruzi* isolate CL and, by extension, in other isolates dependent on gp82 for internalization. The picture is incomplete, with many questions still to be elucidated, such as how gp82, which is GPI-anchored and therefore associates only with the outer leaflet of the MT lipid bilayer, activates PTK, and whether there is a direct connection between PTK or p175 and PLC.

Engagement of gp82 for target cell invasion has been associated with high MT infectivity, a notion reinforced by recent findings with a clone designated CL-14, derived from the CL isolate. MT of clone CL-14, which express low gp82 levels on the surface but otherwise display a surface profile similar to that of the CL isolate, exhibit reduced invasive capacity.[13] In contrast to CL isolate, and consistent with the engagement of molecules other than gp82 to enter host cells, the internalization of clone CL-14 MT is not affected by treatment of parasites with genistein, a specific PTK inhibitor, PLC inhibitor U73122 or drugs that deplete Ca^{2+} from $InsP_3$-sensitive stores.

Gp82, originally defined as the MT surface molecule that reacts with monoclonal antibody 3F6 (MAb 3F6),[14] has been detected in most *T. cruzi* isolates analysed to date but its expression is not ubiquitous. Recently, two isolates were found to be gp82-deficient while expressing MAb 3F6-reactive surface glycoprotein gp30 with Ca^{2+} signaling activity.[3] MT of gp82-deficient *T. cruzi* isolates invade host cells in gp30-mediated manner and are as infective as MT of the CL isolate.[8] As in MT of *T. cruzi* isolate CL, the infectivity of gp82-deficient MT are inhibited by PTK inhibitor genistein and by thapsigargin, which depletes Ca^{2+} from endoplasmic reticulum, implying that a common parasite signaling pathway is activated by gp30 and gp82.[8]

Mucin-like glycoproteins gp35/50, which constitute another group of MT surface molecules with Ca^{2+} signal-inducing activity,[4] are highly glycosylated molecules with glycans O-linked to threonine residues.[15] They are rich in sialic acid and galactose residues, and constitute the main acceptors of sialic acid in a reaction mediated by trans-sialidase (TS),[15] an enzyme that specifically transfers (α2-3)-linked sialic acid from extrinsic host-derived macromolecules preferentially to available galactose acceptors.[16] Sialic acid is not required for gp35/50 binding to host cell. In MT of *T. cruzi* isolate G, which rely on gp35/50 for internalization, the removal of sialic acid by neuraminidase increases the parasite ability to trigger target cell Ca^{2+} response as well as their infectivity.[17] When MT are resialylated by incubation with TS and sialyl lactose, the rate of parasite entry into host cells is restored to levels similar to those before desialylation.[17] Galactofuranose residue may be part of the cell binding site of gp35/50, as deduced from the inhibition of parasite internalization by monoclonal antibody 10D8, which recognizes a galactofuranose-containing epitope.[3]

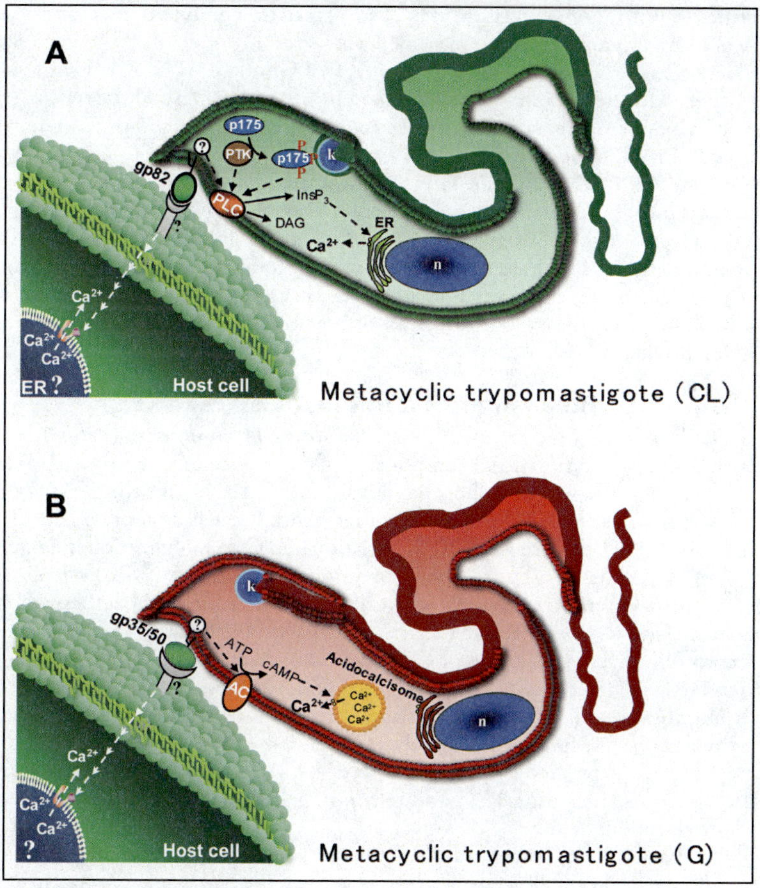

Figure 2. Schematic representation of signaling molecules and pathways that may be activated during *T. cruzi* metacyclic trypomastigote entry into nonphagocytic mammalian cells. A) In metacyclic forms of the CL isolate, which are highly invasive, the binding of the surface glycoprotein gp82 to its receptor induces the activation of protein tyrosine kinase (PTK) that phosphorylates p175. Phospholipase C (PLC) is also activated, generating $InsP_3$ and diacylglycerol (DAG). $InsP_3$ would promote release of Ca^{2+} from endoplasmic reticulum (ER). B) In metacyclic forms of the G isolate, which are poorly invasive, the binding of the mucin-like surface glycoprotein gp35/50 to its receptor triggers the activation of adenylate cyclase and generation of cAMP and Ca^{2+} is mobilized, possibly from acidocalcisomes. Both in CL and G isolates, the connection of gp82 and gp35/50, both of which are GPI-anchored to the plasma membrane, to the downstream signaling components may require adaptor (?) proteins. The routes leading to Ca^{2+} mobilization in host cells upon binding of gp82 or gp35/50 are not known. Dashed arrows represent hypothetical events.

In MT of *T. cruzi* isolates that engage gp35/50 to invade host cells, as is the case of G isolate, a signaling cascade distinct from that induced by gp82 is triggered (Fig. 2B). PTK and PLC are not implicated, whereas the increased parasite infectivity upon treatment with adenylate cyclase activator forskolin indicates the involvement of cyclic AMP.[18] The Ca^{2+} required for cell invasion appears to be mobilized from acidocalcisomes, the vacuoles containing a Ca^{2+}/H^+ exchange system,[19] as inferred from the diminished parasite infectivity upon treatment with ionomycin plus NH_4Cl or nigericin, combinations that release Ca^{2+} from these acidic compartments.[18]

The preferential engagement of gp35/50 by MT of G isolate is associated with low infectivity,[3] as opposed to the high infectivity of *T. cruzi* isolates that enter host cells in gp82- or gp30-mediated manner. Depending on *T. cruzi* isolate, both gp82 and gp35/50 may participate in host cell invasion. Such is the case of the Tulahuen isolate, whose infectivity is intermediate between that of G and CL isolates. Invasion by MT of this isolate is inhibitable by monoclonal antibody to either of these molecules as well as by the native gp82 or gp35/50.[5,7] The involvement of gp82, gp30 and gp35/50 in MT internalization has been established in mammalian cells that are not professional phagocytes. Whether they play a role in MT entry into phagocytic cells was not determined. Experiments with different *T. cruzi* isolates have shown that the rate of infection of macrophages by MT correlates to that of epithelial HeLa cells.

Tissue Culture Trypomastigotes and Host Cell Signaling

Several TCT surface molecules, including TS and the glycoproteins gp85 with adhesive properties towards cytokeratin 18 and components of the extracelullar matrix such as laminin, have been reported to play a role in host cell invasion.[16,20,21] The signaling activity of these molecules has not been determined. On the other hand, there is a report on an undefined soluble factor from TCT that triggers $InsP_3$ formation and Ca^{2+} mobilization in target cells.[22] This Ca^{2+} agonist is generated by the action of an oligopeptidase B (OPB) on an unknown precursor.[23,24] Interaction of TCT with host cells through either of the adhesive molecules expressed on the surface may relay the signal for parasite OPB activation. By itself, OPB, which is a cytosolic enzyme, is devoid of Ca^{2+} signaling effect. Depletion of the gene encoding OPB render TCT defective in mobilizing Ca^{2+} from thapsigargin-sensitive stores in non phagocytic mammalian cells, and in establishing infection in vitro and in vivo.[24] The uptake of TCT by macrophages, however, is largely independent of OPB.[24] It has been proposed that the Ca^{2+} agonist generated by OPB is secreted by TCT and binds to a receptor on the surface of target cells, activating PLC and generating $InsP_3$, which promotes Ca^{2+} release from the endoplasmic reticulum.[24]

Another TCT enzyme that produces a Ca^{2+} agonist is cruzipain, the major *T. cruzi* cysteine protease implicated in host cell invasion.[25] In this case, the Ca^{2+} signaling factor is of host origin. By acting on cell-bound kininogen, cruzipain produces bradykinin that triggers Ca^{2+} mobilization in cells expressing B_2 type of bradykinin receptor, thus contributing for TCT internalization.[26] This kinin-mediated transduction pathway is not ubiquitous, its activation depending on the mammalian cell type and the parasite isolate used.[26] The ability of cruzipain to release kinins from high molecular weight kininogen is modulated by heparan sulfate, which can enhance the enzyme activity up to 35-fold.[27]

To invade host cells, TCT may also use mechanisms other than those mediated by OPB or cruzipain. Ming et al[28] found that TCT attached to transforming growth factor β (TGFβ) receptor-deficient epithelial cell lines, but were unable to penetrate. Susceptibility to TCT infection was restored by transfection with TGFβ receptor genes, and treatment with TGFβ greatly enhanced parasite internalization. As a TGFβ responsive reporter gene is induced in TGFβ sensitive cell lines by TCT, but not by noninvasive epimastigotes, it was postulated that TCT may directly trigger activation of the TGFβ-signaling pathway,[28] but the putative TGFβ-like factor from TCT has never been characterized. In macrophages, TCT internalization may require the involvement of heterodimeric β1 integrins, which belong to a ubiquitous family of integral membrane proteins that link the extracellular matrix to the cortical cytoskeleton. When added to human macrophages, monoclonal antibodies to β1 subunit of VLA integrin family specifically blocked TCT uptake.[29] As that inhibition correlated with the ability to block fibronectin binding to macrophages, it is uncertain whether the parasite interacts with VLA directly or through the binding to fibronectin. Another host cell component reported to participate in TCT adhesion, and possibly invasion, is galectin 3, which increases K-Ras activation and triggers a Ras signal.[30] In experiments with human coronary artery smooth muscle cells, which express galectin-3 on the surface and also secret it, the exogenous galectin-3 enhanced TCT binding and stable transfection of cells with antisense *galectin-3* decreased parasite adhesion.[31]

Diverse signaling routes can be triggered by TCT in target cells, but it is unclear whether they are activated simultaneously and/or to what extent they share common downstream components. Also, despite the considerable number of host cell signaling molecules implicated in TCT invasion, it is difficult to determine which signal transduction pathway they are part of, because the data are fragmentary. In addition to PLC and its product $InsP_3$, which are associated with Ca^{2+} mobilization from intracellular stores, protein kinases, lipid kinases and phosphatases have been reported to participate in TCT internalization.

The role of protein tyrosine kinase (PTK) in *T. cruzi* invasion of macrophages was revealed by treatment of macrophages with genistein, a specific PTK inhibitor.[32] Tyrosine-phosphorylated residues were found to accumulate at the site of parasite association with the macrophage surface, colocalizing with host cell F-actin-rich domains.[33] Activation of macrophage protein kinase C and enhancement of parasite uptake upon incubation with recombinant gp83, a TCT surface ligand, was also reported.[34] *T. cruzi* entry into non professional phagocytes does not require PTK,[22] but may require protein kinase B, as suggests the reduced susceptibility to TCT invasion of transiently transfected 3T3 cells containing an inactive mutant enzyme, when compared to the active mutant-transfected cells.[35]

In addition to protein kinases, protein phosphatases may play a role in TCT internalization. Invasion of TCT, which induces tyrosine dephosphorylation of several proteins in L_6E_9 myoblasts, was found to be greatly reduced in the presence of protein tyrosine phosphatase inhibitors, and in the presence of excess phosphotyrosine, but not of phosphoserine or phosphothreonine.[36] Implication of another phosphatase in TCT internalization comes from observations that, when infected with TCT, human HEp2 tumor cells display altered placental alkaline phosphatase activity and different pattern of actin organization, compared to control cells, and their susceptibility to invasion is decreased by interfering with the enzyme activity before infection.[37]

TCT invasion appears also to require lipid kinases. In macrophages, TCT entry stimulated the formation of lipid products of phosphoinositide (PI) 3-kinases,[38] and the enzyme was localized at the site of parasite interaction with macrophages, rich in F-actin.[33] Upon treatment with PI 3-kinase inhibitor wortmannin, murine and human macrophages, nonphagocytic Vero, L_6E_9 and 3T3 cells became less susceptible to TCT infection.[35]

Multiple mechanisms of mammalian cell invasion are available for TCT. If they use them alternately, sequentially or simultaneously may depend on which *T. cruzi* isolate interacts with which cell type. While the signaling process in TCT is not uncovered, and with many pieces of the various possible signaling cascades in the host cell still lacking, the precise sequence of events following the interaction of the two cells remains undefined. Figure 3 illustrates some of the possibilities TCT may exploit to invade cells that are not professional phagocytes. By attaching to target cells through the surface molecules of gp85 family, for instance, TCT induce the generation of a Ca^{2+} agonist that triggers the signaling cascade involving activation of PLC and Ca^{2+} release in $InsP_3$-dependent manner. Concomitantly or sequentially, the cruzipain secreted by attached TCT exerts its action on kininogen, producing bradykinin that also induces Ca^{2+} response when recognized by its receptor. *T. cruzi* isolates that trigger both signaling pathways may be more invasive, and additional signaling routes, such as those mediated by TGFβ receptors, may further contribute to the process of TCT internalization.

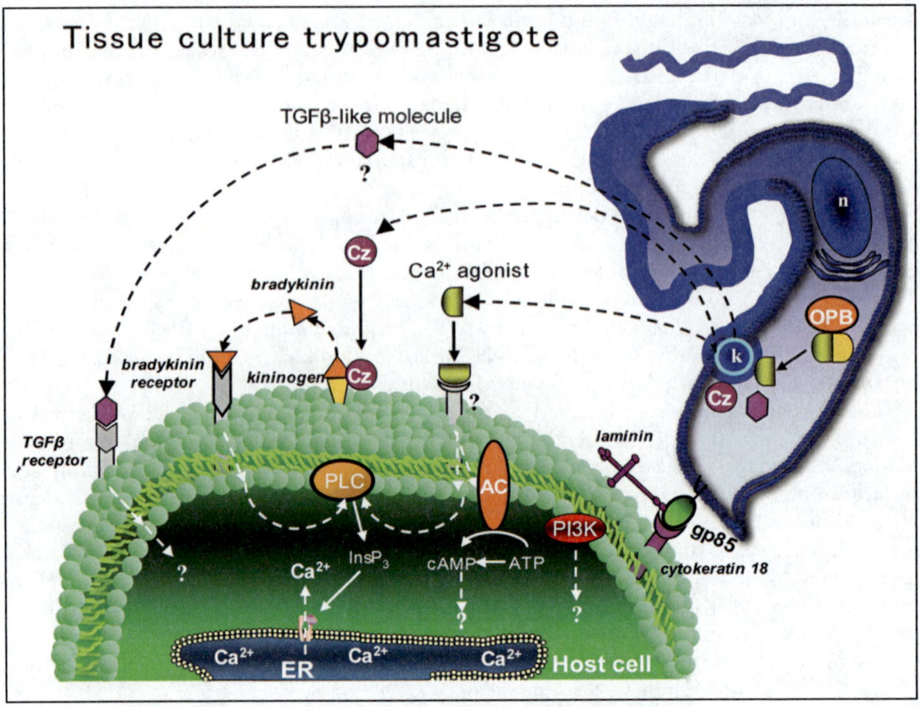

Figure 3. Schematic representation of signaling molecules and pathways that may be activated during penetration of tissue culture trypomastigotes into nonphagocytic mammalian cells. Attachment of the parasite to the host cell, for instance through the laminin-binding surface glycoprotein gp85 that is recognized by cytokeratin 18, may trigger the cytosolic oligopetidase B (OPB)-mediated production of a Ca^{2+} agonist, leading to activation of phospholipase C (PLC) and generation of $InsP_3$ that promotes Ca^{2+} release from endoplasmic reticulum (ER). A Ca^{2+}-signaling factor of host origin, bradykinin, could also be generated by action of *T. cruzi* major cysteine protease cruzipain (Cz) on cell bound kininogen. Among the other possible signaling pathways that may be activated by the parasite is that involving TGF-β receptor. Whether phosphoinositide 3-kinase (PI3K) and adenylate cyclase (AC) are part of any of these signaling routes remains to be defined. Dashed arrows represent hypothetical events.

Host Cell Ca^{2+} Signaling, Actin Cytoskeleton Disassembly and Lysosome Recruitment

Until recently the prevalent notion on TCT invasion of non phagocytic mammalian cells was that, following Ca^{2+} mobilization from intracellular stores and actin cytoskeleton disassembly, lysosomes are recruited to the site of *T. cruzi* penetration and fuse with the plasma membrane, contributing to formation of the parasitophorous vacuole.[39] In that scheme, cAMP potentiated the Ca^{2+}-triggered lysosome exocytosis,[40] which was regulated by synaptotagmin VII, a ubiquitously expressed synaptotagmin isoform that is localized on the membrane of lyososomes in different cell types.[41] That the targeted lysosome exocytosis is the predominant mechanism by which TCT gain access to nonprofessional phagocytic cells has been challenged by the findings that only a minimal fraction of invading TCT associate with host cell lysosomes, whereas the majority of parasites induce plasma membrane invagination and the TCT-containing vacuoles gradually acquire lysosomal markers.[42] According to Wilkowsky et al,[43] the newly forming *T. cruzi* compartments first interact with an early endosome compartment and subsequently with other late endosomes, before interaction with lysosomes.

In the lysosome-independent pathway of *T. cruzi* entry, the invading TCT tightly associated with the host cell plasma membrane is rarely seen colocalized with F-actin and the parasite internalization is facilitated by disruption of target cell actin cytoskeleton.[42] Apparently, regardless of the invasion route, wortmannin sensitive target cell PI 3-kinases are involved in TCT invasion of different nonphagocytic cell types, TCT vacuoles enriched in PI 3-kinase lipid products being detectable following parasite-host cell interaction.[42] Whether PI 3-kinases are associated with Ca^{2+} signaling and actin cytoskeleton rearrangements is not known.

Lysosomal fusion at the site of TCT contact with the target cell may play only a minor role in parasite internalization,[42] but the formation of parasitophorous vacuoles with lysosomal properties, which is also PI 3-kinase-dependent, is essential to prevent the exit of TCT, thus ensuring productive infection.[44] According to Woolsey and Burleigh,[45] actin assembly is required for the fusion of TCT-containing vacuoles with lysosomes as well as for cellular retention of parasites, so that prolonged disruption of actin microfilaments results in significant loss of internalized parasites. Experiments of cytochalasin D pretreatment of cells have shown that lysosome-independent TCT entry is enhanced by rapid and reversible inhibition of actin polymerization, but for the internalized TCT progress to a lysosomal vacuole, which occurs within 60 min for the vast majority of parasites, actin reorganization and assembly is required.[45] In the integrated model proposed by Woolsey and Burleigh,[45] TCT attachment to host cells induces transient depolymerization of cortical actin cytoskeleton, facilitating parasite entry by plasma membrane invagination but not by lysosome-dependent route. Internalized parasites may exit the host cell while the actin polymerization and assembly does not occur and the TCT vacuole maturation, with acquisition of lysosome markers, is not completed.

Concluding Remarks

Activation of signal transduction pathways triggered in the parasite and the host cell, leading to intracellular Ca^{2+} mobilization and Ca^{2+}-induced reorganization of the host cell actin cytoskeleton, constitutes the general mechanism by which *T. cruzi* trypomastigotes invade mammalian cells. With the identification of target cell components involved, plus a plethora of *T. cruzi* molecules that has been identified and characterized structural and functionally, the whole process is beginning to be understood at the molecular level. The infective trypomastigote forms, MT and TCT, engage different molecules to interact with host cells. To enter target cells, MT make use of surface glycoproteins with dual Ca^{2+} signaling activity in a manner dependent of *T. cruzi* isolate. Binding of MT gp82 to its receptor triggers in the parasite a signal transduction pathway involving PTK, PLC and production of $InsP_3$, which is distinct from that induced by mucin-like gp35/50, in which adenylate cyclase, generation of cAMP and Ca^{2+} mobilization from acidocalcisomes are implicated. Preferential activation of either of these routes appears to determine the MT infectivity, gp82- and gp35/50-mediated routes being associated respectively with high and low invasive capacity. Differently from MT, the TCT molecules that bind to host cells as a prelude to invasion, such as the glycoproteins of gp85 family, may not have signaling properties but they could induce TCT enzymes, such as oligopeptidase B and cruzipain, to generate Ca^{2+} signaling factors of parasite or host cell origin. Other signaling routes that may be activated in host cells by TCT include those mediated by TGF-β receptor, integrin family member or galectin 3. Considerable progress has been made towards understanding the mammalian cell invasion by *T. cruzi*, but a lot more work has to be done before we can draw a more complete and detailed picture of that process.

References

1. Docampo R, Moreno SJ. The role of Ca2+ in the process of cell invasion by intracellular parasites. Parasitol Today 1996; 12:61-65.
2. Burleigh BA, Andrews NW. Signaling and host cell invasion by Trypanosoma cruzi. Curr Opinion Microbiol 1998; 1:451-465.
3. Yoshida N. Molecular basis of mammalian cell invasion by Trypanosoma cruzi. Anals Braz Acad Sci 2005; 77:26.
4. Ruiz RC, Favoreto S, Dorta ML et al. Infectivity of Trypanosoma cruzi strains is associated with differential expression of surface glycoproteins with differential Ca^{2+} signaling activity. Biochem J 1998; 330:505-511.
5. Ramirez MI, Ruiz RC, Araya JE et al. Involvement of the stage-specific 82-kilodalton adhesion molecule of Trypanosoma cruzi metacyclic trypomastigotes in host cell invasion. Infect Immun 1993; 61:3636-3641.
6. Ruiz RC, Rigoni VL, Gonzalez J et al. The 35/50 kDa surface antigen of Trypanosoma cruzi metacyclic trypomastigotes, an adhesion molecule involved in host cell invasion. Parasite Immunol 1993; 15:121-123.
7. Yoshida N, Mortara RA, Araguth MF et al. Metacyclic neutralizing effect of monoclonal antibody 10D8 directed to the 35- and 50-kilodalton surface glycoconjugates of Trypanosoma cruzi. Infect Immun 1998; 57:1663-1667.
8. Cortez M, Neira I, Ferreira D et al. Infection by Trypanosoma cruzi metacyclic forms deficient in gp82 but expressing a related surface molecule gp30. Infect Immun 2003; 71:6184-6191.
9. Cardoso de Almeida ML, Heise N. Proteins anchored via glycosylphosphatidylinositol and solubilizing phospholipases in Trypanosoma cruzi. Biol Res 1993; 26:285-312.
10. Manque PM, Eichinger D, Juliano MA et al. Characterization of the cell adhesion site of Trypanosoma cruzi metacyclic stage surface glycoprotein gp82. Infect Immun 2000; 68:478-484.
11. Yoshida N, Favoreto Jr S, Ferreira AT et al. Signal transduction induced in Trypanosoma cruzi metacyclic trypomastigotes during the invasion of mammalian cells. Braz J Med Biol Res 2000; 33:269-278.
12. Favoreto Jr S, Dorta ML, Yoshida N. Trypanosoma cruzi 175 kDa protein tyrosine phosphorylation is associated with host cell invasion. Exp Parasitol 1998; 89:188-194.
13. Atayde VD, Neira I, Cortez M et al. Molecular basis of non virulence of Trypanosoma cruzi clone CL-14. Int J Parasitol 2004; 34:851-860.
14. Teixeira MMG, Yoshida N. Stage-specific surface antigens of metacyclic trypomastigotes of Trypanosoma cruzi identified by monoclonal antibodies. Mol Biochem Parasitol 1986; 18:271-282.
15. Schenkman S, Ferguson MAJ, Heise N et al. Mucin-like glycoproteins linked to the membrane by glycosylphosphatidylinositol anchor are the major acceptors of sialic acid in a reaction catalysed by trans-sialidase in metacyclic forms of Trypanosoma cruzi. Mol Biochem Parasitol 1993; 59:293-304.
16. Schenkman S, Jiang M, Hart GW et al. A novel cell surface trans- sialidase of Trypanosoma cruzi generates a stage-specific epitope required for invasion of mammalian cells. Cell 1991; 65:1117-1125.
17. Yoshida N, Dorta ML, Ferreira AT et al. Removal of sialic acid from mucin-like surface molecules of Trypanosoma cruzi metacyclic trypomastigotes enhances parasite-host cell interaction. Mol Biochem Parasitol 1997; 84:57-67.
18. Neira I, Ferreira AT, Yoshida N. Activation of distinct signal transduction pathways in Trypanosoma cruzi isolates with differential capacity to invade host cells. Int J Parasitol 2002; 32:405-414.
19. Docampo R, Scott DA, Vercesi AE et al. Intracellular Ca^{2+} storage in acidocalcisomes of Trypanosoma cruzi. Biochem J 1995; 310:1005-1012.
20. Giordano R, Chammas R, Veiga SS et al. An acidic component of the heterogeneous Tc-85 protein family from the surface of Trypanosoma cruzi is a laminin binding glycoprotein. Mol Biochem Parasitol 1994; 65:85-94.
21. Magdesian MH, Giordano R, Juliano MA et al. Infection by Trypanosoma cruzi: Identification of a parasite ligand and its host-cell receptor. J Biol Chem 2001; 276:19382-19389.
22. Rodriguez A, Rioult MG, Ora A et al. A trypanosome-soluble factor induces IP_3 formation, intracellular Ca^{2+} mobilization and microfilament rearrangement in host cells. J Cell Biol 1995; 129:1263-1273.
23. Burleigh B, Caler EV, Webster P et al. A cytosolic serine endopeptidase from Trypanosoma cruzi is required for the generation of Ca^{2+} signaling in mammalian cells. J Cell Biol 1997; 136:609-620.
24. Caler EV, Vaena de Avalos S, Haynes PA et al. Oligopeptidase B-dependent signaling mediates host cell invasion by Trypanosoma cruzi. EMBO J 1998; 17:4975-4986.

25. Meirelles MN, Juliano L, Carmona E et al. Inhibitors of the major cysteinyl proteinase (gp57/51) impair host cell invasion and arrest the intracellular development of Trypanosoma cruzi in vivo. Mol Biochem Parasitol 1992; 52:175-184.
26. Scharfstein J, Schmitz V, Morandi V et al. Host cell invasion by Trypanosoma cruzi is potentiated by activation of bradykinin B_2 receptors. J Exp Med 2000; 192:1289-1299.
27. Lima APCA, Almeida PC, Tersariol ILS et al. Heparan sulfate modulates kinin release by Trypanosoma cruzi through the activity of cruzipain. J Biol Chem 2002; 277:5875-5881.
28. Ming M, Ewen ME, Pereira MEA. Trypanosome invasion of mammalian cells requires activation of the TGFβ signaling pathway. Cell 1995; 82:287-296.
29. Fernandez MA, Munoz-Fernandez MA, Fresno M. Involvement of beta 1 integrins in the binding and entry of Trypanosoma cruzi into human macrophages. Eur J Immunol 1993; 23:552-557.
30. Elad-Sfadia G, Haklai R, Balan E et al. Galectin-3 augments K-Ras activation and triggers a Ras signal that attenuates ERK but not phosphoinositide 3-kinase activity. J Biol Chem 2004; 279:34922-34930.
31. Kleshchenko YY, Moody TN, Furtak VA et al. Human galectin-3 promotes Trypanosoma cruzi adhesion to human artery smooth muscle cells. Infect Immun 2004; 72:6717-6721.
32. Vieira MCF, Carvalho TU, Souza W. Effect of protein kinase inhibition on the invasion process of macrophages by Trypanosoma cruzi. Biochem Biophys Res Commun 1994; 203:967-971.
33. Vieira M, Dutra JM, Carvalho TM et al. Cellular signaling during the macrophage invasion by Trypanosoma cruzi. Histochem Cell Biol 2002; 118:491-500.
34. Villalta F, Zhang Y, Bibb KE et al. Signal transduction in human macrophages by gp83 ligand of Trypanosoma cruzi: Trypomastigote gp83 ligand up-regulated trypanosome entry through protein kinase C activation. Mol Cell Biol Res Commun 1999; 2:64-70.
35. Wilkowsky SE, Barbieri MA, Stahl P et al. Trypanosoma cruzi phosphatidylinositol 3-kinase and protein kinase B activation is associated with parasite invasion. Exp Cell Res 2001; 264:211-218.
36. Zhong L, Lu HG, Moreno SN et al. Tyrosine phosphate hydrolysis of host proteins by Trypanosoma cruzi is linked to cell invasion. FEMS Microbiol Lett 1998; 161:15-20.
37. Sartori MJ, Mezzano L, Lin S et al. Role of placental alkaline phosphatase in the internalization of trypomastigotes of Trypanosoma cruzi into HEp2 cells. Trop Med Int Health 2003; 8:832-839.
38. Todorov AG, Einicker-Lamas M, Castro SL et al. Activation of host cell phosphatidylinositol 3-kinases by Trypanosoma cruzi infection. J Biol Chem 2000; 275:32182-32186.
39. Tardieux I, Webster P, Ravesloot J et al. Lysosome recruitment and fusion are early events required for Trypanosoma invasion of mammalian cells. Cell 1992; 71:1117-1130.
40. Rodriguez A, Martinez I, Chung A et al. cAMP regulates Ca^{2+}-dependent exocytosis of lysosomes and lysosome-mediated cell invasion by trypanosomes. J Biol Chem 1999; 274:6754-16759.
41. Caler EV, Chakrabarti S, Fowler KT et al. The exocytosis-regulatory protein sinaptotagmin VII mediates cell invasion by Trypanosoma cruzi. J Exp Med 2001; 193:1097-1104.
42. Woolsey AM, Sunwoo L, Petersen CA et al. Novel PI 3-kinase-dependent mechanisms of trypanosome invasion and vacuole maturation. J Cell Sci 2003; 116:3611-3622.
43. Wilkowsky SE, Barbieri MA, Stahl P et al. Regulation of Trypanosoma cruzi invasion of nonphagocytic cells by the endocytically active GTPases dynamin, Rab5, and Rab7. Biochem Biophys Res Commun 2002; 291:516-521.
44. Andrade LO, Andrews NW. Lysosomal fusion is essential for the retention of Trypanosoma cruzi inside host cells. J Exp Med 2004; 200:1135-1143.
45. Woolsey AM, Burleigh B. Host cell actin polymerization is required for cellular retention of Trypanosoma cruzi and early association with endosomal/lysosomal compartments. Cell Microbiol 2004; 6:829-838.

Chapter 7

Host Cell Actin Remodeling in Response to *Cryptosporidium*

Steven P. O'Hara, Aaron J. Small, Xian-Ming Chen and Nicholas F. LaRusso*

Introduction

Despite sporadic reports of *Cryptosporidium* infection throughout the 1900s, the clinical significance of this parasite in humans was not recognized until the first documented human diagnosis of *C. parvum* in 1976,[1,2] and the subsequent realization that as an opportunistic infectious agent, *Cryptosporidium* infection in AIDS patients was associated with significant morbidity and mortality[3,4] and a causative agent of AIDS-related biliary disease.[5-7] With the advent of highly active antiretroviral therapy, AIDS-related cryptosporidial disease occurs less frequently, yet *Cryptosporidium* is now recognized as a widely dispersed parasite and a significant enteropathogen of immunocompetent and immunocompromised hosts. Two species, *C. hominis* and *C. parvum*, readily infect humans. While *C. parvum* infects a wide range of mammalian hosts, *C. hominis* is believed to infect only humans. Most of the experimental evidence of host cell actin remodeling to date has been accomplished using *C. parvum*.

C. parvum exhibits a monoxenous life cycle, where all developmental stages occur in a single host. The infective cycle begins when a vertebrate host ingests an oocyst. *C. parvum* oocysts excyst in the gastrointestinal tract, releasing infective sporozoites that invade the enteric epithelium. It is these infective sporozoites that have been utilized to look at the initial cascade of events culminating in actin reorganization in cell culture systems. Once internalized at the surface of epithelial cells, the trophozoite undergoes asexual reproduction by merogony forming a Type-1 meront. Six to eight Type-1 merozoites are released from Type-1 meronts when mature; they then invade neighboring enterocytes. Ultrastructurally, internalization of sporozoites and merozoites is similar and it is assumed that both distinct developmental stages utilize the same invasion machinery and drive actin reorganization through similar mechanisms. Less is known about host cell actin reorganization during the sexual stages of parasite development. Therefore, the current brief summary will focus on actin reorganization during zoite internalization and development, focusing primarily on the molecular mechanisms of actin reorganization.

Infection Process and Ultrastructural Observations

Microscopic examination of tissues infected with *Cryptosporidium* reveals dramatic alterations in host cell morphology. Infected cells are typically shortened and the microvilli in the immediate vicinity of the parasite are often branched and dramatically elongated (Fig. 1),

*Corresponding Author: Nicholas F. LaRusso—Miles and Shirley Fiterman Center for Digestive Diseases, Mayo Clinic, 200 First Street, SW, Rochester, Minnesota 55905, USA. Email: larusso.nicholas@mayo.edu

Molecular Mechanisms of Parasite Invasion, edited by Barbara A. Burleigh and Dominique Soldati-Favre. ©2008 Landes Bioscience and Springer Science+Business Media.

Figure 1. Scanning electron micrograph of internalized *Cryptosporidium parvum* in bovine fallopian tube epithelial cells. Several short microvilli are observed on the surface of the cells, while many long, branched microvilli-like structures are observed in the immediate vicinity of the internalized parasites. The base of each parasite reveals a fluted appearence, possibly indicative of the microvillar origin. Bar, 1.5 μm. Reprinted from Forney JR et al. Infection and Immunity 1999; 67:844-852; ©1999 with the permission of authors and the American Society for Microbiology.[16]

however, it has not been determined whether, *C. parvum* preferentially infects in regions where microvilli are more robust or if the infection process initiates the actin based extension of these cellular projections. Immediately following oocyst excystation, the sporozoites exhibit gliding cell motility, a conserved method of motility throughout the apicomplexans. Through sequential gliding, attachment, and internalization, the zoite is ultimately encapsulated in host-derived membrane. The electron microscopic (EM) ultrastructural details of zoite attachment and invasion are well characterized.[8-11] Upon internalization, the zoites are encapsulated in a bi-membrane structure, on the surface of epithelial cells. The bi-membrane structure encapsulating the parasite is comprised of the outer and inner parasitophorous vacuole membranes (PVM), which are derived from the host, and between which is a thin layer of cytoplasm.[8,9,12] The bi-membrane structure, of host origin, links over the top of an invading parasite[13] and surrounds the parasitophorous vacuole that separates the outer parasite pellicle membrane from the host-derived membranes with the exception of the region of membrane fusion at the annular ring (Fig. 2). As the invasion process ensues, a unique structure is formed at the base of the host-parasite interface, containing electron dense material (dense band) with an adjacent filamentous network,[8] later determined to be polymerized actin.[14,15] Therefore, the obligate intracellular parasite proceeds through developmental stages on the surface of epithelial cells in a unique niche that is intramembranous, yet extracytoplasmic.

Accumulation of Actin and Actin-Binding Proteins to Infection Sites

Early ultrastructural observations consistently described the formation of a microfilament "plug" beneath the dense band, suggesting the initiation of actin reorganization within the host cell. Additionally, several researchers described the accumulation of host cell intracellular vesicles immediately beneath the developing parasite[10,16] consistent with cytoskeletal manipulations and similar to observations made in our lab (unpublished). Several groups localized actin to

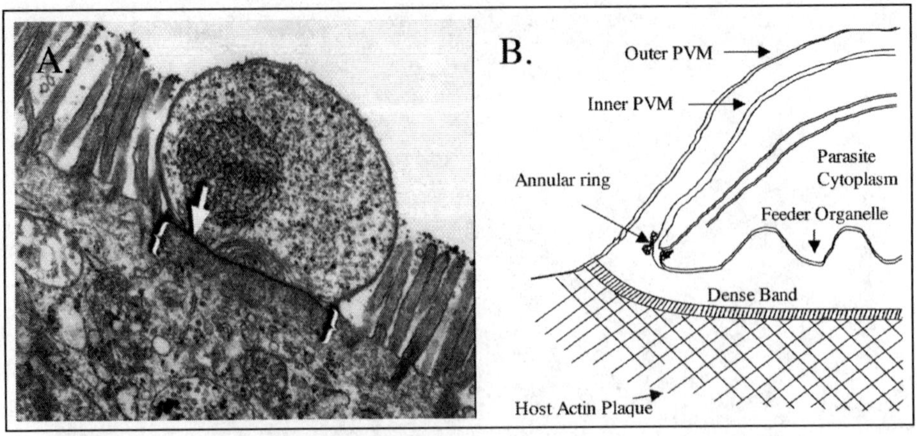

Figure 2. Transmission electron micrograph of a murine small intestinal epithelial cell infected by *Cryptosporidium* and representation of membrane organization. A) Transmission electron micrograph showing a *C. parvum* infected murine small intestinal epithelial cell. The electron dense band (arrow) separates the host cell cytoplasm from the developing parasite. Immediately beneath the dense band, a filamentous network (brackets) is observed. Using various microscopic techniques, including immunofluorescence, phalloidin labeling, immunogold electron microscopy, and β-actin-GFP visualization, it has been demonstrated that the filamentous material is composed of actin. Magnification x15,000. B) Membrane organization of internalized parasite. The outer and inner parasitophorous vacuole membranes (PVMs), derived from the host cell membrane, encapsulate the developing parasite on the surface of epithelial cells. Between the host derived PVMs is a thin layer of host cytoplasm, to which polymerized actin, ezrin and villin have been localized. The host derived inner PVM and parasite derived membranes fuse at a Y-junction to which an electron dense material, the annular ring, has been localized. The feeder organelle membrane, is derived from an anterior vacuole that forms during parasite internalization. Parasite and host derived products have been observed in the region between the feeder organelle and the dense band. Filamentous actin accumulates in the region directly beneath the dense band. Each membrane is drawn with 2 lines to represent a lipid bilayer. Not drawn to scale. A) Reprinted from Elliott DA, Clark DP. Infection and Immunity 2000; 68:2315-2322; 2000 with permission of author and the American Society for Microbiology.[15] B) Figure derived from Tzipori and Griffiths, 1998.[13]

infection sites by immunolocalization or phalloidin staining of infected neonate mouse ileal tissues[14] or in cell culture.[15,16,17] Bonnin et al,[14] using confocal microscopy and phalloidin staining, localized a thin, weakly fluorescent layer of polymerized actin surrounding each internalized parasite suggestive of localization to the PVM. However, using immunogold EM techniques with both β-actin and γ-actin antibodies, the researchers were unable to detect actin in the thin layer of cytoplasm surrounding each parasite, suggesting the sensitivity of this method was insufficient to detect the small amount of actin within the host-derived membranes. Based on the early observations of actin recruitment to infection sites, and the suggestion that the PVM is derived from host membrane, several studies focused on the recruitment of actin binding and remodeling proteins to infection sites. Preliminary data from Forney et al,[16] using immunofluorescent techniques, suggested immediate tyrosine phosphorylation upon inoculation of *C. parvum* sporozoites to bovine fallopian tube epithelial cells with subsequent accumulation of polymerized actin and the actin bundling, microvilli-associated protein, villin, to sites of parasite adherence to these cells. Additionally, the actin binding proteins villin and ezrin, were immunolocalized to the PVM by both immunogold electron microscopy and immunofluorescence.[14] The host-cell origin of the accumulated actin to the cytoplasm subjacent to the infection site was confirmed by the transfection of host cells with a plasmid containing β-actin green fluorescent protein (GFP).[15] Here the researchers used both phalloidin staining and the forced expression of β-actin-GFP to confirm the aggregation of actin filaments to the region of parasite internalization. The studies documented the aggregation of actin to a sharply circumscribed plaque of actin directly beneath the internalized parasite, and suggested that host cell microvilli elongation was not the source of the increased actin fluorescence observed by Forney et al.[16] Additionally, these researchers were unable to immunolocalize tyrosine phosphorylation associated with invading merozoites.

Molecular Pathway of Actin Reorganization

Immunolocalization studies confirmed the rearrangement of actin filaments, and recruitment of actin binding proteins to infection sites, yet the mechanisms, and purpose of this rearrangement remained elusive. Infectivity studies using a series of pharmacological inhibitors of signaling pathways, demonstrated that the protein kinase inhibitors genistein and staurosporine, significantly decreased the number of parasites observed on the surface of cells in a concentration dependent manner. Furthermore, pretreatment of host cells with the Phosphoinositol 3-Kinase (PI3K) inhibitor wortmannin also significantly reduced parasite numbers after 24 hours of incubation, while pretreatment with the G-protein uncoupling agent, suramin, had no effect on the number of parasites detected by microscopy.[16] The accumulation of proteins directly involved in actin nucleation and polymerization including, vasodilator-stimulated phosphoprotein (VASP), the neural Wiskott-Aldrich syndrome protein (N-WASP), and Arp3 of the Arp2/3 complex, to sites of infection, have been detected by immunofluorescence, implying the induction of actin polymerization at sites of infection.[18]

The first mechanistic approaches to determine the molecular pathways of *C. parvum* induced actin aggregation demonstrated that *C. parvum* induced actin polymerization at sites of infection utilizing the actin branching and nucleation machinery of the Arp2/3 complex of proteins. In addition to the accumulation of actin and associated structural components to infection sites, Elliot et al,[18] provided clear evidence of the importance of host actin polymerization during the establishment of an infection site by transfecting the human ileocecal cell line (HCT-8) with a C-terminal fragment of the protein Scar1 (Scar-WA). The C-terminal portion of this protein contains the Arp2/3 binding domain and activates Arp2/3 throughout the cytoplasm, making it less available for processes at the cell periphery. The number of parasites detected in transfected cells, as observed by immunofluorescence microscopy, was dramatically reduced compared to untransfected cells. In addition, cells transfected with a dominant negative form of N-WASP revealed the accumulation of this protein to sites of parasite-host cell interaction, however, since it lacked the Arp2/3 binding domain, actin polymerization at these sites and the number of parasites detected was again dramatically reduced. Neither mutant was able to completely inhibit *C. parvum* invasion; the most dramatic decrease in infection sites detected was observed by transfection of the Scar mutant in which the infection rate was 25% of control cells.[18]

Subsequent work by our group, using a model of biliary cryptosporidiosis, confirmed the importance of N-WASP activation during the development of an infection site and further demonstrated that functional Cdc42 was necessary for N-WASP activation.[19] Several Rho GTPases were analyzed for accumulation at infection sites and activation. It was determined, by both confocal immunofluorescence and immunogold electron microscopy that both Cdc42 and RhoA accumulated at sites of parasite-host cell interaction, yet we could demonstrate the activation of only Cdc42. The functional significance of Cdc42 activation was demonstrated by the dramatic decrease in the number of infection sites detected by immunofluorescence in cells expressing either a dominant negative of Cdc42 (30% of control) or an shRNA vector targeting Cdc42 (20% of control), which effectively suppressed Cdc42 protein levels. Conversely, transfection of a constitutively active Cdc42 significantly increased the number of parasites detected by immunofluorescence following a one-hour infection. Attachment of the parasite to the host cell was not inhibited, and, importantly, when observed by scanning electron

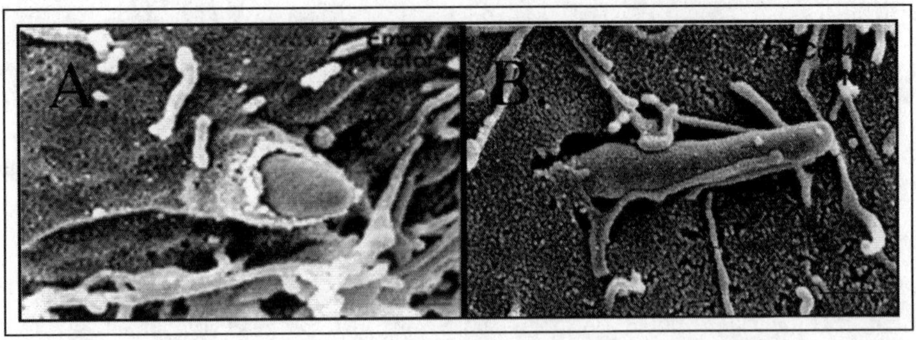

Figure 3. Electron micrographs of *C. parvum* infection of cultured biliary epithelial cells. Inhibition of host cell Cdc42 activation decreases *C. parvum* induced membrane protrusion at the host-parasite interface. A) Electron micrograph showing *C. parvum* attachment and invasion of cells transfected with an empty vector control. The host cell membrane has protruded and covers roughly two-thirds of the invading sporozoite. B) In cells transfected pKR5-Cdc42(17N)-myc (dominant negative Cdc42) the sporozoites maintain the ability to attach to host cells, yet no obvious membrane protrusion was observed. Corresponding transmission electron micrographs revealed that parasites attached to cells expressing the dominant negative Cdc42 failed to induce the dense band structure. Scale bar 0.5 μm. Reprinted from Chen et al. Infection and Immunity 2004;72:3011-3021; ©2004 with permission of authors and the American Society for Microbiology.[19]

microscopy (SEM), parasites attached to cells expressing the dominant negative of or the shRNA targeting Cdc42, revealed limited instances of host membrane remodeling at the region of parasite-host interaction (Fig. 3). Additionally, when these attached parasites were analyzed by transmission electron microscopy (TEM), the actin plaque and dense band typically observed in the host cytoplasm did not form. It was therefore proposed that parasite interactions with the apical surface of host cells initiates a signaling cascade culminating in Cdc42 recruitment and activation, and subsequent N-WASP activation with resultant Arp2/3 nucleation and branching of actin filaments, which facilitates host membrane remodeling and retention of the parasite at the apical surface. In addition to the contribution of the parasite actomyosin system, c-Src, a membrane-associated protein tyrosine kinase important in signal transduction and cortactin, an actin-binding protein known to be a central regulator in cortical actin remodeling and induced cytoskeleton reorganization, were also demonstrated to contribute to *C. parvum* cellular invasion.[20]

The upstream events leading to Cdc42 activation were determined by using both pharmacological inhibitors of host cell PI3K or by cellular expression of a mutant p85 regulatory subunit of class IA PI3K, and assessing activation of Cdc42 following *C. parvum* invasion (<1 hour).[21] It was determined that both the pharmacological inhibitors and mutant PI3K decreased the aggregation of Cdc42 to infection sites and inhibited the activation of this Rho-GTPase. Given that the function of Cdc42 requires guanine nucleotide exchange factors (GEFs), which stimulate the dissociation of GDP from the GDP-bound inactive form, and promotes the formation of the GTP-bound active form, we looked for the known Cdc42-associated GEF, Frabin. Not only did Frabin accumulate at infection sites, the aggregation was dependent on functional PI3K and, Frabin activity was necessary for activation of Cdc42 and subsequent recruitment of actin to infection sites as determined by infection of cells expressing the functionally deficient mutants of Frabin. Therefore, a molecular pathway of actin reorganization has emerged involving a phosphorylation cascade initiated by parasite interaction with the apical domain of epithelial cells, and involves PI3K activation, and the Rho-GTPase, Cdc42, which, interestingly, has been implicated in microvillar as well as filopodial extensions. The phosphorylation cascade culminates in N-WASP activation and subsequent recruitment and activation of the actin nucleation and branching complex of proteins, Arp2/3 (Fig. 4).

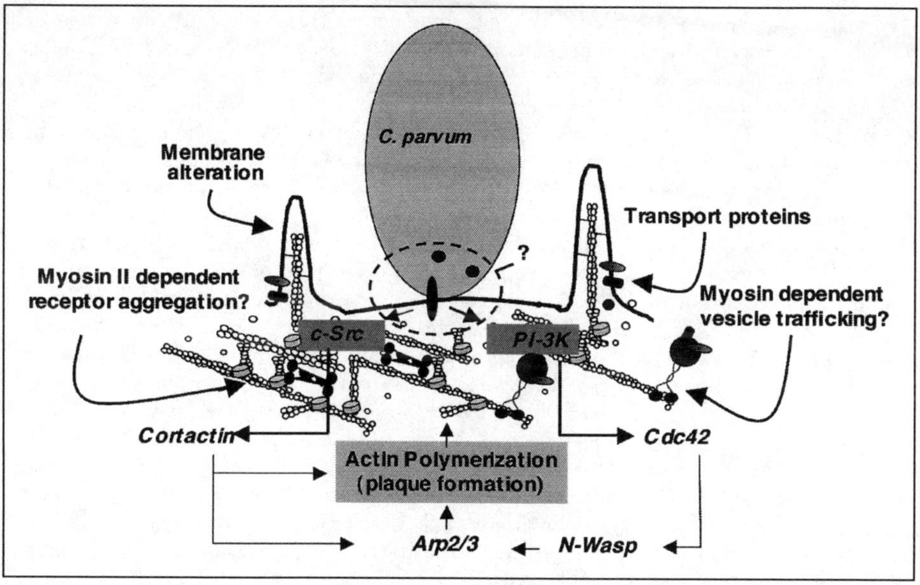

Figure 4. Schematic of *C. parvum*-induced host tyrosine kinase signaling cascades. The mechanism by which *C. parvum* induces host cell tyrosine kinase cascades remains unknown (denoted by question mark). Upon attachment to a host cell, *C. parvum* induces the aggregation (through an obscure mechanism) and activation of c-Src and PI-3K. c-Src dependent activation of cortactin can either directly induce actin polymerization or activate the Arp2/3 complex of proteins. Additionally, activated PI-3K induces Cdc42 activation, which is thought to act through N-Wasp to activate the Arp2/3 complex of proteins. Actin nucleation, branching, and polymerization result in the formation of an actin plaque beneath the internalized parasite and may be involved in membrane alterations involving the myosin dependent trafficking of vesicles and insertion of transport proteins at the site of infection. While actin reorganization is clearly evident, the direct role in the internalization process is still equivocal.

Potential Role of Actin Reorganization

Forney et al and Chen et al[16,17] first suggested an active role for actin polymerization-dependent membrane protrusion in the establishment of infection sites on the surface of epithelial cells. Elliot and Clark,[15] described the formation of an actin plaque, subjacent to the developing parasite within the host cytoplasm, but suggest that the accumulation of actin plays a structural role in maintaining the parasite in its intramembranous, yet extracytoplasmic niche. Previous studies looking at the apicomplexan parasite, *Toxoplasma gondii* demonstrated the active invasion of host cells, independent of cytoskeletal rearrangement or tyrosine kinase activity in the host cell.[22] Subsequent studies demonstrated that the *Toxoplasma* actomyosin cytoskeleton, the force behind apicomplexan gliding cell motility, provided the motive force behind host cell penetration and intracellular localization[23] and a similar mode of entry has been attributed to *plasmodium* spp.[24] *Cryptosporidium* also exhibits actin dependent gliding cell motility,[25,26] however, the initial studies of *C. parvum* motility described limited sporozoite motility, in which motility trails were short and relatively straight.[26,27] Based on the observation that pretreatment of sporozoites with cytochalasin D, which effectively abrogated *C. parvum* motility, yet did not significantly affect parasite numbers after a 24 hour infection, and the unique niche occupied by *C. parvum*, Forney et al[26] proposed that *C. parvum* utilized a unique mechanism of invasion, independent of parasite actin-based motility and membrane penetration. However, more recent investigations by Wetzel et al,[28] which used time-lapse video microscopy to visualize the early events of parasite-host interactions demonstrate that the initial entry into the host cell occurs very rapidly (within 30 seconds) and depends on the conserved penetrative force of the parasite actomyosin cytoskeleton. When an assessment of the initial invasion was performed in the presence of cytochalasin D, using a cytochalasin D resistant cell line, a complete block of rapid internalization was observed, demonstrating the requirement of parasite actin polymerization. Based on the observations by Wetzel et al[28] it seems likely that the observed host actin remodeling and resultant membrane alterations are secondary events that contribute to the establishment of a productive infection site.

The demonstration of the recruitment of actin, actin binding proteins, and actin remodeling proteins to infection sites, and the demonstration of tyrosine phosphorylation, and the recruitment and activation of PI3K, Cdc42, and N-WASP were essential for establishing infection sites within cultured cells strongly suggested the requirement of host actin polymerization for some aspect of the early infection process. Our observations led us to conclude that actin reorganization is intimately involved in parasite retention at the plasma membrane and post-invasion actin dependent processes required for successful parasite development. We have extended our initial studies on signaling cascades responsible for rearrangement of the host actin cytoskeleton to factors that may influence the efficiency of membrane extension and parasite development. Using filopodial and microvillar extensions as models, we have assessed the recruitment of myosins and various transporters, which may serve to expedite membrane extension and serve as a method of nutriment acquisition, to *C. parvum* invasion sites. Localized membrane protrusions such as filopodia are driven by actin polymerization and are often hijacked by intracellular pathogens. The overall rate of membrane protrusion depends on both the actin polymerization rate and the increase in localized cell volume. Interestingly, we determined that the Na+/Glucose Cotransporter, SGLT1, and AQP1, a channel protein selective for the movement of water and other small nonionic molecules[29] accumulate at *C. parvum* invasion sites and participate in efficient membrane protrusion events induced by the parasite.[30] Concordantly, the region of attachment displays localized glucose-driven water influx that is inhibited by either suppression of AQP1 by means of AQP1-small interfering RNA or inhibition of SGLT1 by a specific pharmacological inhibitor, Phlorizin. By inhibiting either of these proteins, we were able to diminish the efficiency of membrane protrusions as determined by both SEM and TEM. Thus, an important role for AQPs and solute transporters in the membrane remodeling associated with the development of the intramembranous/extracytoplasmic niche within host cells by *C. parvum* seems plausible if not likely. Interestingly, pretreatment of host cells with the myosin light chain kinase inhibitor, ML-7 as well as the myosin II ATPase inhibitor, 2,3-butanedione monoxime (2,3-BDM) both exhibited inhibitory effects on infectivity, while parasite attachment was unaffected.[16] Our recent, preliminary studies support the observations of the inhibitory effects of myosin inhibitors on the invasion process. We add that pretreatment of host cells with the myosin II specific inhibitor, blebbistatin, significantly decreases parasite numbers observed at the apical surface by immunofluorescence, and transporter accumulation to infection sites, suggesting a potential role for host cell actomyosin function during the formation of the unique niche occupied by *C. parvum* (Unpublished data). It seems likely; therefore, that *C. parvum* induced actin reorganization, and the recruitment of actin-binding proteins, including myosins, may serve several functions during the development of *C. parvum*.

Summary

Cryptosporidium exhibits a complex strategy to invade and establish productive infection sites, involving complimentary parasite and host cell processes. While the work regarding host cell actin remodeling has greatly enhanced our understanding of the molecular pathways involved in the parasite induced actin reorganization, the specific function of host cell actin remodeling is still equivocal. We contend that host cell actin polymerization contributes to the development of productive *C. parvum* infection sites by generating membrane protrusion events, which may assist in the retention of the parasite at the apical surface within the unique extracytoplasmic niche. With our current understanding of the molecular pathways initiating actin remodeling upon *C. parvum* interactions with host cells, the next logical step is to determine the upstream events resulting in PI3K activation and the specific role of actin remodeling in parasite development, a process that may have implications beyond host-pathogen interactions.

Acknowledgements

Supported by National Institutes of Health grants DK57993, DK24031 and by the Mayo Foundation to N.F.L.

References

1. Meisel JL, Perera DR, Meligro C et al. Overwhelming watery diarrhea associated with Cryptosporidium in an immunosuppressed patient. Gastroenterology 1976; 70:1156-1160.
2. Nime FA, Burek DL, Page M et al. Acute enterocolitis in a human being infected with the protozoan Cryptosporidium. Gastroenterology 1976; 70:592-598.
3. Current WL, Reese NC, Ernst JV et al. Human cryptosporidiosis in immunocompetent and immunodeficient persons. Studies of an outbreak and experimental transmission. New England Journal of Medicine 1983; 308:1252-7.
4. Soave R, Danner RL, Honig CL et al. Cryptosporidiosis in homosexual men. Annals of Internal Medicine 1984; 100:504-11.
5. Cockerill FR, Hurley DV, Malagelada JR et al. Polymicrobial cholangitis and Kaposi's sarcoma in blood product transfusion-related acquired immune deficiency syndrome. American Journal of Medicine 1986; 80:1237-1241.
6. Cello JP. Human immunodeficiency virus-associated biliary tract disease. Seminars in Liver Disease 1992; 12:213-218.
7. Lopez-Velez R, Terazona R, Camacho AG et al. Intestinal and extraintestinal cryptosporidiosis in AIDS patients. European Journal of Clinical Microbiology and Infectious Diseases 1995; 14:677-681.
8. Marcial MA, Madara JL. Cryptosporidium—Cellular-localization, structural-analysis of absorptive cell-parasite membrane-membrane interactions in guinea-pigs, and suggestion of protozoan transport by m-cells. Gastroenterology 1986; 90:583-594.
9. Lumb R, Smith K, Odonoghue PJ et al. Ultrastructure of the attachment of cryptosporidium sporozoites to tissue-culture cells. Parasitology Research 1988; 74:531-536.
10. Aji T, Flanigan T, Marshall R et al. Ultrastructural study of asexual development of Cryptosporidium parvum in a human intestinal cell line. Journal of Protozoology 1991; 38:82s-84s.
11. Huang BQ, Chen XM, LaRusso NF. Mechanisms of attachment and internalization of Cryptosporidium parvum by biliary epithelia: A morphological study. Journal of Parasitology 2004; 90:212-221.
12. Current WL, Reese NC. A comparison of endogenous development of three isolates of Cryptosporidium in suckling mice. Journal of Protozoology 1986; 33:98-108.
13. Tzipori S, Griffiths JK. Natural history and biology of Cryptosporidium parvum. Advances in Parasitology—Opportunistic Protozoa in Humans. 1998:40:5-36.
14. Bonnin A, Lapillonne A, Petrella T et al. Immunodetection of the microvillous cytoskeleton molecules villin and ezrin in the parasitophorous vacuole wall of Cryptosporidium parvum (Protozoa: Apicomplexa). European Journal of Cell Biology 1999; 78:794-801.
15. Elliott DA, Clark DP. Cryptosporidium parvum induces host cell actin accumulation at the host-parasite interface. Infection and Immunity 2000; 68:2315-2322.
16. Forney JR, DeWald DB, Yang SG et al. A role for host phosphoinositide 3-kinase and cytoskeletal remodeling during Cryptosporidium parvum infection. Infection and Immunity 1999; 67:844-852.
17. Chen XM, LaRusso NF. Mechanisms of attachment and internalization of Cryptosporidium parvum to biliary and intestinal epithelial cells. Gastroenterology 2000; 118:368-379.
18. Elliott DA, Coleman DJ, Lane MA et al. Cryptosporidium parvum infection requires host cell actin polymerization. Infection and Immunity 2001; 69:5940-5942.
19. Chen XM, Huang BQ, Splinter PL et al. Cdc42 and the actin-related protein/Neural Wiskott-Aldrich Syndrome Protein network mediate cellular invasion by Cryptosporidium parvum. Infection and Immunity 2004; 72:3011-3021.
20. Chen XM, Huang BQ, Splinter PL et al. Cryptosporidium parvum invasion of biliary epithelia requires host cell tyrosine phosphorylation of cortactin via c-Src. Gastroenterology 2003; 125:216-228.

21. Chen XM, Splinter PL, Tietz PS et al. Phosphotidylinositol 3-kinase and Frabin mediate Cryptosporidium parvum cellular invasion via activation of Cdc42. The Journal of Biological Chemistry 2004; 279:31671-31678.
22. Morisaki JH, Heuser JE, Sibley LD. Invasion of toxoplasma-gondii occurs by active penetration of the host-cell. Journal of Cell Science 1995; 108:2457-2464.
23. Dobrowolski JM, Sibley LD. Toxoplasma invasion of mammalian cells is powered by the actin cytoskeleton of the parasite. Cell 1996; 84:933-939.
24. Field SJ, Pinder JC, Clough B et al. Actin in the merozoite of the malaria parasite, Plasmodium falciparum. Cell Motility and the Cytoskeleton 1993; 25:43-8.
25. Gut J, Nelson RG. Cryptosporidium parvum sporozoites deposit trails of 11A5 antigen during gliding locomotion and shed 11A5 antigen during invasion of MDCK cells in vitro. Journal of Eukaryotic Microbiology 1994; 41:42S-43S.
26. Forney JR, Vaughan DK, Yang S et al. Actin-dependent motility in Cryptosporidium parvum sporozoites. Journal of Parasitology 1998; 84:908-13.
27. Arrowood MJ, Sterling CR, Healey MC. Immunofluorescent microscopical visualization of trails left by gliding Cryptosporidium parvum sporozoites. Journal of Parasitology 1991; 77:315-7.
28. Wetzel DW, Schmidt J, Kuhlenschmidt MS et al. Gliding motility leads to active cellular invasion by Cryptosporidium parvum sporozoites. Infection and Immunity 2005; 73:5379-5387.
29. Agre P, King LS, Yasui M et al. Aquaporin water channels—from atomic structure to clinical medicine. Journal of Physiology (London) 2002; 542:3-16.
30. Chen XM, O'Hara SP, Huang BQ et al. Localized glucose and water influx facilitates Cryptosporidium parvum cellular invasion by means of modulation of host-cell membrane protrusion. Proceedings from the National Academy of Sciences USA 2005; 102:6338-6343.

Chapter 8

Host Cell Actin Remodeling in Response to *Trypanosoma cruzi*:
Trypomastigote Versus Amastigote Entry

Renato A. Mortara,* Walter K. Andreoli, Maria Cecília D.C. Fernandes, Claudio V. da Silva, Adriana B. Fernandes, Carolina L'Abbate, and Solange da Silva

Abstract

Trypanosoma cruzi is the protozoan parasite that causes Chagas' disease, a highly prevalent vector-borne disease in Latin America. Chagas' disease is a major public health problem in endemic regions with an estimated 18 million people are infected with *T. cruzi* and another 100 million at risk (http://www.who.int/ctd/chagas/disease.htm). During its life cycle, *T. cruzi* alternates between triatomine insect vectors and mammalian hosts. While feeding on host's blood, infected triatomines release in their feces highly motile and infective metacyclic trypomastigotes that may initiate infection. Metacyclic trypomastigotes promptly invade host cells (including gastric mucosa) and once free in the cytoplasm, differentiate into amastigotes that replicate by binary fission. Just before disruption of the parasite-laden cell, amastigotes differentiate back into trypomastigotes which are then released into the tissue spaces and access the circulation. Circulating trypomastigotes that disseminate the infection in the mammalian host may be taken up by feeding triatomines and may also transform, extracellularly, into amastigote-like forms.[1] Unlike their intracellular counterparts, these amastigote-like forms,[2] henceforth called amastigotes, are capable of infecting host cells.[3-7] Studies in which the mechanisms of amastigote invasion of host cells have been compared to metacyclic trypomastigote entry have revealed interesting differences regarding the involvement of the target cell actin microfilament system.

Entry of Trypomastigotes and Amastigotes in HeLa and Vero Cells

It has become increasingly evident that many intracellular pathogens specialize in subverting host cell pathways to their benefit. This is particularly well characterized for invasive bacteria such as *Shigella* and *Listeria* and also for enteropathogenic *Escherichia coli* (EPEC).[8-13] For instance, interaction between EPEC and HeLa cells involves aggregation of surface microvilli at the sites of bacterial attachment to the dorsal surface of cultured HeLa cells.[14] In experiments performed in our laboratory some years ago, EPEC were centrifuged onto the cells in order to promote the interaction, and actin aggregation was monitored by staining cells with

*Corresponding Author: Renato A. Mortara—Disciplina de Parasitologia, Departamento de Microbiologia, Imunologia e Parasitologia, UNIFESP, Escola Paulista de Medicina, Rua Botucatu, 862, 6° andar, 04023-062, São Paulo, SP, Brazil.
Email: renato@ecb.epm.br

Molecular Mechanisms of Parasite Invasion, edited by Barbara A. Burleigh and Dominique Soldati-Favre.

fluorescently labeled phalloidin, using what is now known as the FAS (fluorescent actin staining) assay.[14] It was then reasoned that *T. cruzi* amastigotes could, perhaps, interact with the dorsal surface of HeLa cells in a similar fashion. Earlier data in the literature indicated that amastigotes (or amastigote-like forms) could be generated by the extracellular differentiation of trypomastigotes and these forms were capable of invading cultured cells.[3,5,6,15] Using the EPEC-derived protocol, amastigotes were centrifuged onto HeLa cells and shown to promptly aggregate actin filaments by attaching to dorsal surface microvilli.[7,16] Microvillus aggregation was followed by the formation of cup-like structures underneath the parasite (Fig. 1), that resemble the pedestals formed during EPEC attachment/effacing.[17] Extracellular amastigote invasion can be detected within 15 to 30 minutes[7] by several techniques,[2,7,18] including freeze-fracture replicas of recently-infected HeLa cells (Fig. 2).

By contrast, trypomastigotes enter HeLa cells by penetrating at their borders,[7] a characteristic behavior of the highly motile forms that had been described by Schenkman et al.[19] Interestingly, the invasion of HeLa cells by trypomastigotes induced the formation of actin-rich pseudopodial protrusions around the parasites (Fig. 3, see ref. 20). This phenomenon of membrane protrusion was not inhibited by cytochalasin D indicting that the major driving force derived from the parasite, and was detected in HeLa but not in MDCK cells.[20] Later on, other investigators found similar pseudopodial extensions around trypomastigotes invading cardiomyocytes[21] and COS-7 cells (unpublished observations).

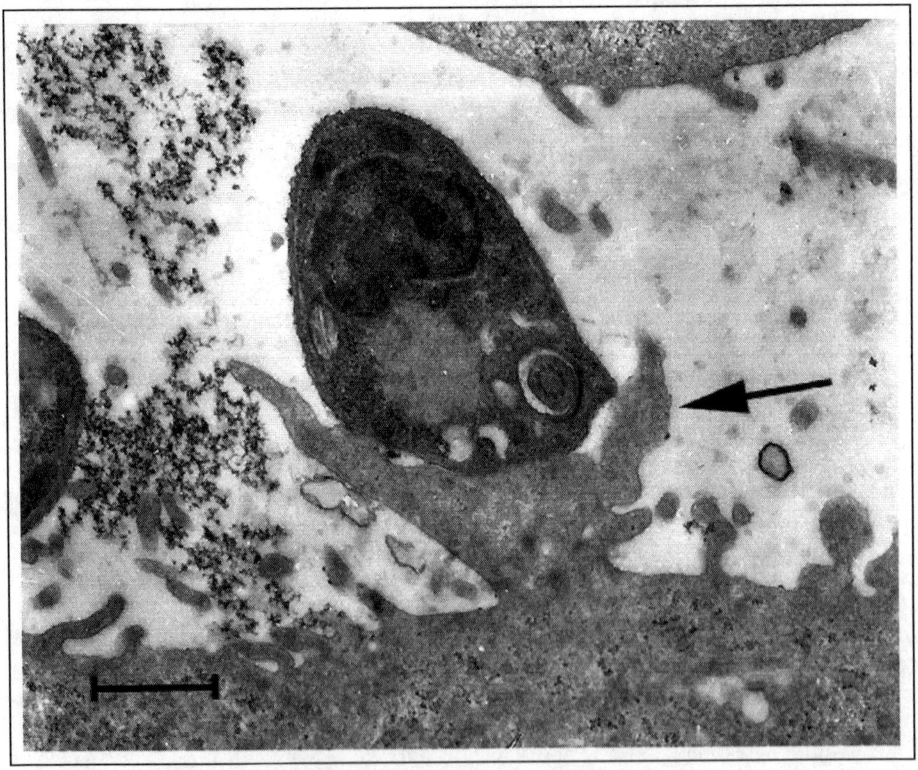

Figure 1. Formation of a cup-like pedestal (arrow) upon invasion of HeLa cells by *T. cruzi* amastigotes. Transmission electron microscopy of HeLa cells recently infected with (G strain extracellular) amastigotes. Bar: 500 nm.

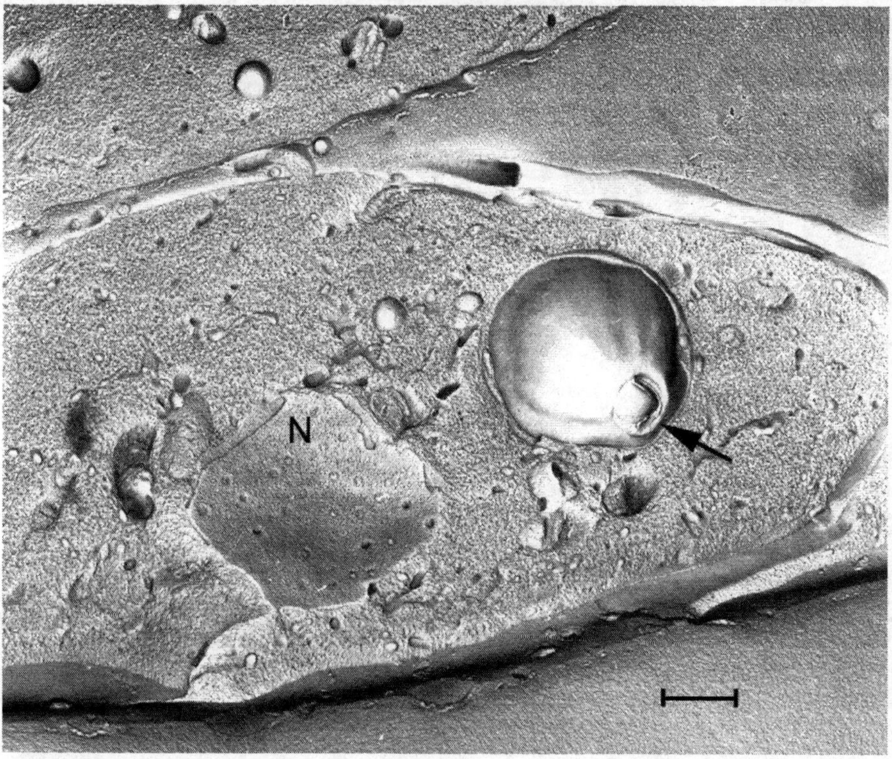

Figure 2. *T. cruzi* amastigotes are highly infective to HeLa cells and can be promptly visualized in the cytoplasm. Freeze-fracture replica of a parasite that has recently invaded a HeLa cell. The parasite flagellar pocket (arrow) can be clearly seen and the nuclear membrane of the host cell (N) identified by the presence of nuclear pores. Bar: 400 nm.

The entry of amastigotes into HeLa and Vero cells was then systematically examined using the centrifugation protocol and compared to cell invasion by metacyclic trypomastigotes.[22] In those studies drugs that interfere with microtubule polymerization (nocodazole) or protein phosphorylation (staurosporine and genistein) were used to evaluate amastigote and trypomastigote invasion of HeLa and Vero cells. The main conclusion was that the effect of a particular inhibitor is characteristic of the host cell-parasite form duet being examined.[22] For instance, trypomastigote invasion of Vero cell invasion was inhibited by staurosporine that dramatically enhanced invasion of HeLa cells.[22]

Besides the formation of actin-rich cups on the surface of HeLa cells, remarkable responses of the cell membrane were also observed upon invasion of Vero cells by extracellular amastigotes. In this fibroblastic cell line, devoid of surface microvilli, protrusive lamellae that formed at the sites of amastigote invasion[16] were markedly similar to the membrane expansions observed during *Shigella flexneri* attaching.[10,23,24]

In order to access the role of actin mobilization in the invasion process, target cells were treated with cytochalasin D, an experimental approach for inhibiting actin polymerization used in several other studies.[21,25-30] Cytochalasin D always inhibited amastigote invasion while systematically enhanced trypomastigote penetration.[22] These observations were consistent with the notion that the trypomastigote motility provides most of the driving force necessary for its

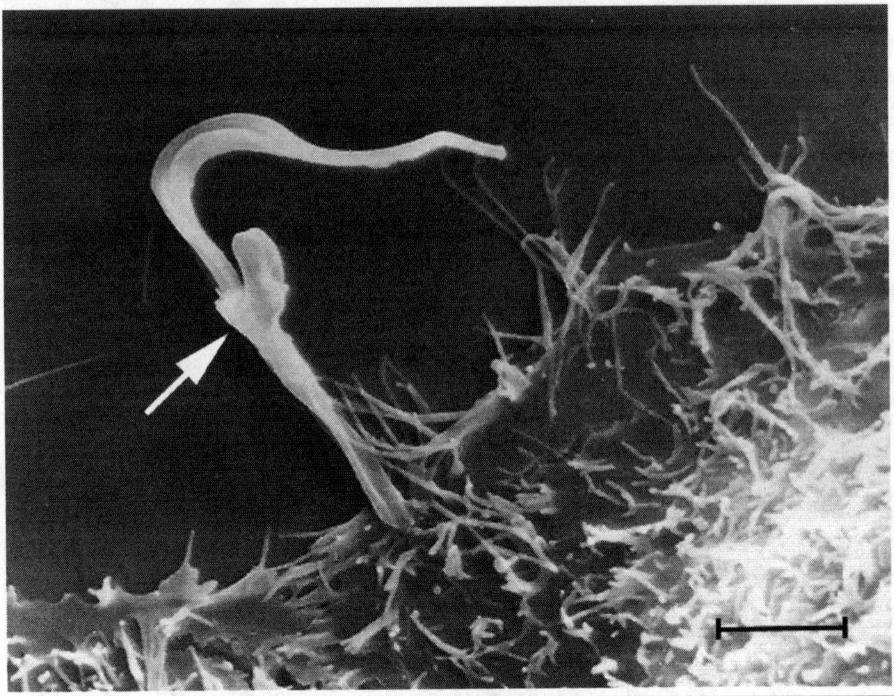

Figure 3. Formation of sleeve-like pseudopodia (arrow) around a trypomastigote invading HeLa cells visualized by scanning electron microscopy.[20] Bar: 2 μm.

internalization.[20,27,28] By contrast, due to their intrinsic motionless nature, amastigote invasion of either HeLa or Vero cells required functional actin microfilaments in order to this parasite form achieve internalization[22] confirming that a major difference between trypomastigote and amastigote invasion regarding the mobilization of host cell actin filaments is the distinct sensitivity to cytochalasin D.

In all the actin-rich membrane extensions formed around invading amastigotes or trypomastigotes, accumulation of cytoskeletal elements (α-actinin, gelsolin, vinculin, talin, tropomyosin and ABP_{280}), integrins or matrix elements could be detected, with some variability observed between the infective forms and target cells.[16] These results were again consistent with the notion that each parasite-host cell pair mobilizes specific interacting components (see Table 1).

T. cruzi Invasion of Cells Transfected with Actin Microfilament Modulators or Rho-GTPases

In the course of these studies, a number of available cell lines with altered expression of cytoskeletal components have been used as targets for *T. cruzi* invasion, in order to appraise their role in the establishment of infection. Cytoskeleton rearrangements at the cell periphery involve the cooperation of several actin microfilament-associated proteins. ABP_{280} (actin-binding protein of 280 kDa), or nonmuscle filamin, cross-links actin filaments into bundles, organizes the cortical network of microfilaments and is also capable of anchoring microfilaments to membrane glycoproteins.[31,32] In order to test the effect of the relative expression of ABP_{280}, transfectants of a melanoma cell line (M2) expressing varying amounts of ABP_{280} were used as

Table 1. Localization of host cell and extracellular matrix components in actin-rich structures formed during invasion of* T. cruzi *infective forms

Host Cell	**HeLa**		**Vero**	
Actin-Rich Structure Component	**Ama[1] 'Cup'**	**Trypo[2] Pseudo-Podium**	**Ama 'Crater'**	**Trypo Pseudo-Podium**
α-actinin	++	++	++	ND
Talin	++	+	+	ND
Vinculin	+	+	++	ND
Gelsolin	++	++	++	ND
Tropomyosin	++	+	++	ND
ABP	++	++	+	ND
Phosphotyrosine	++	+	++	+
α2 Integrin	+	++	+	ND
α3 Integrin	++	ND	+	+
αV Integrin	++	ND	+	+
αVβ5 Integrin	++	++	++	ND
Laminin	++	++	+	+
Fibronectin	++	ND	ND	ND
Vitronectin	+	ND	ND	ND

1. Ama: extracellular amastigote forms; 2. Trypo: metacyclic trypomastigote forms; ++: denotes component accumulation in actin-rich structure; +: denotes component present in the host cell but accumulation not distinguishable in actin-rich structure; ND: denotes component present in the host cell but not detectable in actin-rich structure, or actin-rich structure not detected (from ref. 15).

targets for amastigotes and metacyclic trypomastigotes. The three sub-clones of the M2 line that were used express increasing ABP_{280}:actin molar ratios (A3>A7>A4). Since the ABP_{280}:actin ratio is critical for adequate cortical meshwork rearrangements, these sub-clones exhibit distinct capacities of migration, being maximal in sub-clone A7. The observation that other two clones that have higher (A3) or lower (A4) ABP_{280}:actin ratios displayed hindered motility, confirmed that sub-clone A7 holds the optimum ratio between actin filaments and the accessory protein required for cell migration.[33] Interestingly, trypomastigotes invaded sub-clone A7 better than A3 or A4 whereas amastigotes had the opposite behavior and were capable of greater infection in sub-clones A4 and A3 when compared to A7.[22] These observations suggested that actin-ABP_{280} associations that regulated host cell membrane interactions required for cellular locomotion also modulated, in a selective way, the surface rearrangements involved in the invasion process by each *T. cruzi* infective form.

NIH 3T3 mouse fibroblasts overexpressing variable amounts of another actin assembly regulator protein, gelsolin,[34] were also tested for their susceptibility to invasion by the two infective forms. Clone C3 expresses 25% and clone C5 125% of gelsolin when compared to parental NIH 3T3.[34] Unlike ABP_{280} clones that have a maximum motility at an optimum ABP_{280}:actin molar ratio, the higher the expression of gelsolin, the better the transfected cells migrate.[34] Gelsolin transfectant C3 displayed the higher susceptibility towards metacyclic trypomastigote invasion when compared to either 3T3 or clone C5, suggesting that for this parasite form, there was an optimum level of gelsolin expression that ensureed parasite internalization.[22] By contrast, amastigotes invaded better clone C5 (C5>C3>3T3), clearly indicating that for this parasite form there is an optimum expression of gelsolin that results in better cell invasion.[22] Altogether, these experiments suggested that the requirements for these f-actin modulators during cell invasion by the two *T. cruzi* infective forms was quite distinct, and no direct and simple correlation between their expression (that affects cell motility among other parameters), and susceptibility towards infection could be made.[22]

The observation that amastigotes and trypomastigotes become associated with distinct actin-rich projections upon host cell invasion prompted the examination of the possible role of regulatory Rho GTPases (Rac1, RhoA and Cdc42) in this process. RhoA, Rac1, and Cdc42 are key regulators of independent signal transduction pathways, linking plasma membrane receptors to the assembly of distinct filamentous actin structures. Expression of constitutively active (i.e., GTPase-deficient) mutants of RhoA and Rac1 in cells induces the assembly of contractile actin and myosin filaments (stress fibers) and actin rich surface protrusions (lamellipodia), respectively.[35-37] Constitutively active Cdc42 promotes the formation of actin-rich, finger like membrane extensions (filopodia).[38,39] Invasive bacteria such as *Shigella*, *Salmonella*, *Listeria*, and *Yersinia* can modulate actin microfilament responses required to generate the membrane expansions involved in their internalization[9] by targeting bacterial toxins to the Rho GTPases.[40] Interestingly, *Shigella* and *Salmonella* use type III secretion devices to target toxins to the GTPases that lead to the formation of lamellar expansions at the sites of cell invasion.[9]

To evaluate the relative importance of RhoA GTPases in host cell invasion by different *T. cruzi* infective forms of distinct strains, MDCK cell transfectants that express variants of RhoA, Rac1 and Cdc42 proteins[41] have been used as target cells. Although it has been shown that trypomastigotes modulate Rho1, but not Rac or Cdc42,[30] it was found that metacyclic trypomastigotes *T. cruzi* had no apparent requirement for any of the tested GTPases and their infectivity was apparently only related to the strain virulence.[2] By contrast, *T. cruzi* amastigotes efficiently infected all transfected MDCK cells and invasion was particularly high in constitutively active Rac1V12 cells and specifically reduced in the corresponding dominant negative line Rac1N17, suggesting a key role for Rac1 in this invasion process.[2] In this respect, it is interesting to relate that invasion of these transfected cells by *Salmonella* is also dependent on Rac1,[42] indicating that the two pathogens could exploit similar routes to invade host cells.

Parasite Invasion, Parasite Retention: Role of Actin Filaments

As indicated above, the role of actin filaments in trypomastigote invasion of cultured cells has been extensively studied, mostly by examining the effect of cytochalasin D on the process.[7,16,21,25-30] Until not long ago, the consensus in the field was that trypomastigotes actively invade cells and that the cortical actin filaments hindered cell invasion since their disruption facilitated the process.[26] It had been argued that lysosome recruitment to the sites of trypomastigote invasion could be facilitated if the cortical actin meshwork is disrupted with drugs that interfere with actin polymerization like cytochalasin D.[26] From the studies mentioned earlier, it became clear that host cell actin mobilization is quite different in the process of cell invasion by trypomastigotes and amastigotes. Whereas actin microfilament disruption may even enhance trypomastigote invasion,[26,27] treatment of cells with cytochalasin D always inhibits invasion by amastigotes.[22] The exuberant motility of *T. cruzi* trypomastigotes is believed to be crucial to most of the phenomena observed during cell invasion as well as escape. The notion that parasite components could also induce localized actin depolymerization that could facilitate invasion has also been investigated and it appears that metacyclic trypomastigote surface glycoprotein GP82 might have such an activity.[43] By contrast, amastigotes display scarce movements[44] and rely on as yet poorly characterized ligands[18,45] to invade host cells. Observations using a panel of drugs that disclosed signaling differences in the infectivity by metacyclic trypomastigotes of different strains,[46] indicated that amastigotes engage signaling pathways distinct to trypomastigotes that are also strain-independent.[47] Interestingly though, the efficiency of amastigote invasion in several cell types consistently matches that of metacyclic forms and in some cell types it may even be greater.[2,22,45,48] This would suggest that, although motility may be regarded as a key factor, it does not seem to be required in the case of amastigotes.

Recently it has been confirmed the observations that by treating HeLa and Vero cells with either cytochalasin D or wortmannin for short periods of time, both the invasion and the presence of endosomal and lysosomal markers in recently formed parasitophorous vacuoles was affected. When treated cells were infected from 15 to 90 minutes, it was observed that while both drugs inhibit amastigote invasion,[47] metacyclic trypomastigote invasion was inhibited by wortmannin but enhanced by cytochalasin D (unpublished observations). As noted by others,[30,49-51] treatment of cells with either drug does not block parasite invasion: they only inhibit the process, reinforcing the notion that the parasite is capable of engaging alternative routes to ensure penetration. It was also observed that the proportion of trypomastigote parasitophorous vacuoles carrying either the endossomal marker EEA-1 or LAMP-1 was reduced by cytochalasin D, in agreement to what has been described for trypomastigotes.[30] In spite of the low infectivity of amastigotes in the presence of the drugs, it was possible to observe a similar reduction in parasitophorous vacuole labeling for EEA-1 and LAMP-1when cells were treated with cytochalasin D (unpublished observations). Although preliminary, these data suggest once inside the host cell, the involvement of actin filaments in the maturation of the parasitophorous vacuolar membrane appears to be similar for both parasite forms. Because of the low infectivity of amastigotes in the presence of the drugs and the correspondingly low numbers of parasites per cell, it was not possible to establish whether parasite could exit from infected cell, similarly to what has been described for trypomastigotes.[30,50]

It has become increasingly evident from the growing studies in the literature, that *T. cruzi* has evolved strikingly distinct mechanisms in order to invade host cells. The success of the infection depends on the possibility for the parasite to engage specific and yet efficient mechanisms to gain access to the cytoplasmic milieu of the large variety of host cells it can infect. The notion that different infective forms engage distinct mechanisms and pathways to achieve this goal reinforces the notion that the study of *T. cruzi* invasion of host cells is an area of research that will provide exciting new insights on this parasite-host cell interplay.

References

1. Burleigh BA, Andrews NW. The mechanisms of Trypanosoma cruzi invasion of mammalian cells. Annu Rev Microbiol 1995; 49:175-200.
2. Fernandes AB, Mortara RA. Invasion of MDCK epithelial cells with altered expression of Rho GTPases by Trypanosoma cruzi amastigotes and metacyclic trypomastigotes of strains from the two major phylogenetic lineages. Microbes Infect 2004; 6:460-467.
3. Ley V, Andrews NW, Robbins ES et al. Amastigotes of Trypanosoma cruzi sustain an infective cycle in mammalian cells. J Exp Med 1988; 168:649-659.
4. Pan SCT. Trypanosoma cruzi: In vitro interactions between cultured amastigotes and human skin-muscle cells. Exp Parasitol 1978; 45:274-286.
5. Behbehani K. Developmental cycles of Trypanosoma (Schyzotrypanum) cruzi (Chagas, 1909) in mouse peritoneal macrophages in vitro. Parasitology 1973; 66:343-353.
6. Nogueira N, Cohn Z. Trypanosoma cruzi: Mechanism of entry and intracellular fate in mammalian cells. J Exp Med 1976; 143:1402-1420.
7. Mortara RA. Trypanosoma cruzi: Amastigotes and trypomastigotes interact with different structures on the surface of HeLa cells. Exp Parasitol 1991; 73:1-14.
8. Frischknecht F, Way M. Surfing pathogens and the lessons learned for actin polymerization. Trends Cell Biol 2001; 11:30-38.
9. Cossart P, Sansonetti PJ. Bacterial invasion: The paradigms of enteroinvasive pathogens. Science 2004; 304:242-248.
10. Bourdet-Sicard R, Egile C, Sansonetti PJ et al. Diversion of cytoskeletal processes by Shigella during invasion of epithelial cells. Microbes Infect 2000; 2(7):813-819.
11. Cossart P. Subversion of the mammalian cell cytoskeleton by invasive bacteria. J Clin Invest 1997; 99:2307-2311.
12. Dramsi S, Cossart P. Intracellular pathogens and the actin cytoskeleton. Annu Rev Cell Dev Biol 1998; 14:137-166.
13. Goosney DL, Gruenheid S, Finlay BB. Gut feelings: Enteropathogenic E. coli (EPEC) interactions with the host. Annu Rev Cell Dev Biol 2000; 16:173-189.

14. Silva MLM, Mortara RA, Barros HC et al. Aggregation of membrane-associated actin filaments following localized adherence of enteropathogenic Escherichia coli to HeLa cells. J Cell Sci 1989; 93:439-446.
15. Hudson L, Snary D, Morgan SJ. Trypanosoma cruzi: Continuous cultivation with murine cell lines. Parasitology 1984; 88:283-294.
16. Procópio DO, Barros HC, Mortara RA. Actin-rich structures formed during the invasion of cultured cells by infective forms of Trypanosoma cruzi. Eur J Cell Biol 1999; 78:911-924.
17. Rosenshine I, Finlay BB. Exploitation of host signal transduction pathways and cytoskeletal functions by invasive bacteria. BioEssays 1993; 15:17-24.
18. Silva CV, Luquetti AO, Rassi A et al. Involvement of Ssp-4-related carbohydrate epitopes in mammalian cell invasion by Trypanosoma cruzi amastigotes. Microbes Infect 2006, (in press).
19. Schenkman S, Andrews NW, Nussenzweig V et al. Trypanosoma cruzi invade a mammalian epithelial cell in a polarized manner. Cell 1988; 55:157-165.
20. Schenkman S, Mortara RA. HeLa cells extend and internalize pseudopodia during active invasion by Trypanosoma cruzi trypomastigotes. J Cell Sci 1992; 101:895-905.
21. Barbosa HS, Meirelles MNL. Evidence of participation of cytoskeleton of heart muscle cells during the invasion of Trypanosoma cruzi. Cell Struct Funct 1995; 20:275-284.
22. Procópio DO, Silva S, Cunningham CC et al. Trypanosoma cruzi: Effect of protein kinase inhibitors and cytoskeletal protein organization and expression on host cell invasion by amastigotes and metacyclic trypomastigotes. Exp Parasitol 1998; 90:1-13.
23. Clerc P, Sansonetti PJ. Entry of Shigella flexneri into HeLa cells: Evidence for directed phagocytosis involving actin polymerization and myosin accumulation. Infect Immun 1987; 55:2681-2688.
24. Tran VN, Bourdet-Sicard R, Dumenil G et al. Bacterial signals and cell responses during Shigella entry into epithelial cells. Cell Microbiol 2000; 2:187-193.
25. Andrews NW. From lysosomes into the cytosol: the intracellular pathway of Trypanosoma cruzi. Braz J Med Biol Res 1994; 27:471-475.
26. Tardieux I, Webster P, Ravesloot J et al. Lysosome recruitment and fusion are early events required for trypanosome invasion of mammalian cells. Cell 1992; 71:1117-1130.
27. Schenkman S, Robbins ES, Nussenzweig V. Attachment of Trypanosoma cruzi to mammalian cells requires parasite energy, and invasion can be independent of the target cell cytoskeleton. Infect Immun 1991; 59:645-654.
28. Kipnis TL, Calich VL, Dias da Silva W. Active entry of bloodstream forms of Trypanosoma cruzi into macrophages. Parasitology 1979; 78:89-98.
29. Meirelles MNL, Araujo-Jorge TC, De Souza W. Interaction of Trypanosoma cruzi with macrophages in vitro: Dissociation of the attachment and internalization phases by low temperature and cytochalasin B. Z Parasitenkd 1982; 68:7-14.
30. Woolsey AM, Burleigh BA. Host cell actin polymerization is required for cellular retention of Trypanosoma cruzi and early association with endosomal/lysosomal compartments. Cell Microbiol 2004; 6:829-838.
31. Hartwig JH, Stossel TP. Isolation and properties of actin, myosin, and a new actin binding protein in rabbit alveolar macrophages. J Biol Chem 1975; 250:5696-5705.
32. Hartwig JH, Tyler J, Stossel TP. Actin-binding protein promotes the bipolar and perpendicular branching of actin filaments. J Cell Biol 1980; 87:841-848.
33. Cunningham CC, Gorlin JB, Kwiatkowski DJ et al. Actin-binding protein requirement for cortical stability and efficient locomotion. Science 1992; 255:325-327.
34. Cunningham CC, Stossel TP, Kwiatkowski DJ. Enhanced motility in NIH 3T3 fibroblasts that overexpress gelsolin. Science 1991; 251:1233-1236.
35. Ridley AJ, Hall A. The small GTP-binding protein Rho regulates the assembly of focal adhesions and actin stress fibers in response to growth factors. Cell 1992; 70:389-399.
36. Ridley AJ, Paterson HF, Johnston CL et al. The small GTP-binding protein rac regulates growth factor-induced membrane ruffling. Cell 1992; 70:401-410.
37. Hall A. Small GTP-binding proteins and the regulation of the actin cytoskeleton. Annu Rev Cell Biol 1994; 10:31-54.
38. Nobes CD, Hall A. Rho, Rac, and Cdc42 GTPases regulate the assembly of multimolecular focal complexes associated with actin stress fibers, lamellipodia, and filopodia. Cell 1995; 81:53-62.
39. Kozma R, Ahmed S, Best A et al. The Ras-related protein Cdc42Hs and bradykinin promote formation of peripheral actin microspikes and filopodia in Swiss 3T3 fibroblasts. Mol Cell Biol 1995; 15:1942-1952.
40. Aktories K, Barbieri JT. Bacterial cytotoxins: Targeting eukaryotic switches. Nat Rev Microbiol 2005; 3:397-410.

41. Jou TS, Nelson WJ. Effects of regulated expression of mutant RhoA and Rac1 small GTPases on the development of epithelial (MDCK) cell polarity. J Cell Biol 1998; 142:85-100.
42. Criss AK, Ahlgren DM, Jou TS et al. The GTPase Rac1 selectively regulates Salmonella invasion at the apical plasma membrane of polarized epithelial cells. J Cell Sci 2001; 114:1331-1341.
43. Cortez M, Atayde V, Yoshida N. Host cell invasion mediated by Trypanosoma cruzi surface molecule gp82 is associated with F-actin disassembly and is inhibited by enteroinvasive Escherichia coli. Microbes Infect 2006; 8:1502-1512.
44. Barros HC, Silva S, Verbisck NV et al. Release of membrane-bound trails by Trypanosoma cruzi amastigotes onto modified surfaces and mammalian cells. J Eukaryot Microbiol 1996; 43:275-285.
45. Mortara RA, Procópio DO, Barros HC et al. Features of host cell invasion by different infective forms of Trypanosoma cruzi. Mem Inst Oswaldo Cruz 1999; 94(Suppl 1):135-137.
46. Neira I, Ferreira AT, Yoshida N. Activation of distinct signal transduction pathways in Trypanosoma cruzi isolates with differential capacity to invade host cells. Int J Parasitol 2002; 32:405-414.
47. Fernandes AB, Neira I, Ferreira AT et al. Cell invasion by Trypanosoma cruzi amastigotes of distinct infectivities: Studies on signaling pathways. Parasitol Res 2006, (in press).
48. Stecconi-Silva RB, Andreoli WK, Mortara RA. Parameters affecting cellular invasion and escape from the parasitophorous vacuole by different infective forms of Trypanosoma cruzi. Mem Inst Oswaldo Cruz 2003; 98:953-958.
49. Braga MV, De Souza W. Effects of protein kinase and phosphatidylinositol-3 kinase inhibitors on growth and ultrastructure of Trypanosoma cruzi. FEMS Microbiol Lett 2006; 256:209-216.
50. Andrade LO, Andrews NW. Lysosomal fusion is essential for the retention of Trypanosoma cruzi inside host cells. J Exp Med 2004; 200:1135-1143.
51. Woolsey AM, Sunwoo L, Petersen CA et al. Novel PI 3-kinase-dependent mechanisms of trypanosome invasion and vacuole maturation. J Cell Sci 2003; 116:3611-3622.

Chapter 9

Actin/Myosin-Based Gliding Motility in Apicomplexan Parasites

Kai Matuschewski and Herwig Schüler*

Abstract

Apicomplexan parasites move and actively enter host cells by substrate-dependent gliding motility, an unusual form of eukaryotic locomotion that differs fundamentally from the motility of prokaryotic and viral pathogens. Recent research has uncovered some of the cellular and molecular mechanisms underlying parasite motility, transmigration, and cell invasion during life cycle progression. The gliding motor machinery is embedded between the plasma membrane and the inner membrane complex, a unique double membrane layer. It consists of immobilized unconventional myosins, short actin stubs, and TRAP-family invasins. Assembly of this motor machinery enables force generation between parasite cytoskeletal components and an extracellular substratum. Unique properties of the individual components suggest that the rational design of motility inhibitors may lead to new intervention strategies to combat some of the most devastating human and livestock diseases.

Cast of Characters

Rapid host cell entry is of exquisite importance for virtually every microbial pathogen and is a prerequisite for an intracellular life style. Viruses and bacteria typically induce receptor-mediated endocytosis, escape into the cytosol, and subsequently hijack components of the host cell cytoskeleton. Apicomplexan parasites, such as *Plasmodium* or *Toxoplasma*, have the unique capacity to actively push themselves into a target host cell by utilizing their own actin-myosin motor machinery.[1-4]

Prior to cell invasion the parasites often need to cross a number of biological barriers that they encounter advancing through their life cycles. This function is particularly important during transmission from one host to another, although spread of apicomplexan parasites occurs by at least two fundamentally different routes: Blood parasites, such as *Plasmodium* and *Babesia*, are vector-borne and contracted by infected arthropods—*Anopheles* mosquitoes and ticks, respectively. In the case of *Plasmodium*, motile sporozoites are deposited in the skin tissue upon natural transmission by an infectious mosquito bite.[5,6] In order to eventually commence the pathogenic cycle inside erythrocytes *Plasmodium* first undergoes an obligate liver phase. To reach their final target cells, replication-permissive hepatocytes, sporozoites need to be highly motile and find their way through the entire body of the host.[7] By contrast, tissue parasites such as *Cryptosporidium, Cyclospora, Eimeria* and *Toxoplasma gondii*, are taken up by accidental oral ingestion of oocysts. These release sporozoites, which actively penetrate enterocytes lining the gut lumen. In addition, *T. gondii* can be vertically transmitted to the fetus when tachyzoites

*Corresponding Author: Herwig Schüler—Max Delbrück Center for Molecular Medicine, 13092 Berlin, Germany. Email: schueler@mdc-berlin.de

Molecular Mechanisms of Parasite Invasion, edited by Barbara A. Burleigh and Dominique Soldati-Favre.

penetrate the syncytiotrophoblasts of the placenta. Notably, the transmigratory capacity of *T. gondii* tachyzoites is directly linked to parasite virulence.[8,9]

Regardless of their transmission mode, all apicomplexan sporozoites have to accomplish a number of related tasks: They need to (i) move rapidly out of their site of deposition, (ii) cross, often multiple, biological barriers, (iii) specifically adhere to a suitable host cell, and (iv) invade the target cell under simultaneous formation of a replication-competent organelle, the parasitophorous vacuole (PV). Not surprisingly, these divergent parasites share not only similar activities and a common mode of locomotion but also unique structural features. They contain an apical complex consisting of a set of cytoskeletal structures—apical rings, microtubules and (with the exception of *Plasmodium* species) a conoid—and specialized secretory organelles—micronemes, rhoptries, and dense granules.[1] The highly structured apical complex fulfills functions related to all of the above tasks by sequential release of the secretory organelles.[10] This is arguably the most elaborate invasion machinery in any class of pathogens.

One Continuous Move: Gliding Locomotion, Transmigration and Host Cell Entry

Locomotion of virtually every moving cell is driven by one of three motility modes: (i) crawling motility, as seen in moving fibroblasts, is an amoeboid motion powered by continuous rearrangement of a microfilament meshwork that slowly pushes the cell forward and establishes new cell contacts;[11] (ii) swimming requires specialized axoneme structures, such as flagella or cilia, which are driven by dynein/microtubule motors;[12] (iii) gliding locomotion is a fast substrate-dependent motility that is employed by apicomplexan parasites and other unicellular organisms.[2] This mode of locomotion certainly suits the parasites-life cycle progression best, as it empowers them to move actively, fast, and disseminate to the entire body of the host.

Rearward transport of receptor-ligand interactions can be visualized with glass beads and is thought to translate into forward movement of the parasite.[13,14] Apparently, gliding motility, host cell invasion, and spread between cells are directly linked and driven by a common motor machinery. Gliding motility has now been described for every parasite invasive stage, with the notable exception of *Plasmodium* merozoites (Table 1). Substrate-dependent circular gliding locomotion on preincubated glass slides was first observed in *Plasmodium* sporozoites.[15] Sporozoites perform continuous and extended gliding locomotion without apparent changes in cell shape at a relatively high speed of 1-3 μm/sec.[15] Intriguingly, this capacity gradually develops

Table 1. Motility of apicomplexan parasites

Parasite	Stage	Task	Speed [μm/s]	In Vivo[A]	Ref.
Toxoplasma gondii	sporozoite	transmigration, invasion	1.5	-	75
	tachyzoite	transmigration, invasion	1 - 3	-	18
Plasmodium	sporozoite	accumulation in proboscis	< 2.0	+	17
		skin migration	2.3	+	6,7
		Kupffer cell passage	0.3	+	7
		hepatocyte transmigration	1.6	+	7
	merozoite	erythrocyte invasion	elusive	-	-
	ookinete	midgut penetration	0.04	+	19
E.tenella	sporozoite	enterocyte invasion	6.0	-	13
C. parvum	sporozoite	enterocyte invasion	5.0	-	75

A. '+' indicates that transgene fluorescent parasites have been studied in vivo by intravital imaging techniques during sporozoite maturation after their release from midgut oocysts and is most prominent prior to transmission, i.e., in mosquito salivary glands.[16] Intravital imaging revealed that sporozoites display only residual motility within the mosquito[17] and commence avid locomotion only after transmission to the mammalian skin.[5,6] Induction of fast motility is likely triggered by the heat shock and additional signals, experienced by the sporozoite upon transmission. Once inside the dermis, sporozoites perform a mixture of circular and helical gliding patterns that eventually leads them to a nearby capillary.[6] These two modes of gliding, as well as a third, so-called twirling motility, were first described in *T. gondii* tachyzoites.[18] Helical gliding occurs by forward motility with simultaneous full rotation and probably reflects net translocation in tissues closest. This form of motility was also detected in gliding *Plasmodium* ookinetes that traverse the midgut epithelium of the mosquito.[19] Seemingly, all apicomplexan parasites move through tissues employing similar gliding patterns and, perhaps, also molecular motors. This machinery appears to be very versatile and permits locomotion over a wide range of speeds (Table 1).

During their journey parasites often encounter cellular barriers, typically endothelial cells, that they need to cross in order to reach their ultimate target cell (Fig. 1). An early intriguing observation was that *Plasmodium* sporozoites transmigrate nontarget cells without arresting their gliding behaviour, indicating that they can switch instantaneously from extracellular locomotion on a substrate to intracellular transmigration.[20] Subsequent work has demonstrated that this parasite function is physiologically relevant and required for *Plasmodium* sporozoite entry to the liver[21] and *T. gondii* tachyzoite dissemination.[8] Strikingly, the cellular mechanisms appear to differ substantially (Fig. 1): *Plasmodium* sporozoites traverse the endothelial cells by transmigration, a trans-cellular route, which is accompanied by breaching of the cell plasma membrane.[21] In contrast, *T. gondii* tachyzoites take a para-cellular route and squeeze between cells.[8] Paramigration requires transient breaching of junctional complexes. The underlying cellular and molecular details of either route, including membrane and junction breaching and the nature of the intracellular tracks, remain entirely unsolved.

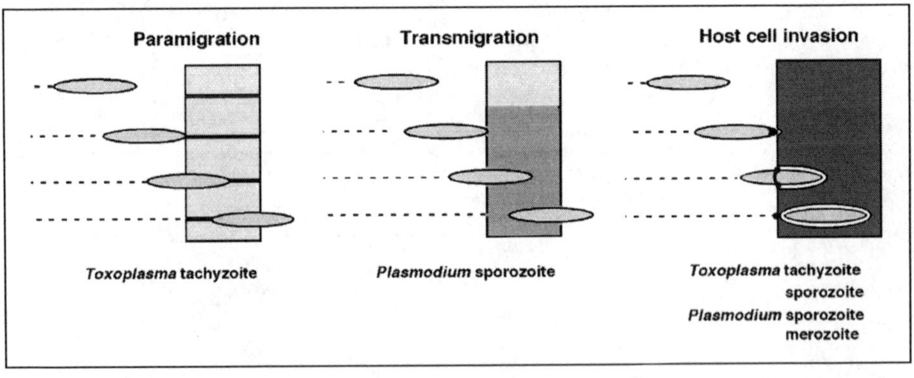

Figure 1. The different tasks for Apicomplexan parasites after transmission to the mammalian host. The basis for all locomotion-related processes is the parasite's own actin-myosin motor machinery that allows continuous forward movement by rearward traction of parasite-substrate interactions (indicated by dashed line). *Toxoplasma gondii* tachyzoites (left) cross cellular barriers on a paracellular route in the absence of cellular damage. *Plasmodium* sporozoites (center) migrate through epithelial cells via membrane breaching (the corresponding cellular damage is indicated by increase in color intensity). Host cell invasion by all apicomplexan parasites (right) is accompanied by the formation of a moving tight junction and, eventually, a parasitophorous vacuole wherein the invasive stage de-differentiates and commences replication. This productive cell entry into a suitable host cell is directed by invasins of the TRAP-family (red). A color version of this figure is available online at www.eurekah.com.

Once a suitable host cell is recognized, it is rapidly invaded by the parasite under simultaneous formation of a moving tight junction and eventually the PV.[22] The PV itself is derived from the host cell plasma membrane and remodeled by sequential secretion of the contents of the apical organelles (Fig. 1). In conclusion, the same motor machinery meets all parasite requirements from initial host cell egress to productive entry of the next target cell and includes extracellular gliding locomotion and endothelial transmigration in a continuous move.

Upside Down: Components of the Motor Machinery

Typically actomyosin-based force generating systems, such as muscle or vesicular transport, are based either on scaffolds of actin and myosin polymers or on soluble monomeric myosins working against F-actin tracks. Therefore in actomyosin-dependent parasite gliding, it was expected that one of the partners must interact with surface molecules that link the motor to the outside while the other must be arranged and fixed so as to integrate the forces and give them directionality. Recent research indicates that during gliding motility transmembrane receptor-bound motile actin stubs are moved along myosin tracks (Fig. 2). The principal constituents needed for motility have been identified (Table 2).

The Backbone: Subpellicular Network and Inner Membrane Complex

A shared feature of all apicomplexan invasive stages is the presence of a unique double membrane layer, the inner membrane complex (IMC). The IMC is in close proximity to the plasma membrane and surrounds the entire cytosol with the exception of the apical tip.[23] Together, the IMC and the parasite plasma membrane delimit a nearly continuous compartment, the submembranous, or cortical, space. The IMC originates from Golgi-derived flattened vesicles and grows by continuous fusion with these vesicles. The IMC is thought to function primarily as a lipid-rich membrane barrier. Additional functions may include storage and regulated release of molecules, e.g., Ca^{2+}, that trigger motility, and a continuous support for the motor machinery. Indeed, a subpellicular network[24] (SPN) underlying the IMC may add to such a supporting function. Together, the SPN and IMC appear to account for the remarkable capacity of apicomplexan parasites to maintain their cell shape while at the same time displaying great torsional flexibility. The major constituents of the SPN are microtubules that grow concomitantly from a microtubule-organizing center at the apical tip.[23] The nature of the interaction between the microtubules and the IMC remains unknown. Electron microscopy studies suggest that the intimate connection is indirect and probably mediated by a layer of dense intra-membranous particles and/or a meshwork of filamentous proteins that ultimately fix the microtubules to the IMC.[25]

Recently, a comprehensive proteome analysis of partially purified cytoskeletal components of *T. gondii* tachyzoites, including the conoid and the apical complex, identified numerous candidate cytoskeletal proteins, including three previously unrecognized β-tubulin isoforms, and an IMC-resident protein, termed IMC4.[26] Thus, we may expect additional proteins to participate in the assembly of the structures that support the molecular motor.

The Immobilized Motor Myosin

The myosin that appears to drive gliding motility in all Apicomplexa is myosin A (MyoA),[27] the founding member of a class of unconventional, tail-less myosins (class XIV) that is restricted to Apicomplexa and ciliates.[28,29] Most members of this protein family are remarkably small, consisting essentially of the globular head domain with ATPase and actin binding activity. Despite its minimal size, purified MyoA behaves like a classical plus end-directed motor and moves at a speed of ~5 μm/sec, in good agreement with the velocity of the motile parasite[30] (Table 1). In other myosins the extended carboxy-terminal tails typically function in regulation and anchoring of the head domain.[31] Such roles appear to be fulfilled by two accessory proteins that were first identified in *T. gondii*, the gliding-associated proteins 45 and 50 (GAP45/GAP50) that link MyoA to the IMC through the integral membrane glycoprotein GAP50.[32] The GAP45/

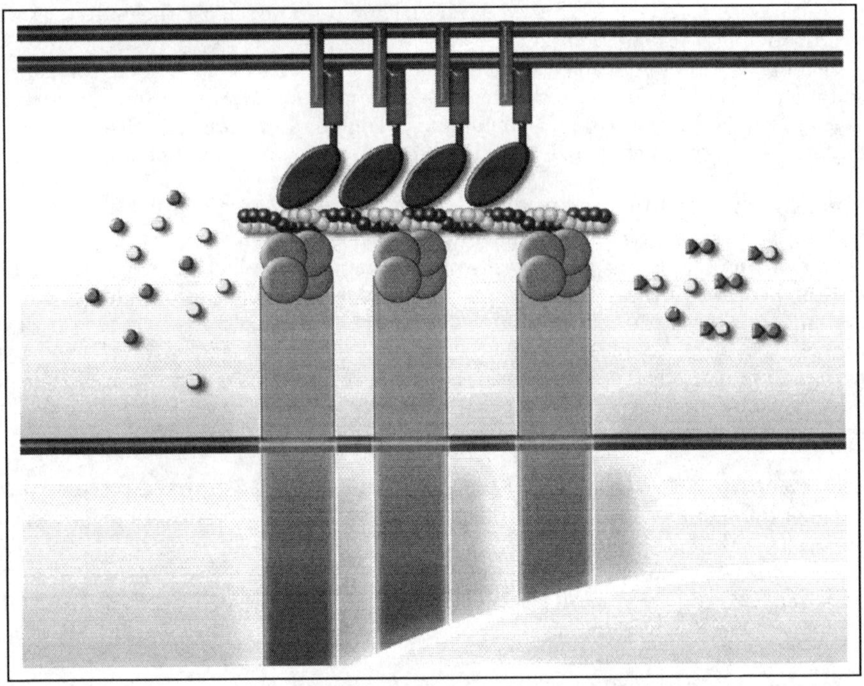

Figure 2. Molecular mechanism of gliding locomotion in Apicomplexan parasites. Parasite transmembrane proteins of the TRAP family (green) in the plasma membrane (blue single line) bind to a target cell or substrate in the extracellular space (sphere emerging below). This binding event is transmitted to the parasite inner membrane space (light brown) and results in recruitment of aldolase tetramers (orange) to the carboxy-terminal domain of the TRAP invasin. This complex binds to short actin filaments (yellow and purple cables) that are formed from actin monomers (yellow and purple spheres) under dynamic regulation by auxiliary proteins (green triangle). Short plus-end directed class XIV myosins (pink) cyclically interact with the F-actin stubs and pass the actin/aldolase/invasin complexes on to the next set of myosins, resulting in net translocation of complexes to the posterior end of the parasite. The myosins are immobilized in the inner membrane complex (blue double line) through a complex including glideosome-associated proteins (cyan) and myosin tail interacting protein (magenta). For a detailed listing of the molecular components of the actin/myosin motor see Table 2. A color version of this figure is available online at www.eurekah.com.

GAP50 pair likely acts in concert with MyoA-tail interacting protein (MTIP), a MyoA-associated protein that is reminiscent of the regulatory light chain.[33] MTIP, which binds directly to MyoA, likely provides the IMC membrane linkage through interaction with GAP50.[32] In *T. gondii*, the MyoA complex is assembled in two consecutive steps that may be triggered by a post-translational modification event: MyoA, the MTIP ortholog MLC-1 and GAP45 first assemble into a soluble complex, which later associates with GAP50 resulting in IMC-anchoring of MyoA.[32] The same complex has recently been shown to be present also in *Plasmodium* merozoites[34-36] further supporting the notion that the intracellular gliding machinery is highly conserved across apicomplexan parasites. Stable immobilization within the IMC likely requires interaction with the SPN. Most importantly, tethering of MyoA to the IMC orients its head into the cortical space and obliges actin to engage the surface receptors (Fig. 2). With the crystal structure of the complex between *Plasmodium* MTIP and the MyoA tail,[37] the architecture of the motor machinery is beginning to be unravelled.

Table 2. Components of the Apicomplexan actin/myosin motor

			Gene[C]			
Protein[A]	Function	Vital[B]	*P. falciparum*	*T. gondii*	*C. parvum*	References
TRAP	sporozoite invasin	+	PF13_0201	-	-	54-56
CTRP	ookinete invasin	+	PFC0640w	-	-	58,59
MTRAP	merozoite invasin	nd	PF10_0181	-	-	34
MIC2	tachyzoite invasin	+	-	20.m0002	-	60
M2AP	MIC2 binding	-	-	33.m0006	-	62
TRAP-C1	sporozoite invasin	nd	-	-	cgd1_3500	76
Aldolase	links invasin to F-actin	nd	PF14_0425	46.m00002	cgd1_3020	46,47
ACT1	formation of short filaments	nd	PFL2215w	25.m00007	cgd5_3160	41-43
ADF1	catalyzes F-actin turnover	+	PFE0165w	39.m00001	cgd5_2800	49
Myo A	motor molecule	+	PF13_0233	46.m00001	cgd7_3790	27
MTIP	binds to MyoA	nd	PFL2225w	55.m00021	cgd6_4160	33
GAP45	protein of the MyoA complex	nd	PFL1090w	42.m08072	1MB.460	32
GAP50	anchors MyoA to IMC	nd	PFI0880c	38.m01061	cgd2_640	32
IMC1	membrane skeletal protein	+	PF11_0431	44.m00004	cgd4_3290	77

A. Abbreviations: ACT1, actin-1; ADF1, actin depolymerising factor-1; CTRP, circumsporozoite and TRAP-related protein; GAP, gliding-associated protein; IMC1, inner membrane complex protein-1; MIC2, micronemal protein-2; MTIP, myosin A tail interacting protein; MTRAP, merozoite-specific TRAP homologue; MyoA, myosin A; M2AP, MIC2-associated protein; TRAP, thrombospondin-associated anonymous protein; TRAP-C1, *Cryptosporidium* TRAP protein-1. B. Vital (+) describes that a loss-of-function parasite can either not be generated or does not progress through the parasite life cycle. Nonvital genes (-) do not perform crucial functions; (nd): no knock-out studies published yet. C. Accession numbers are according to the most recent annotations at www.plasmodb.org, www.toxodb.org, and www.cryptodb.org, respectively.

The Unusual Actin Stubs

A landmark finding in apicomplexan cell biology was the direct role of parasite actin in motility and host cell invasion.[38] Using a genetically engineered *T. gondii* mutant resistant to the microfilament inhibitor cytochalasin-D the authors could functionally separate the roles of parasite versus host cell actin during invasion. This strategy allowed inhibiting host cell actin with no impact on invasion efficiency. This finding formally demonstrated a vital function of the parasite actin in these processes, in agreement with earlier inhibitor studies with *Plasmodium* merozoites.[39] Yet, parasite microfilaments remained elusive, and filamentous structures resembling actin polymers were detectable only in the cortical space between the plasma membrane and the IMC.[40]

Recent work has shown that the ubiquitously expressed actin-1 (ACT1) in *Plasmodium* species and the single actin in *Toxoplasma* both form extremely short, intrinsically unstable polymers[41-43] (see ref. 44 for a review). The intrinsic instability of actin polymers is also reflected by a number of nonhomologous replacements in comparison to conventional actins, which map mainly to putative monomer-monomer interaction sites. These findings explain the inability of researchers to visualize cytosolic F-actin in the parasites. They also fill a gap in parasite motility models, providing a plausible explanation for the functioning of actin in its anticipated cellular location. An attractive hypothesis is that actin polymers may gain stability by incorporation into their correct context within the motor machinery.[45] Interaction of actin polymers with active MyoA itself may thus be a limiting factor that ultimately triggers self-assembly of the actomyosin motor. In addition to transient interaction with myosin, actin polymers also bind to tetramers of aldolase, which in turn bind TRAP-family invasins.[46,47] While this interaction is thought to link surface receptors to the actomyosin motor, stimulation of actin polymerization by aldolase remains to be demonstrated. Recruitment of a key glycolytic enzyme, and perhaps the entire set of glycolytic enzymes, in close proximity to the motor myosins may provide an excellent solution to continuous, high ATP supply.

The apicomplexan genomes encode only a limited repertoire of conventional actin-binding proteins and microfilament regulators.[45,48] This indicates a limited need for fine-tuning microfilament dynamics for diverse purposes, and obviously, intrinsic properties of unconventional parasite actins have evolved so that parasites get by without a multitude of actin regulators. For instance, for actin functioning in the parasites, intrinsic F-actin instability may eliminate the requirement for an actin-depolymerizing activity. Indeed, the universally expressed one of the two *Plasmodium* cofilins, ADF1, lacks F-actin binding capacity, but stimulates nucleotide exchange in G-actin, presumably to accelerate actin polymer turnover by interaction with actin monomers.[49]

The Link to the Outside World: TRAP Family Invasins

Thrombospondin-related anonymous protein (TRAP) is the founding member of apicomplexan transmembrane adhesive proteins that all contain a short carboxy-terminal acidic region exposed to the parasite cytoplasm (cytoplasmic tail domain, CTD), a transmembrane span, and a series of extracellular adhesive domains. TRAP was initially identified and cloned based on sequence similarity of one adhesion module, the thrombospondin repeat (TSR).[50] TRAP orthologs exist in all parasites of the phylum, albeit with considerable sequence variability especially in the composition of the extracellular domains.[51] Common to all of them are copies of two adhesion modules typically found in higher eukaryotes, the TSR type 1 domain (TSR) type 1 domain[52] and the von Willebrand A-domain.[53] Both act as adhesive modules and mediate attachment to heparan sulfate proteoglycans and perhaps other ligands. Reverse genetics showed that *Plasmodium TRAP* is essential for sporozoite motility, and consequently, host cell entry.[54] Mutational studies reproduced the vital role for the CTD[55] and the extracellular adhesion modules, which act in concert.[56] TRAP locates to micronemes from where it is thought to integrate into the plasma membrane via the apical tip of the parasite, and to distribute backwards along the cell periphery in a patchy fashion.[57] A similar direct role in parasite motility and host cell attachment and invasion was demonstrated for the ookinete-specific invasin CTRP[58,59] and for microneme protein-2 (MIC2), the *T. gondii* tachyzoite ortholog of this family.[60] Notably, MIC2 is seen in complex with the *T. gondii*-specific MIC2-associated protein (M2AP) throughout its trafficking to micronemes, shuttling to the parasite surface, and subsequent degradation.[61] The two proteins form a heterohexameric complex that is essential for host cell invasion. Targeted deletion of M2AP impairs but does not abolish host cell entry, demonstrating an auxiliary role in invasion, perhaps by assisting MIC2 secretion.[62]

The first protein identified in the gliding trails of *Plasmodium* sporozoites was circumsporozoite protein (CSP), the major coat protein of the parasite.[63] In analogy, surface antigen 1 (SAG1) of *T. gondii* tachyzoites is deposited into trails. Both proteins contain a GPI lipid anchor and are peripheral membrane proteins that are unlikely to physically interact with the intracellular motor machinery. The role of the surface coat proteins in motility may include signalling a receptor recognition event to the interior. In addition, their continuous release at the anterior pole may provide the parasite with a substrate of its own making, which may facilitate attachment and gliding irrespective of the environment.

Regulation of the Actin/Myosin Motor

Extracellular stimuli and intracellular signalling cascades likely activate and fine-tune gliding motility. In analogy to other motor machineries, work with *T. gondii* tachyzoites indicated a central role for Ca^{2+} in these events.[10,64] The apicomplexan genomes encode for plant-like calcium-dependent protein kinases (CDPKs).[65] Recent gene targeting studies identified a function for CDPK3 in gliding motility of *Plasmodium* ookinetes.[66,67] Gliding motility is initiated by micronemal discharge of the TRAP invasins,[64] which brings in the corresponding CTDs to the cortical space, where all other components are presumably already present. The spectrum of kinase targets that supposedly initiate commitment to motility and invasion remains to be identified. Rearward translocation of the TRAP invasins likely requires additional regulatory steps. For instance, differences in parasite velocity (Table 1) may be triggered by cytoplasmic factors rather than by the TRAP CTDs themselves. A step toward understanding regulation of motility may be the isolation of the molecular targets of small molecules that were identified in a high throughput-screen for *T. gondii* tachyzoite invasion.[68] Intriguingly, this comprehensive analysis revealed a number of motility enhancers in addition to inhibitors suggesting a high degree of regulation to achieve fast locomotion and invasion, yet retain sufficient resources to reach a suitable target cell.

Once the TRAP invasins reach the posterior end of the parasite they need to be shed through proteolytic cleavage within the transmembrane span.[69,70] MIC2 release is essential for *T. gondii* tachyzoite invasion.[60] Of note, mutant *Plasmodium* sporozoites that retain their transmembrane span but lack portions of the CTD display a 'pendulum phenotype', characterized by continuous back and forth movement.[55] This finding indicates that yet unidentified CTD-binding factors contribute to the proteolytic cleavage. Similar to other biological systems, intramembrane cleavage is catalyzed by rhomboid proteases at the posterior pole of the parasite.[71-74] At the same time, the remaining cytoplasmic portions of the TRAP invasins need to be degraded, likely via the ubiquitin/proteasome system, and the actin stubs recycled to the anterior part.

Future Perspectives

In the near future, some remaining basic questions need to be answered, in particular: How is the motile machinery activated, and how is it orchestrated to generate directional motility? Is myosin arranged in tracks with intrinsic polarity, and if so, how are these tracks assembled and regulated? How is actin turnover regulated within the cortical space? Do natural variants of the motor components correlate with parasite virulence?

Understanding the molecular details of parasite motility opens the possibility to generate novel tools that are predicted to be particularly effective, because they target not only multiple functions of a single stage, but also all invasive stages of all apicomplexan parasites, and, hence, hit multiple times during any given Apicomplexan life cycle. In order to translate the findings to drug development an in vitro reconstituted system of the major components, i.e., immobilized MyoA and parasite actin, is a stepping stone towards a high-throughput assay for potential gliding inhibitors. The identification of additional regulators of the parasite actin/myosin motor may also lead to important insights into general principles of cellular motility.

References

1. Fowler RE, Margos G, Mitchell GH. The cytoskeleton and motility in apicomplexan invasion. Adv Parasitol 2004; 56:213-263.
2. Heintzelman MB. Cellular and molecular mechanics of gliding locomotion in eukaryotes. Int Rev Cytol 2006; 251:79-129.
3. Kappe SH, Buscaglia CA, Bergman LW et al. Apicomplexan gliding motility and host cell invasion: Overhauling the motor model. Trends Parasitol 2004; 20(1):13-16.
4. Sibley LD. Intracellular parasite invasion strategies. Science 2004; 304(5668):248-253.
5. Vanderberg JP, Frevert U. Intravital microscopy demonstrating antibody-mediated immobilization of Plasmodium berghei sporozoites injected into skin by mosquitoes. Int J Parasitol 2004; 34(9):991-996.
6. Amino R, Thiberge S, Martin B et al. Quantitative imaging of Plasmodium transmission from mosquito to mammal. Nat Med 2006; 12:220-224.
7. Frevert U, Engelmann S, Zougbédé S et al. Intravital observation of Plasmodium berghei sporozoite infection of the liver. PLoS Biol 2005; 3(6):e192.
8. Barragan A, Sibley LD. Transepithelial migration of Toxoplasma gondii is linked to parasite motility and virulence. J Exp Med 2002; 195(12):1625-1633.
9. Barragan A, Sibley LD. Migration of Toxoplasma gondii across biological barriers. Trends Microbiol 2003; 11(9):426-430.
10. Carruthers VB, Sibley LD. Sequential protein secretion from three distinct organelles of Toxoplasma gondii accompanies invasion of human fibroblasts. Eur J Cell Biol 1997; 73(2):114-123.
11. Small JV, Resch GP. The comings and goings of actin: Coupling protrusion and retraction in cell motility. Curr Opin Cell Biol 2005; 17(5):517-523.
12. Bray D. Cell movements: From molecules to motility. 2nd ed. New York: Garland Publishing, 2000.
13. Russell DG, Sinden RE. The role of the cytoskeleton in the motility of coccidian sporozoites. J Cell Sci 1981; 50:345-359.
14. King CA. Cell motility of sporozoan protozoa. Parasitol Today 1988; 4(11):315-319.
15. Vanderberg JP. Studies on the motility of Plasmodium sporozoites. J Protozool 1974; 21(4):527-537.
16. Vanderberg JP. Development of infectivity by the Plasmodium berghei sporozoite. J Parasitol 1975; 61(1):43-50.
17. Frischknecht F, Baldacci P, Martin B et al. Imaging movement of malaria parasites during transmission by Anopheles mosquitoes. Cell Microbiol 2004; 6(7):687-694.
18. Håkansson S, Morisaki H, Heuser J et al. Time-lapse video microscopy of gliding motility in Toxoplasma gondii reveals a novel, biphasic mechanism of cell locomotion. Mol Biol Cell 1999; 10(11):3539-3547.
19. Vlachou D, Zimmermann T, Cantera R et al. Real-time, in vivo analysis of malaria ookinete locomotion and mosquito midgut invasion. Cell Microbiol 2004; 6(7):671-685.
20. Vanderberg JP, Chew S, Stewart MJ. Plasmodium sporozoite interactions with macrophages in vitro: A videomicroscopic analysis. J Protozool 1990; 37(6):528-536.
21. Mota MM, Pradel G, Vanderberg JP et al. Migration of Plasmodium sporozoites through cells before infection. Science 2001; 291(5501):141-144.
22. Carruthers V, Boothroyd JC. Pulling together: An integrated model of Toxoplasma cell invasion. Curr Opin Microbiol 2006; 9:1-7.
23. Morrissette NS, Sibley LD. Cytoskeleton of apicomplexan parasites. Microbiol Mol Biol Rev 2002; 66(1):21-38.
24. Mann T, Beckers C. Characterization of the subpellicular network, a filamentous membrane skeletal component in the parasite Toxoplasma gondii. Mol Biochem Parasitol 2001; 115(2):257-268.
25. Morrissette NS, Murray JM, Roos DS. Subpellicular microtubules associate with an intramembranous particle lattice in the protozoan parasite Toxoplasma gondii. J Cell Sci 1997; 110(Pt 1):35-42.
26. Hu K, Johnson J, Florens L et al. Cytoskeletal components of an invasion machine-the apical complex of Toxoplasma gondii. PLoS Pathog 2006; 2(2):e13.

27. Meissner M, Schlüter D, Soldati D. Role of Toxoplasma gondii myosin A in powering parasite gliding and host cell invasion. Science 2002; 298(5594):837-840.
28. Heintzelman MB, Schwartzman JD. A novel class of unconventional myosins from Toxoplasma gondii. J Mol Biol 1997; 271(1):139-146.
29. Foth BJ, Goedecke MC, Soldati D. New insights into myosin evolution and classification. Proc Natl Acad Sci USA 2006; 103(10):3681-3686.
30. Herm-Götz A, Weiss S, Stratmann R et al. Toxoplasma gondii myosin A and its light chain: A fast, single-headed, plus-end-directed motor. EMBO J 2002; 21(9):2149-2158.
31. Mermall V, Post PL, Mooseker MS. Unconventional myosins in cell movement, membrane traffic, and signal transduction. Science 1998; 279(5350):527-533.
32. Gaskins E, Gilk S, DeVore N et al. Identification of the membrane receptor of a class XIV myosin in Toxoplasma gondii. J Cell Biol 2004; 165(3):383-393.
33. Bergman LW, Kaiser K, Fujioka H et al. Myosin A tail domain interacting protein (MTIP) localizes to the inner membrane complex of Plasmodium sporozoites. J Cell Sci 2003; 116(Pt 1):39-49.
34. Baum J, Richard D, Healer J et al. A conserved molecular motor drives cell invasion and gliding motility across malaria life cycle stages and other apicomplexan parasites. J Biol Chem 2006; 281(8):5197-5208.
35. Green JL, Martin SR, Fielden J et al. The MTIP-myosin A complex in blood stage malaria parasites. J Mol Biol 2006; 355(5):933-941.
36. Jones ML, Kitson EL, Rayner JC. Plasmodium falciparum erythrocyte invasion: A conserved myosin associated complex. Mol Biochem Parasitol 2006; 147(1):74-84.
37. Bosch J, Turley S, Daly TM et al. Structure of the MTIP-MyoA complex, a key component of the malaria parasite invasion motor. Proc Natl Acad Sci USA 2006; 103(13):4852-4857.
38. Dobrowolski JM, Sibley LD. Toxoplasma invasion of mammalian cells is powered by the actin cytoskeleton of the parasite. Cell 1996; 84(6):933-939.
39. Miller LH, Aikawa M, Johnson JG et al. Interaction between cytochalasin B-treated malarial parasites and erythrocytes: Attachment and junction formation. J Exp Med 1979; 149(1):172-184.
40. Schatten H, Sibley LD, Ris H. Structural evidence for actin-like filaments in Toxoplasma gondii using high-resolution low-voltage field emission scanning electron microscopy. Microsc Microanal 2003; 9(4):330-335.
41. Sahoo N, Beatty W, Heuser J et al. Unusual kinetic and structural properties control rapid assembly and turnover of actin in the parasite Toxoplasma gondii. Mol Biol Cell 2006; 17(2):895-906.
42. Schmitz S, Grainger M, Howell S et al. Malaria parasite actin filaments are very short. J Mol Biol 2005; 349(1):113-125.
43. Schüler H, Mueller AK, Matuschewski K. Unusual properties of Plasmodium falciparum actin: New insights into microfilament dynamics of apicomplexan parasites. FEBS Lett 2005; 579(3):655-660.
44. Schüler H, Matuschewski K. Plasmodium motility: Actin not actin' like actin. Trends Parasitol 2006; 22(4):146-147.
45. Schüler H, Matuschewski K. Regulation of apicomplexan microfilament dynamics by a minimal set of actin-binding proteins. Traffic 2006; 7(11):1433-1439.
46. Jewett TJ, Sibley LD. Aldolase forms a bridge between cell surface adhesins and the actin cytoskeleton in apicomplexan parasites. Mol Cell 2003; 11(4):885-894.
47. Buscaglia CA, Coppens I, Hol WG et al. Sites of interaction between aldolase and thrombospondin-related anonymous protein in Plasmodium. Mol Biol Cell 2003; 14(12):4947-4957.
48. Baum J, Papenfuss AT, Baum B et al. Regulation of apicomplexan actin-based motility. Nat Rev Microbiol 2006; 4(8):621-628.
49. Schüler H, Mueller AK, Matuschewski K. A Plasmodium actin-depolymerizing factor that binds exclusively to actin monomers. Mol Biol Cell 2005; 16(9):4013-4023.
50. Robson KJ, Hall JR, Jennings MW et al. A highly conserved amino-acid sequence in thrombospondin, properdin and in proteins from sporozoites and blood stages of a human malaria parasite. Nature 1988; 335(6185):79-82.
51. Tomley FM, Soldati DS. Mix and match modules: Structure and function of microneme proteins in apicomplexan parasites. Trends Parasitol 2001; 17(2):81-88.
52. Tucker RP. The thrombospondin type 1 repeat superfamily. Int J Biochem Cell Biol 2004; 36(6):969-974.
53. Whittaker CA, Hynes RO. Distribution and evolution of von Willebrand/integrin A domains: Widely dispersed domains with roles in cell adhesion and elsewhere. Mol Biol Cell 2002; 13(10):3369-3387.
54. Sultan AA, Thathy V, Frevert U et al. TRAP is necessary for gliding motility and infectivity of Plasmodium sporozoites. Cell 1997; 90(3):511-522.

55. Kappe S, Bruderer T, Gantt S et al. Conservation of a gliding motility and cell invasion machinery in apicomplexan parasites. J Cell Biol 1999; 147(5):937-943.
56. Matuschewski K, Nunes AC, Nussenzweig V et al. Plasmodium sporozoite invasion into insect and mammalian cells is directed by the same dual binding system. EMBO J 2002; 21(7):1597-1606.
57. Rogers WO, Malik A, Mellouk S et al. Characterization of Plasmodium falciparum sporozoite surface protein 2. Proc Natl Acad Sci USA 1992; 89(19):9176-9180.
58. Dessens JT, Beetsma AL, Dimopoulos G et al. CTRP is essential for mosquito infection by malaria ookinetes. EMBO J 1999; 18(22):6221-6227.
59. Yuda M, Sakaida H, Chinzei Y. Targeted disruption of the Plasmodium berghei CTRP gene reveals its essential role in malaria infection of the vector mosquito. J Exp Med 1999; 190(11):1711-1716.
60. Huynh MH, Carruthers VB. Toxoplasma MIC2 is a major determinant of invasion and virulence. PLoS Pathog 2006; 2(8):e84.
61. Rabenau KE, Sohrabi A, Tripathy A et al. TgM2AP participates in Toxoplasma gondii invasion of host cells and is tightly associated with the adhesive protein TgMIC2. Mol Microbiol 2001; 41(3):537-547.
62. Huynh MH, Rabenau KE, Harper JM et al. Rapid invasion of host cells by Toxoplasma requires secretion of the MIC2-M2AP adhesive protein complex. EMBO J 2003; 22(9):2082-2090.
63. Stewart MJ, Vanderberg JP. Malaria sporozoites leave behind trails of circumsporozoite protein during gliding motility. J Protozool 1988; 35(3):389-393.
64. Carruthers VB, Giddings OK, Sibley LD. Secretion of micronemal proteins is associated with Toxoplasma invasion of host cells. Cell Microbiol 1999; 1(3):225-235.
65. Kappes B, Doerig CD, Graeser R. An overview of Plasmodium protein kinases. Parasitol Today 1999; 15(11):449-454.
66. Ishino T, Orito Y, Chinzei Y et al. A calcium-dependent protein kinase regulates Plasmodium ookinete access to the midgut epithelial cell. Mol Microbiol 2006; 59(4):1175-1184.
67. Siden-Kiamos I, Ecker A, Nybäck S et al. Plasmodium berghei calcium-dependent protein kinase 3 is required for ookinete gliding motility and mosquito midgut invasion. Mol Microbiol 2006; 60(6):1355-1363.
68. Carey KL, Westwood NJ, Mitchison TJ et al. A small-molecule approach to studying invasive mechanisms of Toxoplasma gondii. Proc Natl Acad Sci USA 2004; 101(19):7433-7438.
69. Opitz C, Di Cristina M, Reiss M et al. Intramembrane cleavage of microneme proteins at the surface of the apicomplexan parasite Toxoplasma gondii. EMBO J 2002; 21(7):1577-1585.
70. Zhou XW, Blackman MJ, Howell SA et al. Proteomic analysis of cleavage events reveals a dynamic two-step mechanism for proteolysis of a key parasite adhesive complex. Mol Cell Proteomics 2004; 3(6):565-576.
71. Brossier F, Jewett TJ, Sibley LD et al. A spatially localized rhomboid protease cleaves cell surface adhesions essential for invasion by Toxoplasma. Proc Natl Acad Sci USA 2005; (11)102:4146-4151.
72. Dowse TJ, Pascall JC, Brown KD et al. Apicomplexan rhomboids have a potential role in microneme protein cleavage during host cell invasion. Int J Parasitol 2005; 35(7):747-756.
73. Dowse TJ, Soldati D. Rhomboid-like proteins in Apicomplexa: Phylogeny and nomenclature. Trends Parasitol 2005; 21(6):254-258.
74. Baker RP, Wijetilaka R, Urban S. Two Plasmodium rhomboid proteases preferentially cleave different adhesions implicated in all invasive stages of malaria. PLoS Pathog 2006; 2(10):e113.
75. Wetzel DM, Schmidt J, Kuhlenschmidt MS et al. Gliding motility leads to active cellular invasion by Cryptosporidium parvum sporozoites. Infect Immun 2005; 73(9):5379-5387.
76. Spano F, Putignani L, Naitza S et al. Molecular cloning and expression analysis of a Cryptosporidium parvum gene encoding a new member of the thrombospondin family. Mol Biochem Parasitol 1998; 92(1):147-162.
77. Khater EI, Sinden RE, Dessens JT. A malaria membrane skeletal protein is essential for normal morphogenesis, motility, and infectivity of sporozoites. J Cell Biol 2004; 167(3):425-432.

Chapter 10

Roles of Proteases during Invasion and Egress by *Plasmodium* and *Toxoplasma*

Timothy J. Dowse,[†] Konstantinos Koussis,[†] Michael J. Blackman and Dominique Soldati-Favre*

Abstract

Apicomplexan pathogens replicate exclusively within the confines of a host cell. Entry into (invasion) and exit from (egress) these cells requires an array of specialized parasite molecules, many of which have long been considered to have potential as targets of drug or vaccine-based therapies. In this chapter the authors discuss the current state of knowledge regarding the role of parasite proteolytic enzymes in these critical steps in the life cycle of two clinically important apicomplexan genera, *Plasmodium* and *Toxoplasma*. At least three distinct proteases of the cysteine mechanistic class have been implicated in egress of the malaria parasite from cells of its vertebrate and insect host. In contrast, the bulk of the evidence indicates a prime role for serine proteases of the subtilisin and rhomboid families in invasion by both parasites. Whereas proteases involved in egress may function predominantly to degrade host cell structures, proteases involved in invasion probably act primarily as maturases and 'sheddases', required to activate and ultimately remove ligands involved in interactions with the host cell.

Introduction

"Proteases are enzymes that degrade proteins by hydrolyzing some of their peptide bonds."[1] The above, terminologically correct definition has represented for many years the view we had towards proteases and their biological roles. In recent years though, our knowledge concerning this class of enzymes and their functions has expanded considerably. Once thought of as only nonspecific, "degradative" enzymes, proteases are now also known to be involved in highly specific substrate cleavages that result in changes of protein function. Proteases are involved in a wide range of biological processes such as DNA replication, cell signalling, immunity and apoptosis and from genome sequencing data we know that more than 650 proteases are defined in the human genome and similar numbers are seen in *Drosophila melanogaster* and mouse.[2]

Proteases are divided into five main groups: (1) serine proteases, (2) threonine proteases, (3) cysteine proteases, (4) metalloproteases and (5) aspartic proteases. In the first three groups, an amino acid residue (serine, threonine or cysteine respectively) acts as the catalytic nucleophile that binds to the target peptide bond, whereas in the case of metallo- and aspartic proteases a

[†]These authors contributed equally to this work.

*Corresponding Author: Dominique Soldati-Favre—Department of Biological Sciences, Sir Alexander Fleming Building, Imperial College, London SW7 2AZ, UK; and Département de Microbiologie et Medecine Moléculaire, Falculté de Medecine, Université de Genève, CMU, Rue Michel Servet 1, Genève, Switzerland. Email: Dominique.Soldtati-Favre@medecine.unige.ch

Molecular Mechanisms of Parasite Invasion, edited by Barbara A. Burleigh and Dominique Soldati-Favre. ©2008 Landes Bioscience and Springer Science+Business Media.

Table 1. *Summary of proteases involved in invasion and egress in* Plasmodium *and* Toxoplasma

Class/Name	Domains	Putative Substrate	Localisation	Homologues	Putative Function	Ref.
INVASION						
Serine (rhomboids)						
TgROM1	7 TM	?	Microneme	PfROM1	During invasion	22,24
TgROM4 TgROM5	7 TM	TgMIC2, TgMI6, TgMIC12, TgAMA1	Surface	PfROM4	MPP1?	22,24
Serine (subtilases)						
TgSUB1	Subtilase	TgMIC2, TgM2AP, TgMIC4	Microneme		MPP2?	20
TgSUB2	Subtilase	Rhoptry proteins	Rhoptry		Rhoptry protein maturase	55
PfSUB1	Subtilase	Unknown	Dense granule		?	44
PfSUB2	1 TM subtilase	PfAMA1, PfMSP1	Microneme		MESH	48-50
Other						
MPP3	?	?	Surface	?	?	14
EGRESS						
Cysteine						
Falcipain 2 and 2'	1 TM papain family	Haemoglobin, band 4.1, ankyrin	Food Vacuole	?	Haemoglobin degradation Egress	57,70-72
SERA 5	Papain family	?	PV	?	Egress	80-83
ECP1	Papain family	?	Sporozoites	?	Egress	84
Aspartic				?		
Plasmepsin II	Aspartic protease 1 TM	Haemoglobin	Food Vacuole	?	Haemoglobin degradation Egress	76,77

water molecule performs that role. Each of the main groups is subdivided into clans, which are further divided into families based on sequence homology.[3]

It is not only higher eukaryotes that use proteolytic enzymes. Infectious agents - viruses, bacteria, fungi and protozoa—use proteases for replication, host invasion and metabolism. In the case of the pathogenic protozoa *Plasmodium* and *Toxoplasma* several studies have shown that proteolytic activity is necessary for invasion into and exit from the host cell (egress). Although these studies have provided researchers with invaluable data, the identity, regulation and sub-cellular location of many of the proteases involved in the life cycles of these organisms remain elusive.

The aim of this review is to summarize current knowledge of proteases that are thought to be involved in *Plasmodium* and *Toxoplasma* invasion and egress (see Table 1 for overview).

Apicomplexan Parasite Invasion

Invasion by all apicomplexan parasites is thought to follow a broadly similar mechanism. When a parasite comes into contact with the host cell, it orientates such that the apical pole of the parasite points towards and interacts with the host. Microneme secretion is triggered, as a result of a calcium-mediated signal, and the released proteins are mobilised to the parasite surface. These microneme proteins are thought to mediate high affinity interactions between the parasite and host surfaces. A 'tight junction' forms between the parasite and host, which translocates towards the rear of the parasite via interactions between the microneme protein cytoplasmic tails and a cortical parasite actin-myosin system. This process provides the traction for invasion. Soon after microneme discharge rhoptries, a second group of apical organelle, secrete their contents. Some rhoptry proteins (along with the microneme protein AMA1) have recently been demonstrated to form part of the tight junction, although their precise role is unknown.[4] Other rhoptry material contributes to the formation of the parasitophorous vacuole.

Invasion and egress share common mechanistic features and factors, and in some respects invasion can be thought to begin at the moment of egress, as from this point on the parasite is primed and ready for invasion.[5]

Proteases Involved in Invasion

Many proteins secreted by the parasite during invasion or resident on the surface are proteolytically processed either during their transit through the secretory pathway or on the surface of the parasite whilst it is still intracellular. These processes are important, but are beyond the scope of this review, and have been extensively reviewed recently.[6,7] The topic of this chapter will focus on proteolytic activities and proteases, where known, that are involved in invasion itself. Aside from the insight that studies of these proteases provide into the fascinating fundamental biology of invasion, what makes these proteases a key area of research is their potential as drug targets.[8] Invasion itself is a rapid process, often going to completion within less than a minute following initial interaction between parasite and host cell. However, it has been argued that enzymes involved in the process are likely to be available for long enough to be targeted by drugs, as the synthesis of the components of invasion machinery takes place over a substantial period of time prior to egress. A counter argument, however, suggests that it is precisely the importance of the invasion machinery that has led to its minimal exposure time to extra-cellular milieu, and that any inhibitors will not have access to the well-sequestered enzymes for long enough. The question remains whether drugs targeted to molecules involved in invasion will be successful and will be answered only by further study. The fact that certain protease inhibitors have been demonstrated to inhibit invasion and egress suggests that at least in principal there is enough time for an effective target-drug interaction. Recent studies have led to the identification of proteases involved in essential steps in invasion, and further major breakthroughs can be expected over the next few years.

Shedding of "Invasion Proteins" in Toxoplasma gondii*: Microneme Protein Protease 1*

The Activity

The first studies concerning the processing of the type I transmembrane protein TgMIC2 in *Toxoplasma gondii* (the homologue of *Plasmodium* TRAP) revealed that, following its release from the micronemes and redistribution across the plasma membrane, the protein is shed from the parasite surface. This shedding was found to result from a proteolytic cleavage near the C-terminus of the protein ectodomain, and the hypothetical protease responsible was named Microneme Protein Protease 1 (MPP1).[9] TgMIC2 and other microneme proteins are thought to form the link between the host cell surface and the sub-plasmalemmal actinomyosin motor system, so several investigators suggested that the role of the shedding is to disrupt this link, and thus allow the final steps of invasion to be completed. It has been proposed that MPP1 must therefore be located towards the posterior of the parasite surface.[10]

Interestingly, however, if TgMIC2 redistribution towards the posterior pole is inhibited by cytochalasin D, MPP1 processing still occurs,[9] suggesting that the protease is not exclusively concentrated at the parasite posterior. Additional evidence has shown that MPP1 is constitutively active on the surface of the tachyzoite even when the parasites are intracellular.[11] So if MPP1 is located all over the parasite surface, what is its precise role in invasion? It is possible that host-parasite interactions need to be continually disrupted during the invasion process, but exactly why this would happen is not obvious. Questions such as this highlight the need to identify the MPP1 protease at the molecular level, and since its first description several steps have been made to bring this closer to reality.

Intramembrane Proteolysis

The first development towards discovering the identity of MPP1 was the identification of the cleavage site targeted by the protease. It was observed that other transmembrane TgMIC proteins (TgMIC6 and TgMIC12) are also cleaved near their C-terminus, resulting in their shedding, and this led to the obvious suspicion that MPP1 is responsible for shedding several, if not all transmembrane TgMICs.[11]

Mutagenesis of conserved residues within the transmembrane (TM) domains of TgMIC2, TgMIC6 and TgMIC12 (a homologue of *Eimieria tenella* MIC4;[12]) abrogated their shedding by MPP1, suggesting that the cleavage site was within the TM domain itself. This was confirmed using mass spectrometry, to show that TgMIC6 is cleaved within the TM domain, at the site IA↓GG. This site is conserved in several other microneme proteins from *Toxoplasma* and other apicomplexans, including TgMIC2 and PfTRAP.[11] Importantly, it was also shown that expression of *P. berghei* TRAP in *T. gondii* resulted in MPP1 mediated shedding, suggesting the conservation of MPP1 activity in apicomplexans.[11] Interestingly, another study clearly identified a lysine residue (K692) just upstream of the TM domain of TgMIC2 which, when mutated to alanine (but not arginine), completely abrogated cleavage by MPP1.[13] This could suggest that this region is involved in substrate recognition. Although at the time this result cast doubt on the location of the cleavage site, this issue was finally resolved by mass spectrometric mapping of the TgMIC2 cleavage site, which confirmed its localisation within the TM domain.[14] The mutant analysis by Brossier et al[13] also confirmed the predicted importance of MPP1-mediated cleavage during invasion, as over-expression of the TgMIC2 processing mutant impaired invasion.

Towards Identification of an Enzyme

The location of the MPP1 cleavage site within the TM domain implicated an intramembrane protease in this critical cleavage event. An examination of the list of intramembrane protease families so far discovered immediately suggested the rhomboid-like family as candidate enzymes for MPP1. Rhomboid-1 from *Drosophila*, the founding member of the family, is an intramembrane-cleaving protease (ICliP), which regulates the secretion of an EGF-receptor ligand, Spitz, via cleavage in the Spitz TM domain.[15] Rhomboid-like proteases generally contain six or seven transmembrane domains, with short connecting loops, and the catalytic residues are buried within the lipid-bilayer. Some important characteristics differentiate this family of intramembrane proteases, which are found throughout evolution,[16] from other ICliPs: they are the only intramembrane proteases of the serine class, and the only ones that cleave their substrates in the region of the TM domain nearest the lumen/extracellular region, resulting in shedding of the N-terminal domain. They are also the only class of ICliP that does not require prior cleavage of their substrate by an independent protease.[17]

These features make a rhomboid-like protease a particularly attractive candidate for MPP1, as the cleavage site of TgMIC2 and TgMIC6 is near to the luminal end of the TM domain.

Also importantly, although TgMIC2 is subject to additional proteolytic cleavages on the surface of the parasite (see below), these cleavages are not a prerequisite for MPP1 mediated shedding.[9] When MPP1 was originally described, a number of protease inhibitors were tested in order to identify the class of protease, and thus narrow down the number of candidate genes responsible for the activity. Release of TgMIC2 from the parasite was only inhibited by a serine protease inhibitor, 3,4-dichloroisocoumarin (DCI;[9]), one of only two inhibitors known to inhibit Rhomboid-1 activity (the other being tosyl phenylalanine chloromethyl ketone, TPCK).[15]

Initially this finding was interpreted as being due to inhibition of microneme secretion, as TgMIC4 release was also reduced. The subsequent discovery of the anchoring of TgMIC4 to the membrane via an interaction with TgMIC1/TgMIC6[18] led to a reevaluation of this conclusion, as inhibition of MPP1 would also inhibit TgMIC6 shedding, and therefore that of TgMIC4. Howell et al recently showed that DCI inhibits the intramembrane cleavage of another microneme protein, TgAMA1.[19] MPP1 is thought to be responsible for TgAMA1 cleavage (for details, see below), and thus DCI is a true inhibitor of MPP1. The same authors found that DCI had no effect on secretion of TgSUB1, a subtilisin-like microneme protein which is membrane-bound via a GPI-anchor[20] and cannot therefore be an MPP1 substrate.

Further evidence in support of MPP1 being a rhomboid-like protease was provided by a study by Urban and Freeman.[21] They demonstrated that the substrate specificity of rhomboid-like proteases is governed not by specific sequence motifs, but rather by the presence of helix-destabilising residues near the cleavage site. Analysis of the TM domains of TgMIC2, TgMIC6 and TgMIC12 revealed the presence of such helix-destabilising residues, and the authors went on to show that the TM domains of those microneme proteins could act as substrates for *Drosophila* Rhomboid-1 and the human rhomboid RHBDL2.

Two independent studies then described rhomboid-like proteins in *T. gondii*.[22,23] As MPP1 was indirectly indicated to be constitutively active at the parasite surface,[11] both studies performed analyses of the localisations of some of the *T. gondii* rhomboids (TgROMs). In total, genes for six rhomboid-like proteins have been found in the *T. gondii* genome, called TgROM1-6.[22-24] TgROM1 localises to the micronemes, TgROM2 to the Golgi, and TgROM4 to the parasite surface.[22,23] TgROM3 does not appear to be expressed in tachyzoites, the life-cycle stage where MPP1 was described, and was therefore excluded from further analysis. TgROM6 was predicted from a phylogenetic analysis to be localised to the mitochondrion, and indeed possesses a predicted mitochondrial targeting signal.[24] Recent localisation experiments seem to confirm this prediction (Dowse and Soldati, unpublished). TgROM5 localisation appears to be on the parasite surface, and Brossier et al observed an apparent concentration of this surface-resident rhomboid towards the posterior end of the extracellular parasite. This, combined with data which showed cleavage of TgMIC2 by TgROM5 in a heterologous cell-based assay,[22] led the authors to suggest that TgROM5 is the primary candidate for MPP1. Dowse et al,[23] however, did not observe any cleavage of microneme protein TM domains by TgROM5, despite the fact that this enzyme was active against the TM domain of *Drosophila* Spitz. It should be noted, however, that the substrate TM domains in the two studies were in different contexts in terms of cytoplasmic tails and ectodomains. This is of particular interest as a very recent study demonstrated that TgROM5 appears to have a very flexible substrate specificity, being capable of cleaving many microneme protein TM domains from both *T. gondii* and *P. falciparum*, including TM domains which were not Rhomboid-1 substrates[23a] (see below). However, in spite of this apparent broad specificity, TgROM5 could not cleave microneme protein TM domain-containing substrates, which were nevertheless cleaved by Rhomboid-1 in the Dowse et al study. The reason behind this apparent contradiction remains mysterious.

The story is complicated somewhat further by the phylogenetic distribution of ROMs throughout Apicomplexa. The evidence outlined above suggests that MPP1 is probably conserved throughout the phylum, and indeed target substrates can be predicted in most apicomplexan species where sequence data is available.[24] Whereas orthologues of TgROM4 are found in all *Plasmodium* species examined so far, TgROM5 orthologues are not. Furthermore, in *P. falciparum*, PfROM4 localises to the surface of the merozoite and has recently been shown to cleave the TM domain of the invasion protein EBA-175,[24a] and other invasion proteins[23a] (see below). The significance of this is not yet clear as far as MPP1 in *T. gondii* is concerned, and ultimately only a genetic analysis of MPP1 candidate proteases will enable the identity of MPP1 to be absolutely confirmed.

Both TgROM1 and PfROM1[24a] localise to the micronemes. Since the contents of micronemes are secreted as one of the first steps of invasion,[25] ROM1 might be involved in some aspect of the invasion process. However, as MPP1 activity has been monitored constitutively on the parasite surface in the absence of microneme secretion,[11] ROM1 is unlikely to correspond to MPP1. Furthermore, rhomboids are thought to be constitutively active and therefore need to be spatially separated from their substrate(s) until cleavage is required. Therefore the sub-cellular localization of MPP1 is anticipated to be the parasite plasma membrane. The localisation of ROM1 in the micronemes may be suggestive of a role in shedding of surface proteins encountered during, but not before, invasion and/or egress.

As mentioned above, TgAMA1 is cleaved within its TM domain at a site consistent with rhomboid-mediated cleavage. One significant difference between TgAMA1 and other microneme proteins that have been studied in *T. gondii* is that TgAMA1 can apparently access both sides of the moving junction during invasion,[19] rather than being specifically excluded from the parasite plasma membrane within the forming vacuole as with other microneme proteins. Additionally, TgAMA1 does not contain the conserved hydrophobic tail required for interaction with aldolase, thought to mediate the critical link between microneme-derived adhesins and the actin-myosin motor, and is not required for gliding motility.[26] Depending on the localisation of MPP1 during invasion, this could suggest different processing dynamics of TgAMA1 compared to other TgMICs, and perhaps reflects the role of a second rhomboid-like protease, although it must be noted that TgROM1 was not capable of cleaving PfAMA1 in the study by Baker et al[23a] (see below). This again highlights the fact that many questions still remain unanswered and await functional analyses. Interestingly, *Plasmodium* AMA1 is shed primarily by a different protease (see below).

MPP1 in Plasmodium

As mentioned above, MPP1 is likely conserved throughout Apicomplexa, and *Plasmodium* species contain homologues of TgROM1, TgROM3 and TgROM4.[24] Furthermore, a rhomboid-like activity has been demonstrated at the surface of the *P. falciparum* merozoite,[19] supporting the anticipated conservation of MPP1 throughout the Apicomplexa. But which proteins in *Plasmodium* might be shed by MPP1 mediated cleavage? Certainly, owing to the fact that TRAP is homologous to TgMIC2, predicted to fulfil a similar function, and is cleaved in *T. gondii* in the C-terminal region,[11] one would expect TRAP to be processed in this way. However, a study by Silvie et al[27] showed that in the sporozoite, both TRAP and PfAMA1 appear to be processed by the same enzyme, which has a different inhibition profile to that of the MESH (merozoite sheddase) of merozoites. The shed forms of PfAMA1 from the sporozoites were the same sizes as those shed from merozoites, suggesting that the cleavage took place at the same juxtamembrane location. More work is required to clarify the mechanism of TRAP shedding from the sporozoite.

Two recent papers investigating *P. falciparum* ROMs have shed some light onto MPP1 in this species. O'Donnell et al, investigating PfROM4, demonstrated that this protease localises to the surface of the parasite. They also showed that the EBL family protein, EBA-175, is cleaved within its TM domain, which had previously been predicted, at a site consistent with rhomboid cleavage. They further demonstrated that EBA-175 cleavage is required for invasion to succeed, even if the EBA-175-independent/sialic acid-independent pathway was being used by the parasite. In a cell-based assay, PfROM4 was capable of cleaving the EBA-175 protein, but interestingly this was not a substrate for *Drosophila* Rhomboid-1 or PfROM1.[24a]

A second study elaborated on the findings of these assay. Baker et al[23a] tested several PfROMs, as well as TgROM1 and TgROM5, on the 14 *P. falciparum* adhesins currently thought to be involved in invasion. The substrates tested include the PfRH4 protein recently demonstrated to be upregulated in parasites following a switch from a sialic-acid dependent to a sialic acid independent invasion pathway,[29] and related proteins from the RBL family as well as PfAMA1, proteins from the EBL family, and TRAP family proteins. PfAMA1 has a TM domain resembling other rhomboid substrates, with helix destabilising residues. EBL and RBL family proteins, on the other hand, whilst containing helix-relaxing residues in their TM domains, also contain upstream aromatic residues that create poor rhomboid substrates.

Interestingly, PfROM4 did not cleave Spitz or PfAMA1, also demonstrated by O'Donnell et al. It was however active against EBL, RBL and TRAP family proteins, but Rhomboid-1 could not process EBL and RBL family substrates. PfROM1 on the other hand was highly active against Spitz and PfAMA1, and had weaker levels of activity against the other proteins tested, with some substrates not being cleaved at all by PfROM1. The authors proposed that PfROM1 and Rhomboid-1 have a similar specificity, whereas the specificity of PfROM4 was distinct. Interestingly, TgROM5 was active against all substrates tested, and the authors proposed that this protease has a "dual" specificity.[23a] The significance of this in terms of *T. gondii* is unclear.

Together, these findings suggest that MPP1 in *P. falciparum*, a role most likely filled by PfROM4, is involved in most if not all invasion pathways of *P. falciparum*, and could therefore be an "Achilles heel" of invasion. A recent X-ray crystal structure of a related rhomboid-like protease from *Escherichia coli*, GlpG, reveals the presence of a hydrophilic pocket at the extracellular face of the membrane, which allows access to the water necessary for the hydrolysis reaction performed by the rhomboid to occur.[29a] This pocket could be accessible for potential inhibitors, and this together with the known sensitivity of rhomboids to DCI, proves the principal that drugs could indeed be designed against this "Achilles heel". This structure also demonstrates that in principal it is possible to crystalise rhomboid-like proteases. Structural studies such as this, if performed on apicomplexan ROMs, should reveal at least in part the mechanisms behind this varying substrate specificity observed by Baker et al."

Shedding of Surface Proteins in Plasmodium*: MESH*

MPP1 is thus probably conserved in *Plasmodium* species and involved in shedding of certain proteins during invasion as mentioned above. However, another type of surface shedding has been observed in *Plasmodium*, which has different features to MPP1 mediated shedding.

MSP1 Primary Processing

MSP1 is a large protein complex localised on the surface of *Plasmodium* merozoites. It may be involved in initial low-affinity binding of the parasite to the host cell, and is considered to be a good vaccine candidate. MSP1 is initially expressed as a protein precursor, and is subject to what is called primary processing. This is thought to take place whilst the parasites are developing within the host cell rather than during invasion itself and therefore, while suspected to be important for the function of the protein complex, will not be discussed in depth here. Primary processing has been well reviewed,[8] and in *P. falciparum* results in the full length gene product, after signal peptide removal and GPI anchor modification, being cleaved into four subunits known as $MSP1_{83}$, $MSP1_{30}$, $MSP1_{36}$, and $MSP1_{42}$ (in order from the N- to C-terminus of the original gene product). These fragments are bound together noncovalently in a complex, with $MSP1_{42}$ anchoring the complex in the membrane via its GPI anchor. Other proteins (including MSP6, 7 and 9 in the case of *P. falciparum*,[30-32]) are also bound to this complex. It is the secondary processing of MSP1, which results in the shedding of the vast majority of the complex from the plasma membrane, which we will discuss further.

MSP1 Secondary Processing

During invasion $MSP1_{42}$ is proteolytically cleaved into two fragments (called $MSP1_{33}$ and $MSP1_{19}$) in what is known as secondary processing. This processing results in the release of the MSP1 complex from the parasite surface—an event which appears to be important, as only the post-processing stub ($MSP1_{19}$) appears to be able to penetrate the moving junction and still be localised to the parasite surface after invasion is complete.[33] Other studies have shown that antibodies or small molecules that inhibit MSP1 secondary processing also inhibit invasion.[34,35] The first step towards identification of the MSP-1 shedding protease was the observation that this activity is calcium dependent, sensitive to the serine protease inhibitors PMSF and DFP, and also that the protease responsible is bound to the parasite plasma membrane when the processing event occurs.[36]

AMA1 Processing

PfAMA1, another long-time vaccine candidate in *Plasmodium* is shed during invasion and, as for MSP1, antibodies that inhibit this processing impede invasion.[37,37a] While analysing the activity responsible for PfAMA1 shedding, Howell et al discovered that PfAMA1 is shed by a protease with the same characteristics and inhibition profile as that responsible for MSP1 shedding.[38,39] They concluded that the same protease, named Merozoite Surface Sheddase (MESH,[10]) is responsible for the shedding of the two proteins. Analysis of the MESH cleavage sites of PfAMA1 and PfMSP1 shows no obvious conserved motif, and it has been postulated that MESH may recognise certain sequences by their proximity to the membrane in which it is anchored.[40] This hypothesis is supported by the structural analysis of $MSP1_{19}$, which shows that the tandem EGF domains that comprise this polypeptide are arranged such that the MESH cleavage site is brought close to the plasma membrane, despite being some distance away in terms of primary sequence from the GPI anchor (~96 amino acids).[41,42] Interestingly, shedding of PfAMA1 was not inhibited by cytochalasin D, but neither was the redistribution of the protein over the surface of the parasite, adding weight to the notion that AMA1 is not associated with the parasite's actin-myosin motor system.

Identification of MESH

The characteristics of MESH—a calcium-dependent, surface-bound serine protease—led investigators to search in *Plasmodium* for candidate proteases, and the subtilisin-like family of proteases emerged as primary candidates. The recently completed genome has revealed the presence of three subtilisin-like proteases in *Plasmodium*, named PfSUB1, PfSUB2 and PfSUB3.[43] PfSUB1 was previously identified,[44] and since has been characterised in detail in terms of specificity, although in vivo substrates remain anonymous (see ref. 45 for a detailed review of *Plasmodium* subtilisin-like proteases). Although it may be critical for parasite growth[46] an important study demonstrated that it is not responsible for MESH activity.[47]

PfSUB2, initially described in 1999,[48,49] and identified as a candidate for the MESH protease, was recently identified as the protease most likely responsible for MESH activity. All subtilisin-like proteases have a propeptide, which is generally removed by autocatalytic cleavage during maturation. This propeptide aids folding and is thought to specifically inhibit the enzyme until its removal has occurred, in this way ensuring the enzyme remains inactive until the appropriate time. This phenomenon was taken advantage of by Harris et al,[50] who used recombinant PfSUB2 propeptide to inhibit the activity of mature PfSUB2, which is secreted onto the surface of merozoites, and to demonstrate convincingly that this inhibition results in the inhibition of PfMSP-1 and PfAMA1 shedding. One cannot exclude that PfSUB2 activates the actual protease responsible for MESH activity. Nevertheless this remains the most compelling evidence to date for the identity of a protease responsible for the shedding of an apicomplexan protein during invasion. A recent paper has further demonstrated the importance of the MESH by showing that the cleavage of a merozoite TRAP homologue (PTRAMP) is also PfSUB2 dependent.[51]

Localisation Issues

Harris et al also reported the sub-cellular localisation of PfSUB2, which had until then been thought to reside in the dense granules. In fact, PfSUB2 is a microneme protein, and immunofluorescence and immuno-electron microscopy was used to demonstrate that it is secreted and relocalised to the posterior pole in extracellular merozoites.[50] The authors suggest that this relocalisation is representative of the redistribution of PfSUB2 with the moving junction during invasion under physiological conditions. This could explain how only PfMSP1$_{19}$ and the post-processing stub of PfAMA1 remain on the surface of intracellular post-invasion parasites, as the extracellular domains would be cleaved off as the moving junction passes over the parasite surface.

An obvious advantage of the targeting of MESH to the micronemes is that the protease will not be on the surface until the moment of invasion, thereby ensuring no rogue shedding of critical surface proteins at inappropriate times. However, the presence of PfSUB2 in micronemes could be seen as being somewhat paradoxical, as PfAMA1 is also a microneme protein but is not shed until its release onto the parasite surface. There are several possible explanations for this apparent discrepancy. One possibility is that the propeptide of PfSUB2, whilst known to be cleaved from PfSUB2 in the ER, could remain noncovalently bound to PfSUB2, thus inhibiting its activity, until an environmental change such as a pH shift occurs (for example, following microneme discharge). Alternatively, calcium levels in the micronemes may be too low for optimal MESH activity, and only upon secretion can it become active and fulfil its role as a sheddase. It is additionally possible that the PfAMA1 prosequence, a short hydrophilic sequence that is removed upon translocation of the protein onto the parasite surface, may protect the protein from premature processing in the micronemes, as x-ray crystal studies have shown that this lies adjacent to the PfSUB2 cleavage site in the three-dimensional structure of PfAMA1.[52]

Differences between Toxoplasma *and* Plasmodium *Surface Shedding*

The differences observed between the shedding of PfAMA1 and TgAMA1 raise the obvious question: why did two mechanisms evolve to perform an apparently similar job? It is not absolutely clear, but perhaps the answer lies in the removal of MSP1 during *Plasmodium* invasion. MSP1 is only found in *Plasmodium* species, and being GPI anchored its removal could not be mediated by a rhomboid-like protease. Therefore a separate sheddase had to evolve in order to remove this protein during invasion. Perhaps PfAMA1 is a fortuitous substrate for this enzyme, and thus became shed via this mechanism. This argument is supported by the fact that a rhomboid-like shedding of PfAMA1 can also be observed under certain conditions. However, clearly this rhomboid-like shedding is not sufficient to support invasion, as inhibition of MESH-mediated shedding of PfAMA1 results in inhibition of invasion. It should be noted, however, that PfAMA1 is shed by a sheddase in sporozoites, where MSP1 is absent.[27]

As mentioned above, it has been proposed that MESH-mediated shedding occurs at the moving junction, continuously during invasion, whereas MPP1-mediated shedding occurs only at the posterior pole, during the final stages of invasion.[10] This model remains debatable however, especially in the light of findings that TgAMA1 is shed exclusively by a rhomboid-like activity. If the fundamental function of TgAMA1 and PfAMA1 is similar, why would shedding of PfAMA1 throughout invasion, and the shedding of TgAMA1 at the end of invasion, be equally tolerated? Once again, the answer to this question will be revealed by genetic dissection of MPP1 function during invasion.

Other Microneme Protein Proteases Involved in Invasion in Toxoplasma gondii*: Surface Trimming*

Other microneme protein cleavage events have been characterised in *T. gondii* that occur on the surface, and presumably during invasion. The protease activities mediating these have been called Microneme Protein Protease 2 and 3 (MPP2 and MPP3; refs. 9,14).

Microneme Protein Protease 2

MPP2 is an activity that trims TgMIC2 several times to remove the N-terminal region up to the A-domain.[9] It also trims the TgM2AP C-terminus in three places,[14] the TgMIC4 C-terminus,[53] and TgSUB1.[14] The precise purpose of MPP2 activity is not clear, and may differ in the case of each substrate. It has been recently demonstrated that MPP2 mediated trimming of the TgMIC2 N-terminus activates the binding of ICAM-1 by TgMIC2 - an activity which was proposed to be necessary for transmigrating through tissue barriers.[54] However, it was also observed that ALLN and chymostatin, which inhibit MPP2 activity, did not inhibit host cell invasion,[9] suggesting that MPP2 activity is not essential for host cell invasion in vitro.

As mentioned above, TgSUB1 is a GPI anchored subtilisin-like protease that is shed into culture medium after secretion from micronemes.[20] This protein is currently the best candidate for MPP2. Interestingly, cytochalasin D treatment of parasites, which prevents redistribution of TgMIC2 towards the rear of the parasite, actually results in an enhancement of MPP2 mediated trimming. This is easily understood if MPP2 resides in the micronemes, and is secreted to the parasite surface at the same time as TgMIC2.

Additionally, subtilisins are usually cleaved autocatalytically, and TgSUB1 is a substrate for MPP2 activity. Regulation of TgSUB1 is presumably mediated by calcium levels or by association with the prodomain when TgSUB1 is within the micronemes. Further evidence for the localisation of MPP2 in the micronemes comes from the fact that when TgMIC4 is mistargeted to the parasite surface, MPP2 mediated cleavage does not occur, suggesting a need for this protein to traffic through the micronemes in order to be a substrate for MPP2.[18]

Of the other subtilisin-like proteases in *T. gondii*, only TgSUB2 has been characterised in detail as a potential rhoptry-protein maturase, and does not therefore have a direct role in invasion.[55]

Microneme Protein Protease 3

This activity has been reported by Carruthers and colleagues[14] as a protease that trims TgM2AP at a site just upstream of the three MPP2 cleavage sites. At this point the significance of this activity, or the protease responsible for it is entirely unclear. There are also no data yet to show whether MPP3 mediated cleavage of M2AP is a prerequisite to MPP2 cleavage. No inhibitors have yet been found which impede MPP3 activity, and so the search for the identity of this protease is still in a very early stage. Apart from M2AP there are, as yet, no other determined substrates for MPP3.

Other Potential Proteases Involved in Invasion

Other proteases will presumably be involved in invasion. In *P. falciparum*, the papain-like cysteine protease falcipain1 was reported in an inhibitor study to have a role in invasion.[56] A subsequent genetic disruption of the gene led to a reevaluation of this conclusion, as no defect in invasion could be observed.[57,58]

Also of some interest is a protease described some years ago to modify the red-blood-cell surface during *Plasmodium* invasion. Although the protease, named p76 in *P. falciparum*, has been characterised biochemically, the gene encoding it has yet to be identified, despite the complete genome sequence of *P. falciparum* being available. The protease is sensitive to several serine protease inhibitors,[59] and purified p76 was shown to be able to cleave erythrocyte Band 3 and glycophorin A. Importantly, inhibitors of this Band3 cleavage were also inhibitors of invasion.[60] This activity was hypothesised[61] to be responsible for destabilising the erythrocyte membrane and/or underlying cytoskeleton, in order to facilitate invasion. While this model is tempting, there is no direct evidence that the erythrocyte is proteolytically modified during invasion, and therefore the molecular identity of this protease must be established before its precise role can be revealed.

Other candidate proteases are not currently available, but studies involving inhibition of invasion (see refs. 8 and 62 for review of *Plasmodium* invasion inhibition) clearly show proteases are involved. Whether these proteases are simply the activities described above, or are alternative enzymes also required during invasion remains to be seen.

Egress

Escape from the host cell is a necessity for all apicomplexans in order to progress through their life cycle. In the case of *Plasmodium*, hepatic merozoites have to egress from liver cells, merozoites from red blood cells and sporozoites from oocysts. Proteases appear to play an important role in at least two of these cases. *T. gondii* probably uses similar mechanisms for egress, but the proteolytic events that mediate this process have been less well studied.

Plasmodium

The erythrocytic cycle of this parasite is responsible for the manifestations of clinical malaria and most studies concerning egress have focused on that stage. In addition the released merozoites are potential targets for anti-malarial drugs or vaccines. The proteolytic events and the proteases involved in release of the parasites are less understood than the mechanisms involved in invasion, but nonetheless have received considerable attention. In order to understand the proteolytic mechanisms involved in egress, the first question to be addressed is how the parasite exits the red blood cell.

Escape Pathways

Within the infected red blood cell (RBC), the developing merozoite is surrounded by the erythrocyte membrane (EM) and by the parasitophorous vacuole membrane (PVM). Both these membranes must therefore be ruptured to allow egress, and five models have been proposed regarding the manner and timing of rupture of these membranes.

The first model suggests that the PVM fuses with the EM, and the merozoites egress from the red blood cell (RBC) leaving it intact.[63] Video microscopy was used to support this model, demonstrating that erythrocyte rupture is not an explosive event and that a "parasite ghost"—the remnants of a red blood cell previously occupied by parasites—is observed long after parasites have been released from the cell.

Salmon et al[64] provided evidence for a second model that is somewhat similar to that above. They proposed a two-step egress process, where merozoites still enclosed within the PVM exit from the host erythrocyte without concomitant lysis. The PVM is then rapidly ruptured to complete the process. To dissect the mechanism of egress they used E64, a cysteine protease inhibitor. Examination of E64 treated cultures showed the accumulation of merozoite clusters surrounded by a membrane that was recognised by antibodies against the PVM, suggesting involvement of a cysteine protease in PVM rupture. This protease could act either directly on components of the PVM, or alternatively could act by activating other components involved in the release mechanism.

Wickham et al[65] subsequently presented evidence, again with the aid of protease inhibitors, that parasite release is a two-step event. They proposed a third model in which the PVM membrane undergoes primary rupture and merozoites are released after a secondary rupture of the EM. They found that E64 could inhibit the primary lysis of the PVM. Rupture of the erythrocyte plasma membrane, however, was inhibited by leupeptin and antipain or leupeptin and chymostatin (as was originally demonstrated by Lyon et al[66]).

A fourth model arose from a recent study by Soni et al.[67] These investigators observed that during the ring and trophozoite stages the PVM surrounds the parasite continuously, but during schizogony PVM staining showed a "punctuated pattern". Based on these data the authors suggested that the PVM surrounds individual merozoites and in a primary lysis step the EM is ruptured releasing individual PVM-enclosed merozoites. These are then finally released in a secondary rupture of the PVM. This group also studied the effects of 4 different cysteine protease inhibitors on egress. They demonstrated that leupeptin and E64 treatment resulted in merozoite clusters associated with the PVM, while E64d and MDL blocked any further parasite development irreversibly. This further implicates a role for cysteine proteases in egress.

The last model was proposed by Glushakova et al,[68] and is based on experiments that followed the release of *P. falciparum* merozoites by fluorescence microscopy using differentially labelled membranes. In contrast to all the above mentioned studies, no RBC ghosts were observed and the release of clustered merozoites was interpreted as being an artefact. The authors proposed that an "irregular schizont" is formed a few hours before rupture and an increase of intracellular volume transforms the schizont to a terminal rounded structure that eventually undergoes explosive rupture. According to the same study the PVM ruptures at the same time or shortly before the EM and individual merozoites are released directly into the bloodstream, findings that are consistent with the model of Wickam et al[65] in terms of the temporal events but also suggesting a cascade of regulated events that affect the morphological stages of the parasites.

Role of Proteases in Schizont Rupture

All the above studies provided invaluable information about the events leading up to schizont rupture. Despite the fact that the temporal sequence of events is still a subject of some controversy, protease involvement in these events is a common parameter.[64-67] But what is the identity of these proteases?

Several malarial proteases have been implicated in host cell rupture. These include: falcipain 2, a cysteine protease that cleaves ankyrin and protein 4.1; plasmepsin II, an aspartic protease that cleaves host spectrin, actin and protein 4.1; and members of the cysteine protease-like serine repeat antigen (SERA) family. However, in no case has a precise physiological role been identified.

Falcipain 2, Falcipain 2' and Falcipain 3

Falcipain 2 and the closely related falcipain 2' are cysteine proteases belonging to the papain superfamily. Both have been shown to hydrolyse native haemoglobin and denatured globin, with maximal activity near the pH (~5.2) of the food vacuole,[69] consistent with the proposal that these enzymes are primarily involved in haemoglobin digestion. Gene knock out experiments[70] showed falcipain 2 to be nonessential in blood stages, although the knockout parasites exhibited a defect in haemoglobin hydrolysis, as expected. However, no abrogation of merozoite release was observed in this study.

In contrast with these findings, Dasaradhi et al[71] obtained evidence that falcipain 2 plays a role in egress from schizonts. Following treatment of parasites with siRNAs designed to down-regulate expression of falcipain 2 (*fp-2*), merozoites in late schizogony appeared as clusters enclosed in a thin membrane covering. These clusters were still present 72 h after *fp-2* siRNA treatment, implying that parasite development was inhibited. By using the same transgenic line as Wickham et al,[65] treated with *fp-2* siRNA, they showed that erythrocyte membrane rupture was blocked. Analysis by immunofluorescence assays of these merozoite clusters using antibodies against the erythrocyte membrane protein, band 3, identified the thin membrane surrounding the clusters as the erythrocyte plasma membrane. The authors proposed that falcipain 2 is involved in rupture of the red blood cell membrane.

The difference in findings between these two groups could be explained according to Dasaradhi et al, by the fact that falcipain-2 siRNA may also inhibit falcipain 2′ expression since there is high homology between the two proteases. The above results are strengthened, according to the authors, by previous studies of Hanspal et al[72] showing that falcipain 2 hydrolyses the erythrocyte cytoskeletal proteins band 4.1 and ankyrin at neutral pH, suggesting an additional role for this protease in erythrocyte rupture. It should also be stated, however, that methods of gene knock-down in Apicomplexa using RNAi techniques are frequently met with some scepticism, as many groups have had difficulties with it, and this technique remains controversial in this field.[73]

Falcipain 3 is also a papain-like cysteine protease that has haemoglobinase activity, but it is expressed later in the erythrocytic life cycle than falcipain 2 and 2′.[74] Localisation studies by Dahl et al[75] showed that falcipain 3 exhibits a more widespread localisation pattern in the asexual blood-stage parasite than falcipain 2, although it is more abundant in the food vacuole. These authors speculated that additional further roles in cellular functions other than haemoglobin degradation should not be excluded. Genetic studies should provide invaluable data concerning the requirement for this protease in the erythrocytic cycle of the parasite, and establish whether or not it has a role in egress.

Plasmepsin II

Plasmepsin II is an aspartic protease involved in the degradation of haemoglobin within the food vacuole of the parasite; it is also more active against denatured or fragmented globin. The crystal structure of recombinant plasmepsin II in complex with pepstatin A (a general aspartic protease inhibitor) was the first x-ray crystal structure of a *Plasmodium falciparum* protein.[76] Studies with recombinant plasmepsin II have shown that this protease can cleave native spectrin and selectively cleaves spectrin, actin and protein 4.1 in erythrocyte ghosts at pH 6.8.[77] Confocal microscopy has provided evidence that during late schizogony the protease accumulates in close proximity to the RBC cytoskeleton,[65] while previous studies in trophozoites have shown that plasmepsins are present at the surface of the trophozoites.[78] In mature segmented schizonts, plasmepsin II was found in the parasite periphery in increased amounts compared to early schizonts although at this stage no massive degradation of haemoglobin occurs. The protease is also active at a pH of 6.8, which suggests that it can be active at the neutral pH of the RBC cytoplasm.[77] As in the case of the falcipains discussed above, these recent data therefore suggest that plasmepsin II might be involved in processes other than haemoglobin degradation. However, gene knock out studies have shown that parasites can grow without plasmepsin II although at a slower rate compared to wild type parasites,[79] proving that, in the case that it does play role in egress, it is not essential.

Serine Repeat Antigen (SERA) Family

The *P. falciparum* SERA gene family consists of nine genes, eight located in a gene cluster on chromosome 2 and one (SERA9) on chromosome 9. All SERA proteins possess a central domain that exhibits homology to the papain family of cysteine proteases. In some cases this includes a papain-like catalytic triad (Cys, His, Asn), indicating likely protease activity. However six of them (SERA1-5 and 9) possess a serine residue in place of the active site cysteine, which has raised questions concerning their enzymatic activity. Of particular interest amongst this group has been SERA5 because it is highly abundant, being expressed at much higher levels than the other members, and appears to be essential in blood-stages, while SERA2, -3, -7, and -8 are not essential for erythrocytic growth.[80] SERA5 is synthesized as a 113-126 kDa precursor protein localized to the parasitophorous vacuole. The precursor is processed into an N-terminal 47-kDa fragment, a central 56-kDa and a C-terminal 18-kDa fragment, the last two derived from a 73-kDa intermediate. The 56-kDa domain undergoes further processing to a 50-kDa species.[81] Hodder et al expressed a recombinant form of the SERA5 protease-like domain and showed that this was able to cleave substrates on the C-terminal side of aromatic residues. This activity was inhibited in the presence of the serine protease inhibitor DCI, providing the first evidence of a proteolytic activity for SERA5.[81] Li et al[82] used a recombinant system to study the proteases involved in SERA5 processing. Conversion of the precursor protein to the 47 and 73 kDa fragments was inhibited by diisopropyl fluorophosphate (DFP), another serine protease inhibitor, while conversion of the 56 kDa fragment to the 50 kDa species appeared mediated by a cysteine protease. E64 and leupeptin partially inhibited the conversion of the 73 kDa fragment to the 56 kDa one although this result was not reproducible in vivo. Similar to the findings of Salmon et al,[64] who observed merozoite clusters after treatment of cultures with E64, treatment of middle-stage schizonts with E64 and leupeptin had the same effect, and in parallel inhibited the processing of SERA5. It was speculated by Li et al[82] that the 50 kDa fragment of SERA5 might play a proteolytic role in schizont rupture. Further tests are needed to support this hypothesis. Another important finding that further implicates SERA5 in parasite egress is that antibodies against the 47 kDa N-terminal fragment can inhibit the intraerythrocytic cycle of the parasite, apparently by preventing schizont rupture and release of merozoites.[83]

In 2005, Aly and Matuschewski[84] described complete inhibition of sporozoite egress from the mosquito midgut oocysts following disruption of the *P. berghei* orthologue of SERA8. This putative protease was therefore termed egress cysteine protease 1 (ECP1). Its transcription is specific to mature oocysts, as appears true also of the *P. falciparum* orthologue. Following targeted disruption of the *P. berghei ecp1* gene, sporozoites failed to egress from the midgut oocysts and exhibited an unusual continuous intracellular circular motility that was not observed in wild type oocysts. This elegant study is the first genetic proof that a malarial cysteine protease is necessary for egress. As the authors state these findings "pave the way" for the identification of enzymes that might similarly be involved in merozoite release from liver and blood-stage schizonts.

Toxoplasma

As with *Plasmodium*, *Toxoplasma* egress is a rapid event leading to lysis of the host cell and release of motile parasites that quickly invade neighbouring cells to continue their cycle. By video microscopy it has been observed that escaping parasites pass through the PVM, the host cytoplasm and finally the plasmalemma;[85] parasites rupturing the host cell under induced egress conditions do it in a way that resembles invasion but in an outward direction.[85] Indeed, it has been speculated that there are common mechanisms shared between invasion and egress that are dependent upon both calcium and motility.[5] Unfortunately, there is currently no direct evidence of any protease involvement in *Toxoplasma* egress.

Althought the SERA family have been established as primary candidates for egress enzymes in *Plasmodium*, no clearly direct homologues exist in *T. gondii*. Cysteine proteases related to cathepsins are present however, and have been designated TgCPL, TgCPB and TgCPC (although in older publications are referred to as Toxopains). TgCPB (previously Toxopain-1) has been demonstrated to be a rhoptry protein maturase, but is not involved in proteolysis during invasion or egress itself.[86] There is as yet no published data concerning the other cysteine proteases in *T. gondii*, so any potential role in egress would be purely speculative at this stage.

Egress in Plasmodium *and* Toxoplasma*: Are There Similarities?*

Taking into account the close relationship between *Plasmodium* and *Toxoplasma* and the fact that several proteases have been implicated in egress by *Plasmodium*, it is only reasonable to expect proteolytic activity being involved in *Toxoplasma* released from infected host cells. It remains to be seen whether *T. gondii* will contain the same egress machinery as *Plasmodium* parasites, and perhaps one should consider the differences in intracellular development, and the broader range of host cells that *T. gondii* must egress from. With the complete genome sequences available new data might arise providing investigators with the information needed to characterise the events of egress, a critical stage in the life cycle of these important pathogens that has been relatively neglected.

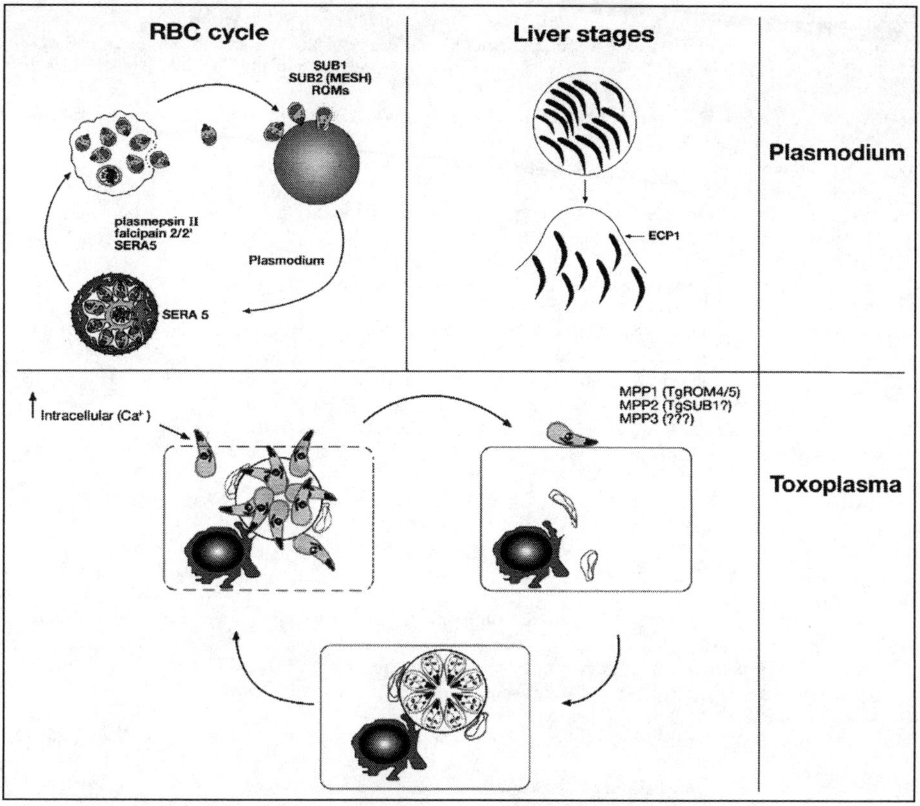

Figure 1. A summary of the involvement of parasite proteases in invasion and egress for *Toxoplasma gondii* and *Plasmodium* parasites. Top left box: the red blood cell (RBC) cycle of *Plasmodium*. The schizont is represented in the bottom RBC, followed by the egress of merozoites, followed by invasion of a new RBC. Top right box: egress of liver cells. Bottom box: a rosette of intracellular *T. gondii* parasites within a vacuole is represented in the bottom cell, followed by egress and then invasion of a fresh host cell. In all boxes, the proteases involved in each particular stage are shown.

Concluding Remarks

Proteases clearly play a critical role in many parasite processes, and their continued study is vital to fully understand the mechanisms of life-cycle progression. The current knowledge of their roles in invasion and egress are summarised in Figure 1. The good chance of utilising proteases as drug targets, together with the fact that they are involved in the absolutely critical processes of invasion and egress, makes this field both intellectually fascinating, and potentially crucial clinically.

References

1. Alberts B. Molecular Biology of the Cell. 3rd ed. New York: Garland Pub., 1994.
2. Lopez-Otin C, Overall CM. Protease degradomics: A new challenge for proteomics. Nat Rev Mol Cell Biol 2002; 3(7):509-519.
3. Barrett AJ, Rawlings ND, Woessner JF. Handbook of proteolytic enzymes. San Diego: Academic Press, 2004.

4. Alexander DL, Mital J, Ward GE et al. Identification of the moving junction complex of Toxoplasma gondii: A collaboration between distinct secretory organelles. PLoS Pathog 2005; 1(2):e17.
5. Hoff EF, Carruthers VB. Is Toxoplasma egress the first step in invasion? Trends Parasitol 2002; 18(6):251-255.
6. Dowse T, Soldati D. Host cell invasion by the apicomplexans: The significance of microneme protein proteolysis. Curr Opin Microbiol 2004; 7(4):388-396.
7. Kim K. Role of proteases in host cell invasion by Toxoplasma gondii and other Apicomplexa. Acta Trop 2004; 91(1):69-81.
8. Blackman MJ. Proteases involved in erythrocyte invasion by the malaria parasite: Function and potential as chemotherapeutic targets. Curr Drug Targets 2000; 1(1):59-83.
9. Carruthers VB, Sherman GD, Sibley LD. The Toxoplasma adhesive protein MIC2 is proteolytically processed at multiple sites by two parasite-derived proteases. J Biol Chem 2000; 275(19):14346-14353.
10. Carruthers VB, Blackman MJ. A new release on life: Emerging concepts in proteolysis and parasite invasion. Mol Microbiol 2005; 55(6):1617-1630.
11. Opitz C, Di Cristina M, Reiss M et al. Intramembrane cleavage of microneme proteins at the surface of the apicomplexan parasite Toxoplasma gondii. EMBO J 2002; 21(7):1577-1585.
12. Tomley FM, Billington KJ, Bumstead JM et al. EtMIC4: A microneme protein from Eimeria tenella that contains tandem arrays of epidermal growth factor-like repeats and thrombospondin type-I repeats. Int J Parasitol 2001; 31(12):1303-1310.
13. Brossier F, Jewett TJ, Lovett JL et al. C-terminal processing of the Toxoplasma protein MIC2 is essential for invasion into host cells. J Biol Chem 2003; 278(8):6229-6234.
14. Zhou XW, Blackman MJ, Howell SA et al. Proteomic analysis of cleavage events reveals a dynamic two-step mechanism for proteolysis of a key parasite adhesive complex. Mol Cell Proteomics 2004; 3(6):565-576.
15. Urban S, Lee JR, Freeman M. Drosophila rhomboid-1 defines a family of putative intramembrane serine proteases. Cell 2001; 107(2):173-182.
16. Koonin EV, Makarova KS, Rogozin IB et al. The rhomboids: A nearly ubiquitous family of intramembrane serine proteases that probably evolved by multiple ancient horizontal gene transfers. Genome Biol 2003; 4(3):R19.
17. Freeman M. Proteolysis within the membrane: Rhomboids revealed. Nat Rev Mol Cell Biol 2004; 5(3):188-197.
18. Reiss M, Viebig N, Brecht S et al. Identification and characterization of an escorter for two secretory adhesins in Toxoplasma gondii. J Cell Biol 2001; 152(3):563-578.
19. Howell SA, Hackett F, Jongco AM et al. Distinct mechanisms govern proteolytic shedding of a key invasion protein in apicomplexan pathogens. Mol Microbiol 2005; 57(5):1342-1356.
20. Miller SA, Binder EM, Blackman MJ et al. A conserved subtilisin-like protein TgSUB1 in microneme organelles of Toxoplasma gondii. J Biol Chem 2001; 276(48):45341-45348.
21. Urban S, Freeman M. Substrate specificity of rhomboid intramembrane proteases is governed by helix-breaking residues in the substrate transmembrane domain. Mol Cell 2003; 11(6):1425-1434.
22. Brossier F, Jewett TJ, Sibley LD et al. A spatially localized rhomboid protease cleaves cell surface adhesins essential for invasion by Toxoplasma. Proc Natl Acad Sci USA 2005; 102(11):4146-4151.
23. Dowse TJ, Pascall JC, Brown KD et al. Apicomplexan rhomboids have a potential role in microneme protein cleavage during host cell invasion. Int J Parasitol 2005; 35(7):747-756.
23a. Baker RP, Wijetilaka R, Urban S. Two plasmodium rhomboid proteases preferentially cleave different adhesins implicated in all invasive stages of malaria. PLoS Pathog 2006; 2(10):e113.
24. Dowse TJ, Soldati D. Rhomboid-like proteins in Apicomplexa: Phylogeny and nomenclature. Trends Parasitol 2005; 21(6):254-258.
24a. O'Donnell RA, Hackett F, Howell SA et al. Intramembrane proteolysis mediates shedding of a key adhesin during erythrocyte invasion by the malaria parasite. J Cell Biol 2006; 174(7):1023-1033.
25. Carruthers VB, Sibley LD. Sequential protein secretion from three distinct organelles of Toxoplasma gondii accompanies invasion of human fibroblasts. Eur J Cell Biol 1997; 73(2):114-123.
26. Mital J, Meissner M, Soldati D et al. Conditional expression of Toxoplasma gondii apical membrane antigen-1 (TgAMA1) demonstrates that TgAMA1 plays a critical role in host cell invasion. Mol Biol Cell 2005; 16(9):4341-4349.
27. Silvie O, Franetich JF, Charrin S et al. A role for apical membrane antigen 1 during invasion of hepatocytes by Plasmodium falciparum sporozoites. J Biol Chem 2004; 279(10):9490-9496.
28. Reed MB, Caruana SR, Batchelor AH et al. Targeted disruption of an erythrocyte binding antigen in Plasmodium falciparum is associated with a switch toward a sialic acid-independent pathway of invasion. Proc Natl Acad Sci USA 2000; 97(13):7509-7514.

29. Stubbs J, Simpson KM, Triglia T et al. Molecular mechanism for switching of P. falciparum invasion pathways into human erythrocytes. Science 2005; 309(5739):1384-1387.
29a. Wang Y, Zhang Y, Ha Y. Crystal structure of a rhomboid family intramembrane protease. Nature 2006.
30. Li X, Chen H, Oo TH et al. A coligand complex anchors Plasmodium falciparum merozoites to the erythrocyte invasion receptor band 3. J Biol Chem 2004; 279(7):5765-5771.
31. Pachebat JA, Ling IT, Grainger M et al. The 22 kDa component of the protein complex on the surface of Plasmodium falciparum merozoites is derived from a larger precursor, merozoite surface protein 7. Mol Biochem Parasitol 2001; 117(1):83-89.
32. Trucco C, Fernandez-Reyes D, Howell S et al. The merozoite surface protein 6 gene codes for a 36 kDa protein associated with the Plasmodium falciparum merozoite surface protein-1 complex. Mol Biochem Parasitol 2001; 112(1):91-101.
33. Blackman MJ, Dennis ED, Hirst EM et al. Plasmodium knowlesi: Secondary processing of the malaria merozoite surface protein-1. Exp Parasitol 1996; 83(2):229-239.
34. Blackman MJ, Scott-Finnigan TJ, Shai S et al. Antibodies inhibit the protease-mediated processing of a malaria merozoite surface protein. J Exp Med 1994; 180(1):389-393.
35. Fleck SL, Birdsall B, Babon J et al. Suramin and suramin analogues inhibit merozoite surface protein-1 secondary processing and erythrocyte invasion by the malaria parasite Plasmodium falciparum. J Biol Chem 2003; 278(48):47670-47677.
36. Blackman MJ, Holder AA. Secondary processing of the Plasmodium falciparum merozoite surface protein-1 (MSP1) by a calcium-dependent membrane-bound serine protease: Shedding of MSP133 as a noncovalently associated complex with other fragments of the MSP1. Mol Biochem Parasitol 1992; 50(2):307-315.
37. Dutta S, Haynes JD, Moch JK et al. Invasion-inhibitory antibodies inhibit proteolytic processing of apical membrane antigen 1 of Plasmodium falciparum merozoites. Proc Natl Acad Sci USA 2003; 100(21):12295-12300.
37a. Dutta S, Haynes JD, Barbosa A et al. Mode of action of invasion-inhibitory antibodies directed against Apical Membrane Antigen 1 of Plasmodium falciparum. Infect Immun 2005; 73(4):2116-2122.
38. Howell SA, Well I, Fleck SL et al. A single malaria merozoite serine protease mediates shedding of multiple surface proteins by juxtamembrane cleavage. J Biol Chem 2003; 278(26):23890-23898.
39. Howell SA, Withers-Martinez C, Kocken CH et al. Proteolytic processing and primary structure of Plasmodium falciparum apical membrane antigen-1. J Biol Chem 2001; 276(33):31311-31320.
40. Blackman MJ. Proteases in host cell invasion by the malaria parasite. Cell Microbiol 2004; 6(10):893-903.
41. Morgan WD, Birdsall B, Frenkiel TA et al. Solution structure of an EGF module pair from the Plasmodium falciparum merozoite surface protein 1. J Mol Biol 1999; 289(1):113-122.
42. Chitarra V, Holm I, Bentley GA et al. The crystal structure of C-terminal merozoite surface protein 1 at 1.8 A resolution, a highly protective malaria vaccine candidate. Mol Cell 1999; 3(4):457-464.
43. Wu Y, Wang X, Liu X et al. Data-mining approaches reveal hidden families of proteases in the genome of malaria parasite. Genome Res 2003; 13(4):601-616.
44. Blackman MJ, Fujioka H, Stafford WH et al. A subtilisin-like protein in secretory organelles of Plasmodium falciparum merozoites. J Biol Chem 1998; 273(36):23398-23409.
45. Withers-Martinez C, Jean L, Blackman MJ. Subtilisin-like proteases of the malaria parasite. Mol Microbiol 2004; 53(1):55-63.
46. Blackman MJ, Corrie JE, Croney JC et al. Structural and biochemical characterization of a fluorogenic rhodamine-labeled malarial protease substrate. Biochemistry 2002; 41(40):12244-12252.
47. Jean L, Hackett F, Martin SR et al. Functional characterization of the propeptide of Plasmodium falciparum subtilisin-like protease-1. J Biol Chem 2003; 278(31):28572-28579.
48. Barale JC, Blisnick T, Fujioka H et al. Plasmodium falciparum subtilisin-like protease 2, a merozoite candidate for the merozoite surface protein 1-42 maturase. Proc Natl Acad Sci USA 1999; 96(11):6445-6450.
49. Hackett F, Sajid M, Withers-Martinez C et al. PfSUB-2: A second subtilisin-like protein in Plasmodium falciparum merozoites. Mol Biochem Parasitol 1999; 103(2):183-195.
50. Harris PK, Yeoh S, Dluzewski AR et al. Molecular identification of a malaria merozoite surface sheddase. PLoS Pathog 2005; 1(3):e29.
51. Green JL, Hinds L, Grainger M et al. Plasmodium thrombospondin related apical merozoite protein (PTRAMP) is shed from the surface of merozoites by PfSUB2 upon invasion of erythrocytes. Mol Biochem Parasitol 2006.
52. Pizarro JC, Vulliez-Le Normand B, Chesne-Seck ML et al. Crystal structure of the malaria vaccine candidate apical membrane antigen 1. Science 2005; 308(5720):408-411.

53. Brecht S, Carruthers VB, Ferguson DJ et al. The Toxoplasma micronemal protein MIC4 is an adhesin composed of six conserved apple domains. J Biol Chem 2001; 276(6):4119-4127.
54. Barragan A, Brossier F, Sibley LD. Transepithelial migration of Toxoplasma gondii involves an interaction of intercellular adhesion molecule 1 (ICAM-1) with the parasite adhesin MIC2. Cell Microbiol 2005; 7(4):561-568.
55. Miller SA, Thathy V, Ajioka JW et al. TgSUB2 is a Toxoplasma gondii rhoptry organelle processing proteinase. Mol Microbiol 2003; 49(4):883-894.
56. Greenbaum DC, Baruch A, Grainger M et al. A role for the protease falcipain 1 in host cell invasion by the human malaria parasite. Science 2002; 298(5600):2002-2006.
57. Sijwali PS, Kato K, Seydel KB et al. Plasmodium falciparum cysteine protease falcipain-1 is not essential in erythrocytic stage malaria parasites. Proc Natl Acad Sci USA 2004; 101(23):8721-8726.
58. Eksi S, Czesny B, Greenbaum DC et al. Targeted disruption of Plasmodium falciparum cysteine protease, falcipain 1, reduces oocyst production, not erythrocytic stage growth. Mol Microbiol 2004; 53(1):243-250.
59. Braun-Breton C, Rosenberry TL, da Silva LP. Induction of the proteolytic activity of a membrane protein in Plasmodium falciparum by phosphatidyl inositol-specific phospholipase C. Nature 1988; 332(6163):457-459.
60. Roggwiller E, Betoulle ME, Blisnick T et al. A role for erythrocyte band 3 degradation by the parasite gp76 serine protease in the formation of the parasitophorous vacuole during invasion of erythrocytes by Plasmodium falciparum. Mol Biochem Parasitol 1996; 82(1):13-24.
61. McPherson RA, Donald DR, Sawyer WH et al. Proteolytic digestion of band 3 at an external site alters the erythrocyte membrane organisation and may facilitate malarial invasion. Mol Biochem Parasitol 1993; 62(2):233-242.
62. Conseil V, Soete M, Dubremetz JF. Serine protease inhibitors block invasion of host cells by Toxoplasma gondii. Antimicrob Agents Chemother 1999; 43(6):1358-1361.
63. Winograd E, Clavijo CA, Bustamante LY et al. Release of merozoites from Plasmodium falciparum-infected erythrocytes could be mediated by a nonexplosive event. Parasitol Res 1999; 85(8-9):621-624.
64. Salmon BL, Oksman A, Goldberg DE. Malaria parasite exit from the host erythrocyte: A two-step process requiring extraerythrocytic proteolysis. Proc Natl Acad Sci USA 2001; 98(1):271-276.
65. Wickham ME, Culvenor JG, Cowman AF. Selective inhibition of a two-step egress of malaria parasites from the host erythrocyte. J Biol Chem 2003; 278(39):37658-37663.
66. Lyon JA, Haynes JD, Diggs CL et al. Plasmodium falciparum antigens synthesized by schizonts and stabilized at the merozoite surface by antibodies when schizonts mature in the presence of growth inhibitory immune serum. J Immunol 1986; 136(6):2252-2258.
67. Soni S, Dhawan S, Rosen KM et al. Characterization of events preceding the release of malaria parasite from the host red blood cell. Blood Cells Mol Dis 2005; 35(2):201-211.
68. Glushakova S, Yin D, Li T et al. Membrane transformation during malaria parasite release from human red blood cells. Curr Biol 2005; 15(18):1645-1650.
69. Shenai BR, Rosenthal PJ. Reducing requirements for hemoglobin hydrolysis by Plasmodium falciparum cysteine proteases. Mol Biochem Parasitol 2002; 122(1):99-104.
70. Sijwali PS, Rosenthal PJ. Gene disruption confirms a critical role for the cysteine protease falcipain-2 in hemoglobin hydrolysis by Plasmodium falciparum. Proc Natl Acad Sci USA 2004; 101(13):4384-4389.
71. Dasaradhi PV, Mohmmed A, Kumar A et al. A role of falcipain-2, principal cysteine proteases of Plasmodium falciparum in merozoite egression. Biochem Biophys Res Commun 2005; 336(4):1062-1068.
72. Hanspal M, Dua M, Takakuwa Y et al. Plasmodium falciparum cysteine protease falcipain-2 cleaves erythrocyte membrane skeletal proteins at late stages of parasite development. Blood 2002; 100(3):1048-1054.
73. Blackman MJ. RNAi in protozoan parasites: What hope for the Apicomplexa? Protist 2003; 154(2):177-180.
74. Sijwali PS, Shenai BR, Gut J et al. Expression and characterization of the Plasmodium falciparum haemoglobinase falcipain-3. Biochem J 2001; 360(Pt 2):481-489.
75. Dahl EL, Rosenthal PJ. Biosynthesis, localization, and processing of falcipain cysteine proteases of Plasmodium falciparum. Mol Biochem Parasitol 2005; 139(2):205-212.
76. Silva AM, Lee AY, Gulnik SV et al. Structure and inhibition of plasmepsin II, a hemoglobin-degrading enzyme from Plasmodium falciparum. Proc Natl Acad Sci USA 1996; 93(19):10034-10039.

77. Le Bonniec S, Deregnaucourt C, Redeker V et al. Plasmepsin II, an acidic hemoglobinase from the Plasmodium falciparum food vacuole, is active at neutral pH on the host erythrocyte membrane skeleton. J Biol Chem 1999; 274(20):14218-14223.
78. Francis SE, Gluzman IY, Oksman A et al. Molecular characterization and inhibition of a Plasmodium falciparum aspartic hemoglobinase. EMBO J 1994; 13(2):306-317.
79. Liu J, Gluzman IY, Drew ME et al. The role of Plasmodium falciparum food vacuole plasmepsins. J Biol Chem 2005; 280(2):1432-1437.
80. Miller SK, Good RT, Drew DR et al. A subset of Plasmodium falciparum SERA genes are expressed and appear to play an important role in the erythrocytic cycle. J Biol Chem 2002; 277(49):47524-47532.
81. Hodder AN, Drew DR, Epa VC et al. Enzymic, phylogenetic, and structural characterization of the unusual papain-like protease domain of Plasmodium falciparum SERA5. J Biol Chem 2003; 278(48):48169-48177.
82. Li J, Matsuoka H, Mitamura T et al. Characterization of proteases involved in the processing of Plasmodium falciparum serine repeat antigen (SERA). Mol Biochem Parasitol 2002; 120(2):177-186.
83. Pang XL, Mitamura T, Horii T. Antibodies reactive with the N-terminal domain of Plasmodium falciparum serine repeat antigen inhibit cell proliferation by agglutinating merozoites and schizonts. Infect Immun 1999; 67(4):1821-1827.
84. Aly AS, Matuschewski K. A malarial cysteine protease is necessary for Plasmodium sporozoite egress from oocysts. J Exp Med 2005; 202(2):225-230.
85. Black MW, Boothroyd JC. Lytic cycle of Toxoplasma gondii. Microbiol Mol Biol Rev 2000; 64(3):607-623.
86. Que X, Ngo H, Lawton J et al. The cathepsin B of Toxoplasma gondii, toxopain-1, is critical for parasite invasion and rhoptry protein processing. J Biol Chem 2002; 277(28):25791-25797.

Chapter 11

Roles of Naturally Occurring Protease Inhibitors in the Modulation of Host Cell Signaling and Cellular Invasion by *Trypanosoma cruzi*

Julio Scharfstein* and Ana Paula C.A. Lima

Abstract

Trypanosoma cruzi trypomastigotes rely on the structural diversity of the cruzipain family of cysteine proteases to infect and multiply in nonprofessional phagocytic cells. Herein, we will review studies demonstrating that the interplay of cruzipain with peptidase inhibitors modulate infection outcome in a variety of experimental settings. Studies with a panel of *T. cruzi* strains showed that parasite ability to invade human smooth muscle cells is influenced by the balance between cruzipain and chagasin, a tight binding endogenous inhibitor of papain-like cysteine proteases. Analysis of *T. cruzi* interaction with endothelial cells and cardiomyocytes indicated that parasite-induced activation of bradykinin receptors drive host cell invasion by $[Ca^{2+}]_I$–dependent pathways. Clues about the mechanisms underlying kinin generation in vivo by trypomastigotes came from analysis of the dynamics of edematogenic inflammation. Owing to plasma extravasation, the blood-borne kininogens accumulate in peripheral sites of infection. Upon diffusion in peripheral tissues, kininogens (i.e., type III cystatins) bind to heparan sulphate chains, thus constraining interactions of the cystatin-like inhibitory domains with cruzipain. The cell bound kininogens are then turned into facile substrates for cruzipain, which liberates kinins in peripheral tissues. Subjected to tight-regulation by kinin-degrading metallopeptidases, such as angiotensin converting enzyme, the short-lived kinin peptides play a dual role in the host-parasite balance. Rather than unilaterally stimulating pathogen infectivity via bradykinin receptors, the released kinins potently induce dendritic cell maturation, thus stimulating type 1 immune responses. In conclusion, the studies reviewed herein illustrate how regulation of parasite proteases may affect host-parasite equilibrium in the course of *T. cruzi* infection.

Abbreviations

ACE, angiotensin converting enzyme; BK, bradykinin; bradykinin receptor; CPM/N, carboxypeptidase M/N; chagasin, endogenous cysteine protease inhibitor of *T. cruzi*; cruzipain, the major cysteine protease of *T. cruzi*; HK/LK, high-/low molecular weight kininogen; LBK, Lysyl-bradykinin; TCTs, tissue culture trypomastigotes; tGPI-mucin, trypomastigote-derived glycosylphosphatidylinositol-anchored mucin; TLR, Toll-like receptor; VSPh, methylpiperazine-Phe-homoPhe-vinylsulfone-benzene.

*Corresponding Author: Julio Scharfstein—Instituto de Biofisica Carlos Chagas Filho, UFRJ, Rio de Janeiro, Brazil. Email: scharf@biof.ufrj.br

Molecular Mechanisms of Parasite Invasion, edited by Barbara A. Burleigh and Dominique Soldati-Favre.

Introduction

Invasion of nonprofessional phagocytic cells by *T. cruzi* involves multiple activation pathways, depending on the nature of the host cell, parasite strain and developmental form (metacyclics or tissue-culture derived trypomastigotes).[1] Analysis of $[Ca^{2+}]_i$ fluxes in both pathogen and host cells revealed the existence of a bi-directional flow of communication that persists throughout the penetration process.[2,3] Early studies on the Ca^{2+}-dependent responses which tissue-culture trypomastigotes (TCTs)[4] evoke via activation of GPCRs[5] led to the concept that the pathogen usurps a natural cellular wound repair mechanism involving Ca^{2+}-regulated lysosomal delivery and fusion to host cell surfaces by a synaptotagmin VII-dependent pathway.[6] More recently, an alternative penetration pathway involving PI3-kinase was described.[7] Accordingly, *T. cruzi* is also able to trigger invaginations of plasma membrane, forming nascent parasitophorous vacuoles whose maturation is critically depend on actin remodelling.[7,8] Although the nature of activation signals that drive the lysosomal-independent versus the Ca^{2+}-regulated lysosomal exocytosis responses were not characterized, they are not necessarily mediated by the same receptors, and furthermore, may likely differ from one cell system to another.

How precisely parasite proteases participate in each of these two well-defined stages of the invasion processes is far from clear. Historically, the first suggestion that proteases might play a crucial role in the invasion process can be traced back to the pioneering work by Piras and coworkers,[9] who reported on the effects which serum "priming" imparted on the phenotypic maturation of tissue culture trypomastigotes. Later on, efforts to delineate the role of proteolysis in the invasion of non-professional phagocytic cells focused on particular classes of *T. cruzi* peptidases. For example, the prolyl-oligopeptidase designated as Tc80[10] drew initial attention owing to its collagenase activity, presumably required for tissue invasion. More recently, studies performed with selective Tc80 inhibitors indicate that this prolyl-oligopeptidase is also critically involved in host cell invasion.[11] Given indications that parasite-host cell adhesion involves interactions with various extracellular matrix components,[12-13] it is possible that Tc80 may strengthen cell surface adhesive interactions required for formation of synapses with host cells.

The first example of a TCT-derived protease that triggers intracellular Ca^{+2} mobilization through G-protein coupled receptors (GPCRs) came from studies with an oligopeptidase B-type of enzyme present in *T. cruzi* extracts.[14] Since this peptidase is found in the cytoplasm of the parasites, it was proposed that Ca^{+2} inducing signaling molecules were generated upon processing of an inactive polypeptide precursor, the newly formed agonists being subsequently transported from the cytoplasm to extracellular spaces.[15,16] Another activation pathway leading to intracellular Ca^{+2} mobilization through the triggering of GPCRs emerged from our studies on cruzipain, the major lysosomal cysteine protease of *T. cruzi*. As discussed later in this chapter, cruzipain (and/or cruzipain isoforms) generate signaling molecules that enhance parasite uptake by a broad range of nonprofessional phagocytic cells (e.g., endothelial cells, neonatal cardiomyocytes, and smooth muscle cells).[17-20]

Structural and Functional Diversity of the Cruzipain Family

For over a decade, research on seemingly unrelated *T. cruzi* antigens and papain-like cysteine proteinases evolved in parallel before they converged, in the early 1990s.[21-25] The cathepsin L-like peptidase, named cruzipain,[23] was extensively characterized at the biochemical, structural and molecular level[26] while less is known about the functional properties of the *T. cruzi* cathepsin B-like enzyme.[27] Assigned to the clade A of the C1 papain-like branch, the cathepsin L-like cysteine proteinases from the cruzipain family are encoded by a multigene family arranged in tandem arrays on several chromosomes.[28-29] Members of the cruzipain family share with type 1 cysteine proteases of *Trypanosoma brucei* and *Leishmania* a long glycosilated[30] and highly antigenic[22,31] C-terminal extension of unknown function which is absent in the thiol cathepsin counterparts from lysosomes of mammalian cells. The C-terminal extension is linked to enzyme's catalytic (central) domain through a polythreonine hinge susceptible to autocatalytic cleavage.[22,24,32] Cruzipain is synthesized as a precursor that undergoes maturation by proteolytic excision of the N-terminal pro domain, a feature shared by all papain-like enzymes. Data concerning the kinetic properties and substrate specificity of natural cruzipain derived mostly from studies using the enzyme purified from epimastigotes,[21,33-35] later complemented by analysis of recombinant enzymes.[36] This knowledge was refined by structure-activity studies based on the X-ray structure of cruzain,[37] name designated for a recombinant form of cruzipain truncated at its C-terminal extension. Collectively, these studies revealed that cruzipain (i) displays dual cathepsin L and cathepsin B-like specificity[33,38] (ii) unlike mammalian counterparts, cruzipain has the capability of accommodating Proline in the P_2 position[34,35] (iii) is modulated by substrate binding to a temperature-sensitive site[33,39,40] (iv) is stable and active at a broad pH range, indicating that it is not strictly dependent on an acidic environment as is often the case of lysosomal-like proteases.[24,33,36]

Initial analyses of the structural organization of the cruzipain genes, performed with different parasites strains, appointed a variable copy number (14 to-100) arranged in tandem and distributed in at least 4 chromosomes.[28,41] The recent completion of the genome shows at least 14 copies of cruzipain genes in the CL Brener strain.[42] At the present time, there are only 4 full-length cruzipain sequences annotated in the *T. cruzi* genome database, sharing over 97% identity. This notwithstanding, earlier studies using RT-PCR revealed cruzipain sequences presenting 80-95% of amino acids identity in the Dm28 strain.[29] Noteworthy, cruzipain transcripts found in epimastigotes are dominated by sequences showing more than 95% amino acid identity with the major cruzipain isoform (cruzipain 1), while the sequences encountered in trypomastigotes and amastigotes share 80% amino acid identity, suggesting that some cruzipain isoforms are developmentally regulated.[29] Since the majority of the nonconserved substitutions among these isoforms are concentrated in the central domain, we anticipated[29] that they might have distinct substrate preferences, and possibly, different biological roles. Indeed, we subsequently demonstrated that the cruzipain 2 isoform is preferentially expressed by Dm28c trypomastigotes and amastigotes.[43] Of note, the substrate preference of recombinant cruzipain 2 is narrower than cruzain,[43] the main differences being ascribed to S_2, the S_1' and the S_2' sites.[43b] Consistent with this, we found that Mu-F-hF-VSPh, a potent inhibitor of recombinant cruzain[44] (i.e., cruzipain 1), is also able to inactivate recombinant cruzipain 2, albeit at reduced efficiency (Lima, APL and Scharfstein, J, unpublished data). Thus, it appears that cruzipain gene polymorphism has equipped *T. cruzi* with an armamentarium of functionally diverse isoforms. During cellular growth, the parasites (epimastigotes and amastigotes) may rely on isoforms with broad specificity to promote lysosomal protein catabolism and/or growth factors.[17,44] In addition, the metabolically quiescent TCTs may depend on a specific isoforms to process particular subsets of precursor proteins, generating signals that enhance cellular invasion.[18-20]

Cysteine Proteases Are Factors of Parasite Virulence

For many years, the evidence linking trypomastigotes' capacity to invade cells to the activity of cysteine proteases was at best circumstantial. Invasion assays performed in different culture systems revealed that polyclonal antibodies directed against the C-terminal extension (GP25) partially protected primary cultures of smooth muscle human cells.[45] Uncertainties concerning the fine specificity of the polyclonal antibodies used in these early studies precluded a definitive assessment of the molecular mechanism underlying their protective effects. In the early 90´s, using primary cultures of cardiomyocytes to study the role of cruzipain in cellular invasion, we found that Z-(SBz)Cys-Phe-CHN_2, an irreversible inhibitor designed against cathepsin L, partially blocked cellular invasion and arrested intracellular parasite development when tested at the low μM range.[17] Assays performed with a related radiolabeled probe, Z-Phe-Tyr^{125}-CHN_2, confirmed that these first-generation inhibitors targeted cruzipain in monolayers of infected heart cells,[17] raising the proposition that cruzipain was a therapeutic target. After the elucidation of the X-ray structure of this protease,[37] cruzipain inhibitors containing the vinyl-sulfone reactive group became an indispensable tool in functional studies, in view of the technical difficulties posed by targeted deletion of the multigene cruzipain family, and the obstacles in employing RNA interference techniques in *T. cruzi*. Some of these cruzipain inhibitors were of low toxicity, proving capable of eradicating infection in animal models of Chagas' disease.[44]

Years later, we reported that Dm28c TCTs invade primary human endothelial cells[18] and neonatal cardiomyocytes[19] by triggering Ca^{2+} fluxes via bradykinin receptors, and implicated cruzipain as the protease generating "kinins", i.e., the cognate peptide ligand for these GPCRs.[18] Alternative pathways, i.e., independent of kinin receptor signalling, but likewise involving cruzipain-mediated proteolysis, were identified as we performed invasion assays with human smooth muscle cells.[20] Using several *T. cruzi* strains, we found that parasite competence to invade smooth muscle cells was directly correlated with activity levels of secreted cruzipain.[20] For example, the poorly infective *T. cruzi* G-strain expresses low levels of active cruzipain enzyme as compared to Dm28c strain.[20] Notably, the infectivity of G-TCTs was drastically enhanced by the addition of Dm28c TCT conditioned medium and this effect was abolished by the broad spectrum inhibitor E-64. At least two functionally distinct cruzipain isoforms (Lima, AP, unpublished data) were detected in the Dm28c conditioned medium, but the lack of isoform-specific inhibitors has precluded an assessment of their individual roles in protease-mediated enhancement of G strain infectivity. Of note, the effects of Dm28 conditioned medium were cancelled by thapsigargin, indicating that factors steering invasion mobilized Ca^{2+} from host cell intracellular stores.[20] In addition to the critical requirement of cysteine protease activity, the effect of TCTs (Dm28c) conditioned medium depends on an as yet unidentified membrane-bound substrate(s), possibly displayed on lipid vesicles shed by TCTs.[46] Although the nature of these lipid-bound protein substrate(s) shed by TCTs is not known, proteomic analysis has recently revealed the presence of members of the trans-sialidase (TS/Tc85) family and of glycophosphatidylinositol linked mucins (tGPI-mucins) (Torrecilhas et al, unpublished data). It remains to be determined if cruzipain may process such polymorphic surface glycoproteins, perhaps liberating peptides that stimulate endocytic uptake by nonprofessional phagocytic cells through the triggering of hitherto unknown receptors[13] (Fig. 1).

Modulation of Cruzipain by Endogenous Inhibitors Has Impact on Parasite Infectivity

The mechanisms underlying the control of endogenous cysteine protease activity in lower eukaryotes remained elusive for years. Protein inhibitors of the cystatin family are ubiquitous in metazoa, thus would seem to act as natural candidates as endogenous modulators of papain-like enzymes in protozoa. However, such genes are so far absent from the genomes of trypanosomatids. Evidences of a novel family of endogenous inhibitors of cysteine proteases came from our studies on a *T. cruzi* protein of 11 kDa protein, designated as chagasin, characterized as a potent tight-binding reversible inhibitor of papain-like cysteine proteases.[47] Chagasin is encoded by a single gene and is expressed at higher levels in trypomastigotes and amastigotes as compared to epimastigotes.[47] Intriguingly, chagasin is mostly associated with membranes in trypomastigotes while in epimastigotes it is expressed at low levels, mostly as a soluble protein.[47] A striking physical-chemical property of chagasin, its thermo-resistance, facilitated its dissociation from complex with endogenous enzymes upon boiling, enabling the detection of its inhibitory activity in parasite lysates.[47] Although chagasin shares several biochemical properties with cystatins, it has no sequence similarity with other known cysteine protease inhibitors, suggesting that it is the prototype of a new family of inhibitors, and recently classified in the MEROPS database as family 142, belonging to clan IX. Several chagasin-like genes sharing low similarity were rapidly identified in other protozoa and even in bacteria.[48-50,50b] In agreement with earlier in silico predictions,[48,51] the high-resolution solving of the crystal structure of chagasin[52b] or that of chagasin bound to falcipain (Wang et al., submitted) and the NMR structure of the chagasin protein in solution[52] revealed the presence of an immunoglobulin-like domain scaffold projecting loops that interact with the active site of target cysteine proteases. Similar features were described for a chagasin homologue of *L. mexicana*, designated as inhibitor of cysteine peptidases (ICP).[53]

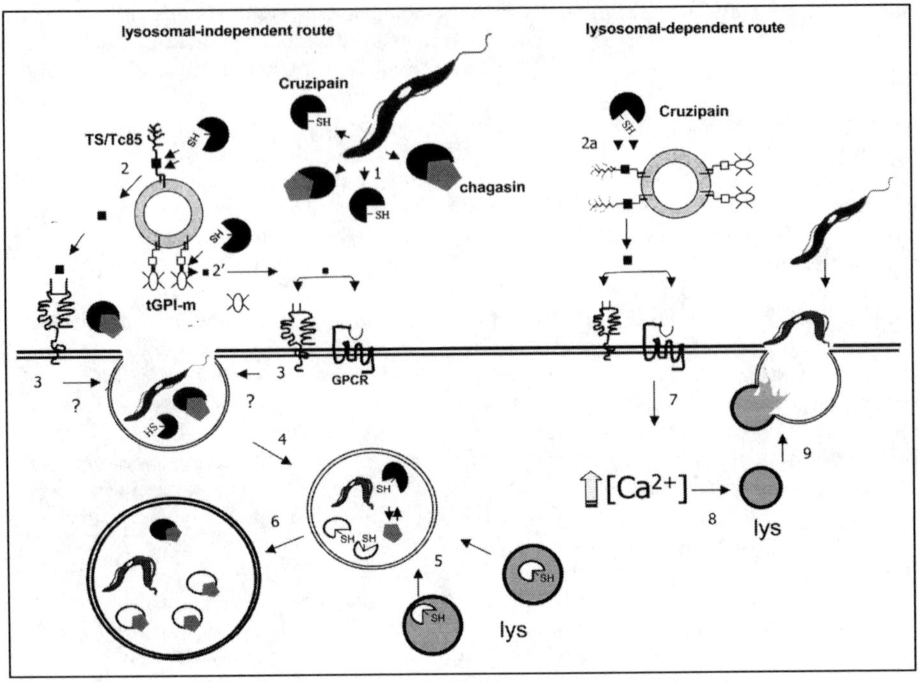

Figure 1. Cruzipain mediates invasion of smooth muscle cells by processing molecules associated with shed trypomastigote membranes. Two hypothetical mechanisms involving the participation of *T. cruzi* cysteine proteinases (for simplicity, represented by cruzipain) are illustrated. On the left side of the figure, TCT secrete soluble cruzipain and cruzipain-chagasin complex[54] (arrow 1) while at the same time shedding lipid vesicles.[46] Members of the trans-sialidase family (TS/Tc85)[12,13] (arrow 2), GPI-anchored mucins[47] (arrow 2′) or other molecules displayed on the shed vesicles (Torrecilhas et al, unpublished data) may serve as substrates for soluble cruzipain. Following ligand binding to as yet uncharacterized host cell receptors, the peptides may induce membrane invaginations (arrow 3) via the lysosome-independent pathway. Once internalized, TCTs may keep on secreting cruzipain and cruzipain-chagasin complex in the nascent parasitophorous vacuole (arrow 4). As vacuole maturation proceeds through fusion with endosomes or lysosomes, free chagasin may inactivate host cell cysteine proteases discharged in the parasitophorous vacuole (arrow 5). On the right side of the figure, the parasites may invade smooth cells by alternatively engaging the lysosomal entry route (right side of panel). Parasite strains expressing high molar ratios of cruzipain over chagasin may initiate this response by proteolytically generating peptide ligands (arrow 2, right) for G-protein coupled receptors that trigger lysosomal fusion to the plasma membrane via $[Ca^{2+}]_i$-dependent mechanisms (arrows 7-9, right panel).

The observation that expression levels of chagasin are low in epimastigotes and inversely correlated with those of cruzipain[47] led us to investigate whether these molecules intersect during trafficking. Chagasin colocalized with cruzipain both in the Golgi complex and reservosomes of epimastigotes, where the inhibitor was mostly present as tight-binding molecular complexes with cruzipain.[54] Interestingly, the quantification of chagasin and of cruzipain levels in several *T. cruzi* strains revealed that the inhibitor/enzyme molar ratio in epimastigotes is rather constant (~1:50) despite strain-dependent variability in the expression of the individual proteins.[54] As mentioned, the G strain displayed a drastic reduction in cysteine protease activity, while expression of the protease inhibitor remain normal, i.e., chagasin/cruzipain ratio of 1:5.[54] Using Dm28 parasites engineered to express 4-fold more chagasin than wild type, we observed that chagasin impaired several biological functions typically associated with cruzipain activity.[54] Biochemical and ultrastructural studies suggested that chagasin may intersect with pro-cruzipain trafficking pathways, most likely in the Golgi complex. By inhibiting the autocatalytic removal of the pro-segment from pro-cruzipain zymogen(s), chagasin may tightly control a limiting step of cruzipain maturation. Thus, differences in cruzipain activity among *T. cruzi* strains and/or isolates may reflect alterations in chagasin levels and/or trafficking pathways of zymogens. The complexity of this regulatory mechanism is also highlighted by the findings that cruzipain isoforms can trans-activate each other, for example, mature forms of cruzipain 2 are able to process pro-cruzipain 1 (Reis, FC and Lima, AP, unpublished data), thus recapitulating the trans-activation mechanisms originally described in studies with cysteine proteases of *L. mexicana.*[55]

Although the evidence so far clearly appoints to a role of chagasin as a negative modulator of cruzipain activity in epimastigotes, we still lack information regarding its role in the trypomastigotes and amastigotes. Intriguingly, and in striking contrast with epimastigotes, lysates of TCTs display elevated contents of chagasin molecules as compared to epimastigotes.[49] Moreover, chagasin has a different sub-cellular distribution in TCT, being observed at the cell surface and in the flagellar pocket.[47] The possibility that variable expression levels of chagasin and cruzipain could influence *T. cruzi* infectivity in vitro was confirmed by examining the phenotype of TCTs overexpressing chagasin. Indeed, these parasites displayed reduced infectivity for human smooth muscle cells in vitro.[54] Of note, recombinant chagasin reduced the extent of cellular uptake of wild-type TCTs in a dose-dependent manner, thus recapitulating the phenotypic properties (i.e., reduced infectivity) of TCTs overexpressing chagasin.[54] By analogy to the results obtained with poorly infective G-strain TCTs, cellular invasion of the chagasin overexpressors was enhanced upon addition of conditioned medium derived from wild type Dm28c TCTs. Further, the reconstitution response required active cysteine protease,[54] again supporting the notion that balance in the ratio between chagasin/cruzipain modulates the efficiency of smooth cell invasion by *T. cruzi.*

At the present time, it is unclear if the balance between cruzipain/chagasin may have impact on the engagement of lysosomal-independent pathway of cellular invasion.[7] In the absence of synapses, parasite strains may critically depend on high-level secretion of cruzipain isoforms (and conversely, on low level secretion of chagasin molecules) to efficiently process substrates either displayed on shed vesicles,[54] and/or cell surfaces (Fig. 1).[18,19] Considering that chagasin inhibitory activity on cruzipain is significantly (40%) reduced at acid pH (Lima AP, unpublished data), the acidification resulting from Ca^{2+}-induced lysosomal fusion with host cell plasma membrane may cause partial dissociation of cruzipain from chagasin complexes, leading to increased accumulation of active protease within "synapses". As discussed latter in this chapter, the acidification within such secluded spaces may facilitate cruzipain-dependent generation of GPCR agonists (e.g., "kinins")[18,19] thus further enhancing the cellular invasion process. Although providing a rationale to understand how alternative invasion pathways may eventually converge, the above mechanisms do not offer clues as to how chagasin may contribute to the dynamics of the penetration process. Development of chagasin nulls in *T. cruzi* is being currently pursued to evaluate if this cysteine inhibitor may dampen potentially deleterious activities of mammalian cathepsin L/B/S, following lysosomal fusion/discharge to the parasitophorous vacuole (Fig. 1).

Role of Cruzipain in Cellular Invasion Mediated by Kinin Receptors

The realization that cruzipain and tissue kallikrein display similar substrate specificity suggested that the parasites may be able to process kininogens, liberating lysyl-bradykinin (LBK).[56] The term "kinin" refers to a small group of vasoactive metabolites structurally related to the nonapeptide bradykinin (BK), released from an internal moiety of high or low molecular weight kininogens (HK/LK) by the action of plasma or tissue kallikrein.[57] In chronic inflammation, the concerted action of neutrophil elastase and mast cell tryptase on oxidized forms of kininogens may also liberate a slightly larger kinin, Met-LBK.[58] As discussed latter in this chapter, kinins can be directly liberated from HK/LK by the action of a wide range of microbial cysteine proteases, such as those expressed by parasitic such as *T. cruzi*[18,56] and by promastigotes belonging to the *L. donovani* complex.[59] Once released, the short-lived kinins (half life of <15 sec in the plasma) exert their biological effects by the paracrine mode, through the activation of distinct sub-types of heterotrimeric GPCRs, B_2R or B_1R.[60-62] Conditions leading to excess formation of the intact kinins (BK or LBK) typically result in the down-modulation B_2R,[63] thus attenuating host cell signalling. Further underscoring the importance of kinin regulation, the long-range effects on bradykinin receptors localized at the vascular lining are prevented by the metabolic action of kinin-degrading peptidases,[57] e.g., the angiotensin converting enzyme (ACE/kininase II). While intact kinins (BK or LBK) are the agonists for the constitutively expressed B_2R, excision of the C-terminal Arg of BK/LBK by carboxypeptidase N/M (kininase I)[57,60,62] generates high-affinity ligands for the B_1R, which are upregulated during inflammation.[62]

Kinins were traditionally viewed as classical mediators of acute inflammation (e.g., inducers of oedema formation, vasodilation and pain sensations), but these short-lived peptide hormones are currently regarded as general modulators of circulatory homeostasis.[57,60] Of potential interest for research on the pathogenesis of Chagas' disease, studies in other disease models suggest that B_1R upregulates expression of connective tissue growth factor and collagen I production, thus may play a role in the pathogenesis of chronic fibrosis.[64] Of further interest, we recently reported that bradykinin potently stimulates dendritic cell maturation by triggering the constitutively expressed B_2 kinin receptor subtype, ultimately driving Th1 polarization via the IL-12 pathway.[65]

Although usually described as a prototypical $G\alpha_{q/11}$-coupled receptor leading to PLC-β-mediated generation of second messangers inositol-1,4,5-triphosphate, diacylglycerol and calcium,[61] B_2R may also trigger the PI3-kinase pathway,[66] or modulate intracellular cAMP levels and PKA through coupling to either G_s or G_i.[61] Further, by activating small G proteins Rho, Rac and or Cdc42 via coupling to $G\alpha_{12/13}$, B_2R signaling modulates cytoskeleton organization, and consequently affects cell shape and motility.[61] In addition, kinin receptor signaling can activate ERK/MAPK cascade via G_i.[63] In vascular cells, the increase in $[Ca^{2+}]_i$ can activate the nitric oxide/cGMP pathway that induces vasodilation. In some cells, stimulation of kinin receptors also generate prostaglandins via activation of phospholipase A2 and D.[61]

At first sight, the findings that natural cruzipain slowly liberates LBK from purified kininogens in test tube conditions[56] was paradoxical because kininogens are members of the cystatin family of cysteine protease inhibitors, hence are able to inactivate papain-like enzymes such as cruzipain, in vitro.[67] To visualize how cruzipain may liberate the internal kinin moiety (domain D4) from cell-bound kininogens during the cellular invasion process, one must bear in mind that HL and LK also share the N-terminal D1-3 domains, all of which are homologous to cystatins, although only the last two (D2-3) are functionally active inhibitors of cysteine proteinases.[68] Of note, the D3 domain of kininogens displays a binding site for endothelial cells that overlaps with the cystatin domain.[69] At the C-terminal end of the bradykinin (D4) sequence of HK is a histidine-rich motif (D5H) that binds to heparan sulphate chains.[70] Significantly, HK is also tethered to the cell surfaces by binding to chondroitin sulphate proteoglycans.[71]

Given indications that glycosaminoglycans serve as platforms for the cell-surface accumulation of HK in a variety of mammalian cells,[70,71] we reasoned that interactions with heparan sulphate could impair its cystatin-like inhibitory functions, perhaps allowing for increased proteolytic processing by cruzipain. Indeed, studies in solution revealed that heparan sulphate redirects the substrate specificity of cruzipain, generating multiple HK break-down products.[72] Interestingly, peptide bonds localized in the cystatin-like segments of HK were more efficiently hydrolysed by cruzipain when we added heparan sulphate to the reaction mixture.[72] Moreover, as predicted, the cysteine inhibitory activity of HK was drastically reduced (~10 fold increase of K_{iapp}).[72] Kinetic data obtained with a fluorogenic peptide substrate spanning the N-terminal flanking side of bradykinin indicated that the catalytic efficiency of cruzipain is enhanced up to

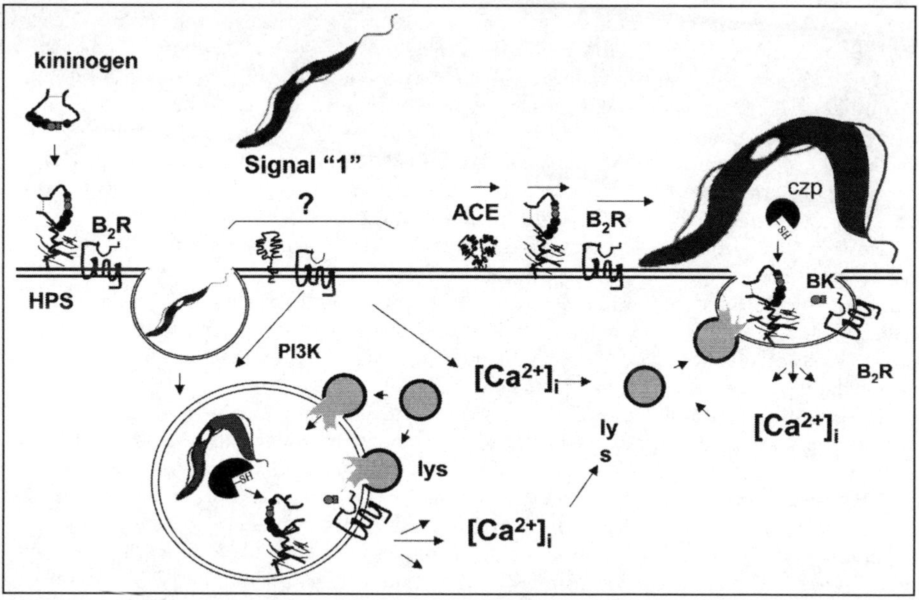

Figure 2. Cruzipain promotes parasite invasion of cardiovascular cells by triggering the B_2R signaling pathway. The scheme illustrates two hypothetical mechanisms leading to kinin generation in secluded sites formed by juxtaposed parasite and host cell membranes. At the left side of the panel, stimuli originating from the parasite or host cell origin, here generically defined as signal "1", induce plasma membrane invaginations.[7,8] Formation of the nascent parasitophorous vacuole may lead to internalization of high-molecular weight kininogen (docked to heparan sulphate chains, HPS) and B_2R. PI3K-dependent maturation of the parasitophorous vacuole occurs through membrane fusion either with endosomes (not represented) or lysosomes (low pH indicated by grey color). Next, active cruzipain diffuses from the flagellar pocket into this secluded/slightly acidic microenvironment. The protease liberates intact kinins (e.g., BK) from kininogen molecules docked to HPS. BK-induced triggering of B_2R mobilizes $[Ca^{2+}]_i$, ultimately driving lysosomal recruitment and fusion to plasma membranes (right side of panel). Cruzipain benefits from the secluded/acidic environment formed in this "synapse", releasing at high efficiency BK from HPS-bound kininogens. BK-induced triggering of B_2R intensifies the $[Ca^{2+}]_i$ fluxes, thus further increasing parasite uptake through the lysosomal dependent entry pathway. Alternatively, B_2R-engagement may occur independently of the PI3K route, provided that stimuli eliciting weak Ca^{2+}-lysosomal fusion to plasma membranes are generated at very early stages of interaction with cardiovascular cells.

6-fold in the presence of heparan sulphate.[72] We then demonstrated that heparan sulphate increased the efficiency of the kinin-releasing reaction, albeit only at relatively narrow concentration range.[72] Collectively, these studies indicated that optimal cooperative interactions between cruzipain, HK and heparan sulphate proteoglycans may significantly enhance the kinin-releasing activity of TCTs (Fig. 2).

Cruzipain and ACE/Kininase II Plays Opposite Roles in the Modulation of Host Cell Invasion through the B_2R-Dependent Pathway

Studies performed with monolayers of primary human umbilical vein endothelial cells (HUVECs) or of Chinese Hamster Ovary cells transfected with the rat-B_2 bradykinin receptor gene (CHO-B_2R) confirmed the cell-bound kininogens served as substrates for cruzipain.[18]

Using $[Ca^{2+}]_i$ responses as a read-out, we showed that purified cruzipain (preactivated) triggered potent activation responses in CHO-B_2R or HUVECs.[18] The $[Ca^{2+}]_i$ response was blocked by HOE-140, a specific antagonist of the B_2R subtype, or by E-64, an irreversible inhibitor of papain-like cysteine proteases, hence confirming that cruzipain liberated the kinin agonist from cell-bound forms of kininogens.[18] Assays performed with living TCTs revealed that the pathogen evoked strong repetitive $[Ca^{2+}]_i$ transients in CHO-B_2R, but not in mock-transfected CHO cells, and furthermore, the responses were reflected as changes in cellular invasion indexes.[18] Interestingly, parasite uptake by CHO-B_2R was enhanced upon addition of purified HK or alternatively, by addition of physiological concentration of the B_2R peptide agonist, i.e., BK, into the serum-free medium.[18] At high doses (~100 nM) the BK agonist prevented host cells from being overinfected due to B_2R down-regulation of the kinin receptor.[18,63] Furthermore, mAbs directed to kininogens blocked invasion on CHO-B_2R, but did not interfere with the baseline levels of infection of CHO mock, suggesting that cell-bound HK/LK serve as precursors for the B_2R agonist(s) released by cruzipain (Fig. 2).

Studies with synthetic inhibitors of cruzipain unexpectedly revealed that membrane-permeable cruzipain inhibitors reduced extent of parasite invasion via the B_2R, while addition of soluble inhibitors such as cystatin C or E-64 did not interfere at all with parasite infectivity.[18] Considering that trypomastigotes are poorly endocytic[73] and that these flagellates accumulate cruzipain in the flagellar pocket,[24] the failure of hydrophilic inhibitors in preventing cellular invasion was interpreted as evidence that the kinin-releasing reaction most likely occurs in enclosed areas formed by juxtaposition of host cell and parasite plasma membranes.[18] We therefore proposed a model whereby cruzipain molecules diffuse from the parasites' flagellar pocket into this intercellular space (Fig. 2). In this secluded microenvironment, perhaps equivalent to a "synapse", active forms of cruzipain may be possibly spared from physiological inactivation by soluble forms of plasma protease inhibitors (e.g., cystatins, kininogens, α_2-macroglobulin).[18] Although not directly demonstrated, this notion also implies that surface-bound kininogens, along with bradykinin-receptors, are actively recruited to such signalling centres.[18]

The findings of residual infection levels in cultures (CHO-B_2R, HUVECs or neonatal cardiomyocytes) supplemented with HOE-140 (B_2R antagonist)[18,19] revealed that, as expected, the kinin signalling pathway is not the sole signalling route(s) driving parasite uptake in such cells. The possibility that the parasites mobilize cooperative pathways of activation to optimally invade cardiovascular cells is under investigation. As discussed before, it is not obvious how the lysosomal-independent entry mechanism may facilitate engagement of the kinin signalling pathway. Although entirely speculative, it is possible that plasma membrane invaginations induced via the PI3-kinase or other "priming" stimuli (Fig. 2) may induce the internalization of surface-bound kininogens along with kinin receptors and their cognate heterotrimeric G proteins into the nascent parasitophorous vacuole. Pending on the ratios of cruzipain/chagasin secreted in the tight luminal space of the vacuole, the kinin hormones may be then excised from kininogens, thus triggering Ca^{2+}-regulated lysosomal exocytosis through B_2R from within the nascent vacuole (Fig. 2). Alternatively, low-grade generation of kinins by cruzipain secreted at early stages of host-parasite contact may initiate Ca^{2+}-dependent exocytosis. Lysosomal fusion induced by this "priming" response may suffice to generate a secluded/acidic environment which, as previously discussed, may cause partial dissociation of cruzipain from chagasin complexes. Owing to the increased efficiency of the kinin-releasing reaction, cruzipain may further stimulate the endocytic uptake of the parasites (Fig.2).

Another interesting aspect that emerged from these in vitro studies concerned the modulatory activity of kinin-degrading peptidases (e.g., ACE/kininase II). Using ACE inhibitors, such as captopril, we found that TCT infectivity in cultures of CHO-B_2R was enhanced over 4-fold (no effect in assays with CHO mock).[18] These results were consistent with the notion that the ACE inhibitor increased the half-life of intact kinins (BK or LBK) released by cruzipain, thus enhancing B_2R stimulation. Although the ACE inhibitor potentiation is less accentuated in assays with endothelial cells or cardiomyocytes, parasite uptake is markedly reduced upon addition of HOE-140. Studies in progress aim to determine why the parasites fail to compensate B_2R blockade of cardiomyocytes by engaging alternative (i.e., kinin-independent) mechanisms, e.g., the TGF-β-dependent route.[74,75] Thus, analysis of the mechanisms underlying generation and degradation of ligands for bradykinin B_2 receptors indicated that cruzipain and ACE/kininase II play opposite roles in this process, respectively acting as positive and negative modulators of host cell invasion.

Regulation of the B_2R Signaling Pathway Modulates Pathogenic Outcome in Experimental *T. cruzi* Infection

During the asymptomatic stage of the disease, endogenous kinins generated in infected tissues are likely kept at low levels by ACE and other kinin-degrading peptidases. In principle, it is possible that dysfunctional states leading to down-modulation of ACE by cardiovascular cells may generate transient windows of opportunity for B_2R signaling, with some incremental gain in parasite infectivity. Although some *T. cruzi* variants may be able to inactivate ACE, the natural selection of such powerful pro-inflammatory pathogens may be too detrimental to the host, thus may not likely contribute to the establishment of host-parasite equilibrium in the settings of a chronic infection. More recently, we developed a subcutaneous model of infection to investigate if parasite-induced activation of kinin receptors modulates pathogenic outcome.[19] Analysis of the dynamics of inflammation evoked by TCTs revealed that these infective forms rapidly induce plasma extravasation through capillary beds by activating neutrophils via a TLR2-dependent mechanism.[75b] The plasma leakage, although incipient at early stage of inflammation, leads to accumulation of plasma-borne kininogens in the peripheral tissues, thereby allowing for cruzipain-mediated liberation of kinins.[75b] As a result of B_2R-dependent increases in vascular permeability, the levels of kinins generated in peripheral tissues raise sharply, stimulating DC maturation via B_2R.[65,75b] As predicted, the activated DCs migrate to draining lymph nodes where they activate virgin T cells, promoting IL-12 dependent differentiation of type 1 immunity.[75b] In summary, our in vivo studies suggest that rather than unilaterally benefiting the pathogen, the activation of the cruzipain/kinins/B_2R pathway stimulates host defenses,[75b] hence may contribute to the maintenance of host-parasite equilibrium.

Carboxypeptidase M/N (kininase I) Is a Positive Modulator of Cardiovascular Cell Invasion by the Inducible B_1R Pathway

The possibility that alterations in kinin homeostasis might be compromised during the chronic stage of the disease is under current investigation. Confronted with intense selective pressure by the immune system, parasites may take advantage of the availability of plasma-borne proteins (i.e., including kininogens) that permeate the inflamed heart to invade cardiovascular cells through the upregulated B_1R. This possibility is supported by in vitro studies showing that Dm28c TCTs aptly invade LPS-stimulated endothelial cells, but not resting cells, by engaging the B_1R (TLR4/NFk-B-induced). These effects were blocked by kininase I inhibitors, implicating this peptidase in the conversion of intact kinins into B_1R agonists[19] (Fig. 3). More recently, we demonstrated that tGPI anchors, i.e., TLR2 ligands expressed by TCTs,[76] can likewise induce B_1R in primary endothelial cells, rendering them susceptible to B_1R/kininase I-dependent pathways of cellular invasion (Andrade, D. and Scharfstein, J., in preparation). At present, it is unknown if the B_1R/kininase I-dependent pathway of invasion contributes to the low grade but persistent infection observed in chronically infected mice. For example, it would be interesting to know if *T. cruzi* may rely on their own metalloproteinases, such as the recently described GP63 homologues,[76b] to generate ligands for B_1R. Of interest, preliminary studies suggest that chronic myocardial fibrosis is significantly attenuated in $B_1R^{-/-}$ animals. Ongoing studies should determine if the attenuation of myocardium fibrosis observed in $B_1R^{-/-}$ mice is an indirect consequence of reduced

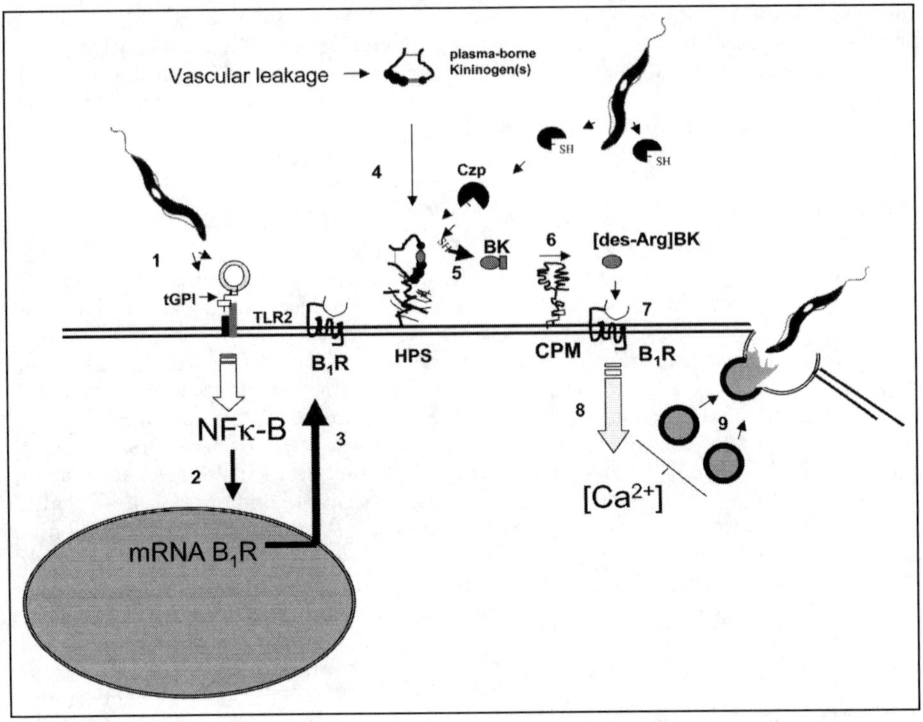

Figure 3. Kininase I is a positive modulator of cardiovascular cell invasion by the inducible B_1R. At the left side of the panel, TLR2 ligands (e.g., tGPI-mucin) associated with shed vesicles (arrow 1) stimulate NFk-B (arrow 2), thus upregulating expression of B_1R (arrow 3) on cardiovascular cells. Triggering of B_1R depends on the following sequence of events: first, plasma-borne kininogens permeating inflamed tissues bind to host cell surfaces via interactions with HPS (arrow 4). The secreted forms of cruzipain then promote the excision of intact kinin peptides (e.g., BK) from kininogen molecules bound to HPS (arrow 5). The released BK is then converted into the B_1R agonist ([des-Arg]BK) by the action of kininase I (carboxypeptidase M) (arrow 6). Invasion by the B1R-dependent route (7) involves signaling by the Ca^{2+}/lysosomal dependent entry route.[19]

tissue parasitism. Alternatively, the attenuated fibrosis observed in $B_1R^{-/-}$ mice may reflect impairment of B_1R-dependent immunopathology, such as deficient leukocyte recruitment to sites of infection[77] and/or to reduced deposition of collagen, a B_1R-dependent response which in other disease models is associated with increased stability of mRNA for connective tissue growth factor.[64] Additional studies are required to determine if overt stimulation of B1R might aggravate IL-1β-dependent hyperthophy in cardiomyocytes, perhaps contributing to the pathological response that TCTs initiate through the activation of TLR2.[78]

Concluding Remarks

In this chapter we have reviewed studies showing that cellular invasion by *T. cruzi* is finely modulated by interplay between multiple proteases (i.e., cruzipain, ACE, and carboxypeptidase M/N), with naturally occurring peptidase inhibitors (chagasin and kininogens). Research focused on the mechanisms regulating generation and degradation of ligands for GPCRs implicated kinin receptors as drivers of cardiovascular cell invasion by the Ca^{2+}-dependent lysosomal pathway. Lessons taken from these relatively simple in vitro studies opened new avenues for research on the immunopathogenesis of Chagas' disease.

Acknowledgements

The research work of our group was supported by CNPq, FAPERJ, WHO-TDR, VW Foundation and the Wellcome Trust (72349/Z/03/Z).

References

1. Burleigh BA, Woosley AM. Cell signaling and Trypanosoma cruzi invasion. Cell Microbiol 2002; 4:701-11.
2. Moreno SN, Silva J, Vercesi AE et al. Cytosolic-free calcium elevation in Trypanosoma cruzi is required for cell invasion. J Exp Med 1994; 180(4):1535-40.
3. Dorta ML, Ferreira AT, Oshiro ME et al. Ca^{2+} signal induced by Trypanosoma cruzi metacyclic trypomastigote surface molecules implicated in mammalian cell invasion. Mol Biochem Parasitol 1995; 73(1-2):285-9.
4. Tardieux I, Webster P, Ravesloot J et al. Lysosome recruitment and fusion are early events required for trypanosome invasion of mammalian cells. Cell 1992; 71(7):1117-30.
5. Leite MF, Moyer MS, Andrews NW. Expression of the mammalian calcium signaling response to Trypanosoma cruzi in Xenopus laevis oocytes. Mol Biochem Parasitol 1998; 92:1-13.
6. Caler EV, Chakrabarti S, Fowler KT et al. The exocytosis-regulatory protein synaptotagmin VII mediates cell invasion by Trypanosoma cruzi. J Exp Med 2001; 193:1097-04.
7. Woolsey AM, Sunwoo L, Petersen CA et al. Novel PI 3-kinase-dependent mechanisms of trypanosome invasion and vacuole maturation. J Cell Sci 2003; 116:3111-22.
8. Woolsey AM, Burleigh BA. Host cell actin polimerization is required for cellular retention of Trypanosoma cruzi and early association with endosomal/lysosomal compartments. Cell Microbiol 2004; 6(9):829-38.
9. Piras MM, Piras R, Henriquez D et al. Changes in morphology and infectivity of cell culture-derived trypomastigotes of Trypanosoma cruzi. Mol Biochem Parasitol 1982; 6(2):67-8.
10. Santana JM, Grellier P, Schrével J et al. A Trypanosoma cruzi-secreted 80 kDa proteinase with specificity for human collagen types I and IV. Biochem J 1997; 324:129-37.
11. Grellier P, Vendeville S, Joyeau R et al. Trypanosoma cruzi prolyl-oligopeptidase Tc80 is involved in nonphagocytic mammalian cell invasion by trypomastigotes. J Biol Chem 2001; 276:47078-86.
12. Giordano R, Fouts DL, Tewari D et al. Cloning of a surface membrane glycoprotein specific for the infective form of Trypanosoma cruzi having adhesive properties to laminin. J Biol Chem 1999; 274:3461-8.
13. Magdesian MH, Giordano R, Ulrich H et al. Infection by Trypanosoma cruzi. Identification of a parasite ligand and its host cell receptor. J Biol Chem 2001; 276:19382-9.
14. Burleigh BA, Andrews NW. A 120-kDa alkaline peptidase from Trypanosoma cruzi is involved in the generation of a novel Ca^{2+}-signaling factor for mammalian cells. J Biol Chem 1995; 270(10):5172-80.
15. Burleigh BA, Caler EV, Webster P et al. A cytosolic serine endopeptidase from Trypanosoma cruzi is required for the generation of Ca^{2+} signalling in mammalian cells. J Cell Biol 1997; 136(3):609-20.
16. Caler EV, Vaena de Avalos S, Haynes PA et al. Oligopeptidase B-dependent signalling mediates host cell invasion by Trypanosoma cruzi. EMBO J 1998; 17:4975-86.
17. Meirelles MN, Juliano L, Carmona E et al. Inhibitors of the major cysteinyl proteinase (GP57/51) impair host cell invasion and arrest the intracellular development of Trypanosoma cruzi in vitro. Mol Biochem Parasitol 1992; 52(2):175-84.
18. Scharfstein J, Schmitz V, Morandi V et al. Host cell invasion by Trypanosoma cruzi is potentiated by activation of bradykinin B(2) receptors. J Exp Med 2000; 192(9):1289-300.
19. Todorov AG, Andrade D, Pesquero JB et al. Trypanosoma cruzi induces edematogenic responses in mice and invades cardiomyocytes and endothelial cells in vitro by activating distinct kinin receptor (B1/B2) subtypes. FASEB J 2003; 17:73-5.
20. Aparicio IM, Scharfstein J, Lima APCA. A new cruzipain-dependent pathway of human cell invasion by Trypanosoma cruzi requiring trypomastigote membranes. Infec and Immun 2004; 72:5892-02.
21. Cazzulo JJ, Cazzulo Franke MC, Martinez J et al. Some kinetic properties of a cysteine proteinase (cruzipain) from Trypanosoma cruzi. Biochim Biophys Acta 1990; 1037(2):186-91.
22. Scharfstein J, Schechter M, Senna M et al. Trypanosoma cruzi: Characterization and isolation of a 57/51,000 m.w. surface glycoprotein (GP57/51) expressed by epimastigotes and bloodstream trypomastigotes. J Immunol 1986; 137(4):1336-41.
23. Cazzulo JJ, Couso R, Raimondi A et al. Further characterization and partial amino acid sequence of a cysteine proteinase from Trypanosoma cruzi. Mol Biochem Parasitol 1989; 33(1):33-41.

24. Murta AC, Persechini PM, Padron TD et al. Structural and functional identification of GP57/51 antigen of Trypanosoma cruzi as a cysteine proteinase. Mol Biochem Parasitol 1990; 43(1):27-38.
25. Eakin AE, Mills AA, Harth G et al. The sequence, organization, and expression of the major cysteine protease (cruzain) from Trypanosoma cruzi. J Biol Chem 1992; 267(11):7411-20.
26. Cazzulo JJ. Proteinases of Trypanosoma cruzi: Potential targets for the chemotherapy of Chagas disease. Curr Top Med Chem 2002; 2:1261-71.
27. Nobrega OT, Santos Silva MA, Teixeira AR et al. Cloning and sequencing of tccb, a gene encoding a Trypanosoma cruzi cathepsin B-like protease. Mol Biochem Parasitol 1998; 97(1-2):235-40.
28. Campetella O, Henriksson J, Aslund L et al. The major cysteine proteinase (cruzipain) from Trypanosoma cruzi is encoded by multiple polymorphic tandemly organized genes located on different chromosomes. Mol Biochem Parasitol 1992; 50(2):225-34.
29. Lima AP, Tessier DC, Thomas DY et al. Identification of new cysteine protease gene isoforms in Trypanosoma cruzi. Mol Biochem Parasitol 1994; 67(2):333-8.
30. Aslund L, Henriksson J, Campetella O et al. The C-terminal extension of the major cysteine proteinase (cruzipain) from Trypanosoma cruzi. Mol Biochem Parasitol 1991; 45(2):345-7.
31. Scharfstein J, Rodrigues MM, Alves CA et al. Trypanosoma cruzi: Description of a highly purified surface antigen defined by human antibodies. J Immunol 1983; 131(2):972-6.
32. Hellman U, Wernstedt C, Cazzulo JJ. Self-proteolysis of the cysteine proteinase, cruzipain, from Trypanosoma cruzi gives a major fragment corresponding to its carboxy-terminal domain. Mol Biochem Parasitol 1991; 44(1):15-21.
33. Lima AP, Scharfstein J, Storer AC et al. Temperature-dependent substrate inhibition of the cysteine proteinase (GP57/51) from Trypanosoma cruzi. Mol Biochem Parasitol 1992; 56(2):335-8.
34. Nery ED, Juliano MA, Meldal M et al. Characterization of the substrate specificity of the major cysteine protease (cruzipain) from Trypanosoma cruzi using a portion-mixing combinatorial library and fluorogenic peptides. Biochem J 1997; 323(2):427-33.
35. Serveau C, Lalmanach G, Juliano MA et al. Investigation of the substrate specificity of cruzipain, the major cysteine proteinase of Trypanosoma cruzi, through the use of cystatin-derived substrates and inhibitors. Biochem J 1996; 313(3):951-6.
36. Judice WAS, Cezari MHS, Lima APCA et al. Comparison of the specificity, stability and individual rate constants with respective activation parameters for the peptidase activity of cruzipain and its recombinant form, cruzain, from Trypanosoma cruzi. Eur J Biochem 2001; 268:6578-86.
37. McGrath ME, Eakin AE, Engel JC et al. The crystal structure of cruzain: A therapeutic target for Chagas' disease. J Mol Biol 1995; 247(2):251-9.
38. Gillmor SR, Craik CS, Fletterick RJ. Structural determinants of specificity in the cysteine protease cruzain. Prot Sci 1997; 6:1603-11.
39. Stoka V, McKerrow JH, Cazzulo JJ et al. Substrate inhibition of cruzipain is not affected by the C-terminal domain. FEBS Lett 1998; 429:129-33.
40. Salvati L, Mattu M, Polticelli F et al. Modulation of the catalytic activity of cruzipain, the major cysteine proteinase from Trypanosoma cruzi, by temperature and pH. Eur J of Biochem 2001; 268(11):3253-58.
41. Tomas AM, Kelly JM. Stage-regulated expression of cruzipain, the major cysteine protease of Trypanosoma cruzi is independent of the level of RNA. Eur J Biochem 1996; 76(1-2):91-103.
42. El Sayed NM, Myler PJ, Bartholomeu DC et al. The genome sequence of Trypanosoma cruzi, etiologic agent of Chagas disease. Science 2005; 309:409-15.
43. Lima APCA, Reis FCG, Serveau C et al. Cysteine protease isoforms from Trypanosoma cruzi, cruzipain 2 and cruzain, present different substrate preference and susceptibility to inhibitors. Mol Biochem Parasitol 2001; 114(1):41-52.
43b. dos Reis FC, Judice WA, Juliano MA et al. The substrate specificity of cruzipain 2, a cysteine protease isoform from Trypanosoma cruzi. FEMS Microbiol Lett 2006; 259(2):215-20.
44. Engel JC, Doyle PS, Hsieh I et al. Cysteine protease inhibitors cure an experimental Trypanosoma cruzi infection. J Exp Med 1998; 188(4):725-34.
45. Scharfstein J, Mendonça-Previato L, Borojevic R. Human antibody response to a surface glycoprotein isolated from Trypanosoma cruzi. Proc Pontificia Academia Scientiarvm 1982; 51:150-4.
46. Goncalves MF, Umezawa ES, Katzin AM et al. Trypanosoma cruzi: Shedding of surface antigens as membrane vesicles. Exp Parasitol 1991; 72(1):43-53.
47. Monteiro ACS, Abrahamson M, Lima APCA et al. Identification, characterization and localization of chagasin, a tight-binding cysteine proteinase inhibitor in Trypanosoma cruzi. J Cell Sci 2001; 114(21):3933-42.
48. Rigden DJ, Mosolov VV, Galperin MY. Sequence conservation in the chagasin family suggests a common trend in cysteine proteinase binding by unrelated protein inhibitors. Protein Sci 2002; 11:1971-77.

49. Sanderson SJ, Westrop GD, Scharfstein J et al. Functional conservation of a natural cysteine peptidase inhibitor in protozoan and bacterial pathogens. FEBS Lett 2003; 542:12-6.
50. Riekenberg S, Witjes B, Saric M et al. Identification of EhICP1, a chagasin-like cysteine protease inhibitor of Entamoeba histolytica. FEBS Lett 2005; 579:1573-8.
50b. Pandey KC, Singh N, Arastu-Kapur S et al. Falstatin, a cysteine protese inhibitor of Plasmodium falciparum, facilitates erythrocyte invasion. PLoS Pathog 2006; 2(11):e117.
51. Rigden DJ, Monteiro ACS, De Sá MFG. The protease inhibitor chagasin of Trypanosoma cruzi adopts an immunoglobulin-type fold and may have arisen by horizontal gene transfer. FEBS Lett 2001; 504:41-4.
52. Salmon D, Aido-Machado R, Diehl A et al. Solution structure and backbone dynamics of the Trypanosoma cruzi cysteine protease inhibitor chagasin. J Mol Biol 2006; 357:1511-1521.
52b. Figueiredo da Silva AA, Carvalho Vieira LD, Krieger MA et al. Crystal structure of chagasin, the endogenous c ysteine-protease inhibitor from Trypanosoma cruzi. J Struct Biol 2006, in press.
53. Smith BO, Picken NC, Westrop GD et al. The structure of Leishmania mexicana ICP provides evidence for convergent evolution of cysteine peptidase inhibitors. J Biol Chem 2006; 281:5821-5828.
54. Santos CC, Santanna C, Terres A et al. Chagasin, the endogenous cysteine-protease inhibitor of Trypanosoma cruzi, modulates parasite differentiation and invasion of mamalian cells. J Cell Sci 2005; 118:901-15.
55. Brooks DR, Tetley L, Coombs GH et al. Processing and trafficking of cysteine proteases in Leishmania mexicana. J Cell Sci 2000; 113:4035-41.
56. Del Nery E, Juliano MA, Lima AP et al. Kininogenase activity by the major cysteinyl proteinase (cruzipain) from Trypanosoma cruzi. J Biol Chem 1997; 272:25713-18.
57. Bhoola KD, Figueroa CD, Worthy K. Bioregulation of kinins: Kallikreins, kininogens, and kininases. Pharmacol Rev 1992; 44:1-80.
58. Kozik A, Moore RB, Potempa J et al. A novel mechanism for bradykinin production at inflammatory sites. Diverse effects of a mixture of neutrophil elastase and mast cell tryptase versus tissue and plasma kallikreins on native and oxidized kininogens. J Biol Chem 1998; 273:33224-33229.
59. Svensjö E, Batista PR, Brodskyn CI et al. Interplay between parasite cysteine proteases and the host kinin system modulates microvascular leakage and macrophage infection by promastigotes of the Leishmania donovani complex. Microbes Infect 2006; 8(1):206-20.
60. Farmer SG, Burch RM. Biochemical and molecular pharmacology of kinin receptors. Annu Rev Pharmacol Toxicol 1992; 32:511-536.
61. Blaukat A. Structure and signalling pathways of kinin receptors. Andrologia 2003; 35:17-23.
62. Marceau F, Bachvarov DR. Kinin receptors. Clin Rev Allergy Immunol 1998; 6:385-401.
63. Blaukat A, Micke P, Kalatskaya I et al. Downregulation of bradykinin B2 receptor in human fibroblasts during prolonged agonist exposure. Am J Physiol Heart Circ Physiol 2003; 284:H1909-16.
64. Ricupero DA, Romero JR, Rishikof DC et al. Des-Arg(10)-kallidin engagement of the B1 receptor stimulates type I collagen synthesis via stabilization of connective tissue growth factor mRNA. J Biol Chem 2000; 275(17):12475-80.
65. Aliberti J, Viola JP, Vieira-de-Abreu A et al. Cutting edge: Bradykinin induces IL-12 production by dendritic cells: A danger signal that drives Th1 polarization. J Immunol 2003; 170:5349-53.
66. Xie P, Browning D, Hay N et al. Activation of NFk-B by bradykinin through a Gaq- and Gbg-dependent pathway that involves phosphoinositide 3-kinase and Akt. J Biol Chem 2000; 275:24907-14.
67. Stoka V, Nycander M, Lenarcic B et al. Inhibition of cruzipain, the major cysteine proteinase of the protozoan parasite, Trypanosoma cruzi, by proteinase inhibitors of the cystatin superfamily. FEBS Lett 1995; 370:101-4.
68. Kaufmann J, Haasemann M, Modrow S et al. Structural dissection of the multidomain kininogens. Fine mapping of the target epitopes of antibodies interfering with their functional properties. J Biol Chem 1993; 268:9079-91.
69. Herwald H, Hasan AAK, Godovac-Zimmermann J et al. Identification of an endothelial cell binding site on kininogens' domain. J Biol Chem 1995; 270:14634-42.
70. Renné T, Dedio J, David G et al. H-Kininogen utilizes heparan sulfate proteoglycans for accumulation on Endothelial Cells. J Biol Chem 2000; 275:33688.
71. Renne T, Muller-Esterl W. Cell surface-associated chondroitin sulfate proteoglycans bind contact phase factor H-kininogen. FEBS Lett 2001; 500(1-2):36-40.
72. Lima APCA, Almeida IL, Tersariol I et al. Heparan sulfate modulates kinin release by Trypanosoma cruzi through the activity of cruzipain. J Biol Chem 2002; 277:5875-81.

72b. Barboza M, Duschak VG, Fukuyama Y et al. Structural analysis of the N-glycans of the major cysteine proteinase of Trypanosoma cruzi. Identification of sulfated high-mannose type oligosaccharides. FEBS J 2005; 272(15):3803-15.
73. de Souza W. Structural organization of the cell surface of pathogenic protozoa. Micron 1995; 26:405-30.
74. Ming M, Ewen ME, Pereira ME. Trypanosome invasion of mammalian cells requires activation of the TGF beta signaling pathway. Cell 1995; 82(2):287-96.
75. Waghabi MC, Keramidas M, Feige JJ et al. Activation of transforming growth factor beta by Trypanosoma cruzi. Cell Microbiol 2005; 7(4):511-7.
75b. Monteiro ACS, Schmitz V, Svensjo E et al. Cooperative activation of TLR2 and Bradykinin B2 receptor is required for induction of type-1 immunity in mouse model of subcutaneous infection by Trypanosoma cruzi. J Immunol 2006; 177(9):6325-35.
76. Campos MA, Almeida IC, Takeuchi O et al. Activation of Toll-like receptor-2 by glycosylphosphatidylinositol anchors from a protozoan parasite. J Immunol 2001; 167:416-423.
76b. Cuevas IC, Cazzulo JJ, Sanchez DO. gp63 homologues in Trypanosoma cruzi: surface antigens with metalloprotease activity and a possible role in host cell infection. Infect Immun 2003; 71(10):5739-49.
77. McLean P, Ahluvalia A, Perretti A. Association between kinin B1 receptor expression and leukocyte trafficking across mesenteric postcapillary venules. J Exp Med 2001; 192:367-80.
78. Petersen CA, Krumholz KA, Burleigh BA. Toll-like receptor 2 regulates interleukin-1beta-dependent cardiomyocyte hypertrophy triggered by Trypanosoma cruzi. Infect Immun 2005; 73(10):6974-80.

Chapter 12

Biogenesis of and Activities at the *Toxoplasma gondii* Parasitophorous Vacuole Membrane

Anthony P. Sinai*

Abstract

Apicomplexan parasites like *Toxoplasma gondii* are distinctive in their utilization of parasite encoded motor systems to invade cells. Invasion results in the establishment of the parasitophorous vacuole (PV) within the infected cell. Most apicomplexans complete their intracellular tenure within the infected cell in the PV that is demarcated from the host cytoplasm by the parasitophorous vacuole membrane (PVM). In this chapter I focus on the events surrounding the formation of the PVM and selected activities attributed to it. Its central role as the interface between the parasite and its immediate environment, the host cytoplasm, is validated by the diversity of functions attributed to it. While functions in structural organization, nutrient acquisitions and signaling have been defined their molecular bases remain largely unknown. Several recent studies and the decoding of the *Toxoplasma* genome have set the stage for a rapid expansion in our understanding of the role of the PVM in parasite biology.

Toxoplasma gondii, like all apicomplexan parasites are obligate intracellular pathogens. This family of parasites utilize their own actin-myosin based motor systems to gain entry into susceptible cells[1] establishing themselves, in some cases transiently (e.g., *Theileria spp*[2]) in specialized vacuolar compartment, the parasitophorous vacuole (PV). The *T. gondii* PV is highly dynamic compartment defining the replication permissive niche for the parasite.[3] The delimiting membrane defining the parasitophorous vacuole, the parasitophorous vacuole membrane or PVM is increasingly being recognized as a specialized "organelle" that in the context of the infected cell is extracorporeal to the parent organism, the parasite. A systematic study of this enigmatic organelle has been severely limited by several issues. Primary among these is the fact that it is formed only in the context of the infected cell thereby limiting the amount of material. Secondly, unlike other cellular organelles that can often be purified by conventional approaches, the PVM, cannot be purified away from host cell organelles[4] (see below). In spite of these significant obstacles considerable progress has been made in recent years toward understanding the biogenesis of the PVM, identification of its protein complement and the characterization of activities within it. These studies demonstrate that the PVM, on its own and by virtue of its interactions with cellular components, plays critical functions in the structural integrity of the vacuole, nutrient acquisition and the manipulation of cellular functions.[3,5] In addition it appears that the repertoire of activities at the PVM is likely to be plastic reflecting temporal changes associated with the replicative phase of parasite growth.[5] Finally, the PVM likely forms the foundation for the cyst wall as the parasite differentiates in the establishment of latent infection.[6] As the critical border crossing between the parasite and invaded cell the study of the

*Anthony P. Sinai—Department of Microbiology Immunology and Molecular Genetics. University of Kentucky College of Medicine, 800 Rose St., Lexington, Kentucky 40536, U.S.A. Email: sinai@uky.edu

Molecular Mechanisms of Parasite Invasion, edited by Barbara A. Burleigh and Dominique Soldati-Favre.

PVM provides a fertile area for new investigation aided by the recent decoding of the *Toxoplasma* genome (available at wwww.ToxoDB.org) and the application of proteomic analyses[7-9] to basic questions in parasite biology.

Biogenesis of the PVM

Unlike a typical phagosome membrane formed by conventional phagocytosis or macropinocytosis[10,11] the *T. gondii* PVM is formed during the active invasion of the host cell by the parasite.[1] *Toxoplasma* invasion is a rapid and temporally staged (Fig. 1). Productive attachment via the apical end to the plasma membrane results from the secretion of the micronemes.[12,13] In response to as yet undetermined signals the parasite corkscrews itself into the host cell advancing before it the host plasma membrane that forms the nascent PVM. This rapid (5-30 seconds) invasion is temporally accompanied by the discharge of the club shaped rhoptries that deliver a number of critical proteins in a cloud of lipid vesicles into the newly forming vacuole[14,15] (Fig. 1). The physical force drives the parasite completely within the confines of the host cell and the entry scar at the plasma membrane is sealed.

Several features of this invasion process directly impact the protein composition of the vacuolar membrane. The early freeze-etch microscopic studies of Porchet-Hennere demonstrated that the early vacuolar membrane was devoid of intramembrane particles suggesting the selective exclusion of membrane proteins in the course of invasion.[16,17] A significant and important refinement to the model came from the studies of Mordue et al who demonstrated that the exclusion of membrane proteins at the time invasion was in fact selective, by excluding proteins containing Type I- single spanning trans membrane domains while permitting inclusion of lipid-anchored proteins[18] (Fig. 1). Signals for exclusion from the PVM were not encoded in the transmembrane domain itself but rather the cytoplasmic tail suggesting sorting occurs at the level of the host cytoskeleton.[18] A refinement of this original study revealed that the lipid environment did not contribute to the sorting of proteins into the nascent vacuole as both lipid raft associated and nonraft Type I transmembrane domain containing proteins were generally excluded from the vacuole[19] (Fig. 1). Interestingly, while proteins with single transmembrane domains were excluded, polytopic membrane proteins in the lipid raft fractions gained access to the PVM.[19] These studies demonstrate that the *T. gondii* PVM is not in fact devoid of host proteins as originally surmised[16,17] suggesting the vacuole is not entirely segregated from cellular functions. The function of stripping the nascent vacuole of excluded proteins is mediated by the "moving junction", a transient structure formed exclusively at the site of parasite invasion (Fig. 1).

First characterized morphologically in the context of *Plasmodium* and *Toxoplasma* entry the "moving junction" appeared as an area increased electron density at the interface between the parasite surface and the host plasma membrane (Fig. 1).[20,21] Additional studies revealed the moving junction to be critical in parasite invasion resulting in the significant deformation of the transiting organism,[21-24] exclusion of host markers[18,25] as well as antibodies opsonizing the parasite surface.[26] The first insights into the molecular composition of the moving junction have emerged recently in publications from the Boothroyd[27] and Dubremetz[28] laboratories. Over the last couple of years the availability of the *T. gondii* genome information and large scale organelle proteomic projects have resulted in the identification of a slew of novel activities in the micronemes[8] and rhoptries.[7] Immunofluorescence and immunoelectron microscopic analyses using antibodies against the rhoptry fraction revealed a degree of compartmentalization within the rhoptry itself and the identification of a subset of proteins restricted to the rhoptry necks (RON proteins).[7] These studies demonstrate that the parasite delivers a complex of RON proteins[27,28] into the host plasma membrane establishing the moving junction. The complexed RON proteins together with the microneme derived AMA1[29] released onto the parasite surface serve as the anchor points permitting the parasite to squeeze through the moving junction.[27,28] The secreted RON proteins however do not become a part of the new PVM but remain at the moving junction where they can be detected as "rings" at the constriction of invading parasites that demarcate the intracellular and extracellular portions of the parasite.[27] As the parasite

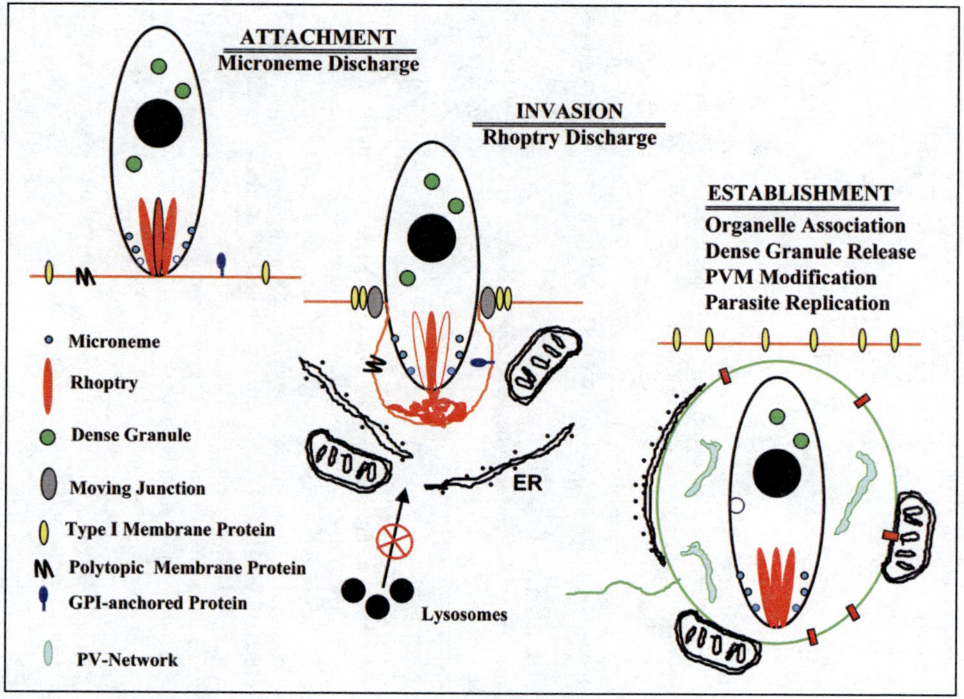

Figure 1. Schematic representation of *Toxoplasma gondii* invasion and the establishment of the parasitophorous vacuole. Internalization is initiated by productive attachment of the apical end of the parasite facilitated by microneme secretion. Invasion driven by parasite derived motor function occurs following the establishment of the "moving junction" in the host plasma membrane. The host plasma membrane forms the leading edge of the nascent vacuole with limited if any contribution from the discharge of rhoptries. The moving junction selectively excludes Type I membrane proteins from the nascent vacuole while permitting access to both polytopic membrane proteins and GPI-anchored proteins. The exclusion of Type I proteins removes signals for endocytic maturation preventing lysosome fusion and maturation. The vacuole is established by the delivery of *T. gondii* proteins from the rhoptries. This is followed by the constitutive secretion of proteins from dense granules. PVM-organelle association, interactions with the cytoskeleton, the delivery of mechanisms for nutrient scavenging and manipulation of host functions (not depicted) serve to establish the replication permissive niche.

invades, the host plasma membrane is forced through the moving junction to form the nascent PVM. Upon internalization of the parasite the PVM is sealed at the posterior end of the parasite and a transient scar is detected at the plasma membrane.[27]

Stripped of most host proteins, the lipid bilayer forming the nascent vacuole appears to be derived predominantly from the host plasma membrane.[25] This was elegantly demonstrated by the Ward laboratory using real-time electrophysiological analyses of invading parasites.[25] Their studies convincingly show that the parasite contributes little if any lipid to the new vacuole,[25] This despite the discharge of lipid by the rhoptries.[15,30,31] Of note, both raft and nonraft localizing lipid probes were incorporated in the nascent vacuole at equivalent rates.[19]

The newly formed PVM is rapidly modified by the temporally and spatially staged release of proteins from the parasite rhoptries and dense granules[24,32-34] (Fig. 1). The first proteins to appear in the PVM are derived from the rhoptries.[33,34] Among the proteins secreted are ROP1,[35] ROP2 (and ROP2 family members)[36] as well as quite likely a number of more recently identified rhoptry proteins.[7]

Within minutes of complete internalization, a large scale secretion event, at the posterior end of the parasite results in the release of a number of dense granule proteins.[37] Several of these GRA proteins including GRA3,5,7, 8,9 and 10 appear at the PVM[38-44] (reviewed in ref. 45). Unlike rhoptries, where the prevailing wisdom holds that discharge is limited to parasite invasion, dense granules form the basis of the constitutive and default secretory pathway.[45,46] What is not known is whether the relative concentrations of GRA proteins loaded into the secretory granules varies at different phases in the parasites intracellular residence. One might expect such changes to occur at the time of the developmental switch from a tachyzoite to bradyzoite[6,47] when components required for the formation of the cell wall[48] would be required.

The Physical Organization of the PV and PVM

The PVM is a dynamic entity. To better understand the organization of the PVM it is important to consider the structural basis of the PV. Electron microscopic studies indicate that the vacuolar space contains a proliferation of membrane tubules referred to the as the PV network (reviewed in ref. 45) (Fig. 1). The network is structurally[49,50] and antigenically[45,50-54] distinct from the PVM but does exhibit some degree of connectivity to it. The fact that dense granule proteins exhibit differential localization patterns between the network and PVM while rhoptry proteins appear to localize to the PVM alone suggests a degree of organization.[45,50-54] Recent studies have examined the organization of the *T. gondii* PV using scanning and transmission electron microscopy of thick preparations. The application of new sample preparation techniques coupled with high resolution electron microscopy reveal a remarkably detailed picture of vacuolar organization.[49,55] In these stunning images the parasites appear to be decorated with the network structure which form a mesh like pattern with connections to the PVM.[49,55,56] This organization has been suggested to provide a degree of mechanical support from within the lumen of the vacuole.[49] These structures may in fact be elements of the host microtubule cytoskeleton that have recently been shown to project into the vacuole.[57]

On the outer leaflet of the PVM are associated organelles[4,24,55,58-60] and elements of the cytoskeleton.[61-65] The growth of the vacuole and its resident parasites is under a high degree of regulation that must be responsive to signals from within the lumen of the vacuole and the host cytoplasm. One generally observes the PV packed with parasites and an absence of spacious vacuole suggesting vacuolar expansion and parasite replication are intimately linked. The changes implicit in regulating the growth of the vacuole are dynamic as morphometric analyses demonstrate that a typical vacuole increases its surface area four fold and its volume eight fold during a 24 hour period post infection.[49]

The PVM itself is not a uniform membrane. This heterogeneity is evident on a gross level by the differences in the distribution and extent of PVM-organelle association.[4,58] In addition there may be functional subdomains within the vacuolar membrane as well. Another "modification" of the PVM are membranous extensions emanating from many but not all vacuoles

(Fig. 1). These PVM-extensions were first described for a number of dense granules proteins including GRA3[38] (Fig. 2D), GRA5,[41] GRA7[57] and 8.[43] How the level of organization in the PVM impacts function remains an open question.

Activities Associated with the Early PVM

Organelle Association

Morphologically, a feature defining the *T. gondii* PVM is the intimate association of host mitochondria and the endoplasmic reticulum[4,24,55,58-60] (Fig. 1). A key element of this interaction is the tight apposition of the host organelle membranes with the PVM but the absence of membrane fusion. The extent of the PVM-organelle association, and the relative distribution has been quantified morphometrically and appear to vary, particularly for the extent of PVM-mitochondrial association depending on the cell type and approach to quantification.[4,58] This variation very likely reflects the relative concentration of these organelles in the cell, particularly in the juxtanuclear area where the vacuoles tend to reside.[4] All indications point to the PVM-organelle association being a highly stable and in effect irreversible interaction.[4] This property prevents one from purifying the PVM for biochemical analysis as even in "PVM-enriched" fractions the PVM itself is akin to a contaminating fraction of a mitochondrial prep.

The kinetics of PVM-organelle association are rapid and detectable within a minute of parasite entry.[66] This properties correlate with those for the discharge of rhoptries at the time of parasite invasion.[32,33] The topology of ROP2 in the PVM was elucidated in the elegant experiments of Beckers and colleagues as having the N-terminus exposed to the host cytoplasm.[36] What drew attention to ROP2 was that the examination of the N-terminus of mature ROP2 (ROP2 is processed in the secretory pathway en route to the rhoptry such that the N-terminus is at or close to aa98[67]), were the features encoded in aa98-127.[66] These features include the presence of a positively charged amphipathic helix, a high incidence of Ser and Thr and a poor representation of prolines and bulky amino acids.[36,66] Together these features are remarkably reminiscent of a mitochondrial import signal[68,69] suggesting the subversion of the mitochondrial import as a potential mechanism for the recruitment of mitochondria to the PVM.

The capacity of the host cytoplasm exposed domain of ROP2 (ROP2hc) to be partially translocated across the mitochondrial outer membrane in a manner dependent on the putative N-terminal signal sequence (aa98-127) coupled with the targeting of ROP2hc to mitochondria following expression in mammalian cells strongly argue for ROP2 as the mediator of PVM-mitochondrial association.[66] Interestingly, fusion of the 30aa N-terminal signal to GFP directs the chimeric protein to the mitochondria following the transient transfection of host cells.[66] In addition, deletion of aa98-127 results in redirection of ROP2hc to the endoplasmic reticulum[66] although its significance is unclear. Furthermore ROP2hc interacts with organelles with very high affinity essentially mimicking what is observed for the PVM in infected cells. The role of ROP2 in PVM-mitochondrial association is further strengthened by the marked reduction in the extent of mitochondrial association in parasites expressing a ROP2 antisense construct.[70]

The mechanism for PVM-ER association remains elusive. While deletion of aa98-127 from ROP2hc results in localization and high affinity binding to the ER,[66] the question of whether this represents an actual interaction in infected cells remains an open question. The experiments of Nakaar et al,[70] showed no reduction in the extent of ER-recruitment following the suppression of ROP2 expression using the antisense approach.

Resistance to Lysosome Fusion

The seminal studies of Jones and Hirsch[71,72] were the earliest to demonstrate that the *T. gondii* vacuole failed to fuse with lysosomes (Fig. 1). This observation confirmed and refined in multiple subsequent studies indicate that the *T. gondii* PVM is nonfusogenic with regard to components of the endocytic and exocytoic pathways of membrane traffic[73-76] and fails to acidify.[75,77] The ability to prevent lysosomal fusion is critically dependent on active parasite invasion as internalization by phagocytosis results in rapid lysosome fusion and the killing of the parasite.[74] The property of resistance to lysosome fusion is established at the time of invasion by the exclusion of signals for entry into the endocytic cascade by the moving junction.[3,18]

The capacity to establish and maintain the nonfusogenic state appears to be intrinsic to the PVM as demonstrated by the elegant experiments of Hakansson and colleagues.[78] In this study the authors used the microfilament depolymerizing agent cytochalasin D to block the entry of partially invaded parasites. While the parasites were frozen at the host plasma membrane in the process of invasion, rhoptry discharge was not affected. PVM vesicles defined by the presence of rhoptry proteins and termed E-vacuoles (empty vacuoles) were found to have retained their nonfusogenic properties indicating that the PVM and PVM alone is responsible for this property.[78]

Reorganization of the Host Cytoskeleton

Several studies have examined the effect on *T. gondii* development on the host cytoskeleton. Studies have focused on microtubules[61-63,65] and intermediate filaments,[64,65] both of which appear to form a scaffold or overcoating of the vacuole. The distortion of the host cytoskeleton in infected cells is actually dynamic as evidenced from recent studies from the Coppens laboratory.[57,61]

The extent of manipulation of the microtubule cytoskeleton by *T. gondii* was recently revealed in a groundbreaking study from the Coppens lab. As noted above, the fact *T. gondii* PVM resists lysosomal fusion was taken as evidence for the complete lack of interaction with this organelle system (reviewed in ref. 3). This is no longer believed to be true. In studies examining cholesterol utilization by *T. gondii* it was noted that the bulk of parasite cholesterol comes from exogenous sources along the lysosomal LDL pathway.[61,79] Perturbation of this pathway resulted in the sequestration of cholesterol and the inhibition of parasite growth.[79] In this recent study Coppens et al executed painstakingly detailed electron microscopic analyses to reveal that intact lysosomes are delivered to the lumen of the vacuole by microtubule supported invaginations of the PVM.[57] Thus while the lysosomes themselves do not fuse with the PVM (see above) their contents including cholesterol and hydrolytic enzymes are safely delivered to the parasite.[57] The presence of secreted proteinase inhibitors in the lumen of the vacuole,[80] and the pore activity of the PVM[81] (see below) that would prevent acidified environment likely contributes to the neutralization of the lysosomes degradative power.

In addition to the microtubules involved in the delivery of lysosomes a far more profound reorganization of the microtubule cytoskeleton is observed. The microtubules overcoating the PVM are short and disordered.[57] Remarkably, the microtubule organizing center (MTOC) relocated to a location adjacent to the vacuole, potentially drawn in by the thread like PVM-extension that appear to extend along microtubules.[57]

Nutrient Acquisition

The central tenet of parasitism is the acquisition of nutrients from the host. The *T. gondii* PVM appears to be adapted to this scavenging function allowing for both building blocks of intermediary metabolism and lipids transit. The *T. gondii* PVM is porous by virtue of a nonspecific pore activity allowing free bidirectional access to compounds under 1300 daltons.[81] While described over a decade ago, the molecular basis of this activity remains elusive. With regard to lipidic nutrients, PVM-associated organelles are the likely source at the site of membrane-membrane contact[3,4] as has been described in diverse systems for inter-organellar lipid transfer.[82,83] Sterols on the other hand are actively scavenged from the lysosomal pathway as described above.[61,79] Additionally, lipids are also scavenged from lipid storage droplets in the host cell[57,84] presumably using the cytoskeleton as conduits for delivery to the PVM.

Manipulation of Signaling

The *Toxoplasma* infected cell is fundamentally remodeled at the level of transcription[85] and by extension the proteome.[86] Many of these changes are due to the parasite-mediated subversion of critical cellular transcriptional machineries. Work in our laboratory[87,88] and that of others[85,89-91] has found that the activation of the host transcription factor NFkappaB accompanies infection in several cell types. The activation of NFκB is mediated by the phosphorylation of its inhibitor I kappaBα at 2 critical serine residues (reviewed in ref. 92). Remarkably, in infected cells, this event happens at the PVM and has been linked to a parasite encoded I kappa B kinase activity (TgIKK).[88,93] The PVM therefore functions as a signaling platform from which diverse cellular functions may be manipulated.

Identification of Novel Activities at the PVM

Over the last few of years there has been a burst of interest in the study of the PVM addressing its structural, nutrient scavenging and signaling interactions with the host cell. As noted throughout the chapter many questions remain to be answered providing a fertile area of investigation.

The recent completion of the *Toxoplasma* genome (http://www.toxodb.org/) as well as proteomic studies defining the secretory organelles[7,8] have and will continue to advance our understanding of PVM. We have recently undertaken a proteomic analysis of the PVM using a set of unique reagents. By purifying PVM-containing host organelle fractions in rabbit fibroblasts as immunogens we have successfully generated a pair of high titre multivalent polyclonal antisera against the *T. gondii* PVM with minimal host reactivity (Martin, Fentress and Sinai, manuscript in preparation). These reagents are being used to identify PVM proteins by both MALDI-TOF and multi-dimensional protein identification technology (MuDPIT).[94]

The recent application of new techniques and resources the study of the PVM will undoubtedly chip away at the difficulties in the study of this unique organelle. Recent and ongoing studies indicate we are in fact on the threshold of a new era of investigation of the PVM which will provide new molecular insights at the pathogen-host interaction.

References

1. Dobrowolski JM, Sibley LD. Toxoplasma invasion of mammalian cells is powered by the actin cytoskeleton. Cell 1996; 84:933-939.
2. Shaw MK. The same but different: The biology of Theileria sporozoite entry into bovine cells. Int J Parasitol 1997; 27(5):457-474.
3. Sinai AP, Joiner KA. Safe haven: The cell biology of nonfusogenic pathogen vacuoles. Annu Rev Microbiol 1997; 51:415-462.
4. Sinai AP, Webster P, Joiner KA. Association of host cell endoplasmic reticulum and mitochondria with the Toxoplasma gondii parasitophorous vacuole membrane: A high affinity interaction. J Cell Sci 1997; 110:2117-2128.
5. Lingelbach K, Joiner K. The parasitophorous vacuole membrane surrounding Plasmodium and Toxoplasma: An unusual compartment in infected cells. J Cell Sci 1998; 111:1467-1475.
6. Weiss LM, Kim K. The development and biology of bradyzoites of Toxoplasma gondii. Front Biosci 2000; 5:D391-405.
7. Bradley PJ, Ward C, Cheng SJ et al. Proteomic analysis of rhoptry organelles reveals many novel constituents for host-parasite interactions in Toxoplasma gondii. J Biol Chem 2005; 280(40):34245-34258.
8. Zhou XW, Kafsack BF, Cole RN et al. The opportunistic pathogen Toxoplasma gondii deploys a diverse legion of invasion and survival proteins. J Biol Chem 2005; 280(40):34233-34244.
9. Hu K, Johnson J, Florens L et al. Cytoskeletal components of an invasion machine-the apical complex of Toxoplasma gondii. PLoS Pathog 2006; 2(2):e13.
10. Jutras I, Desjardins M. Phagocytosis: At the crossroads of innate and adaptive immunity. Annu Rev Cell Dev Biol 2005; 21:511-527.
11. Desjardins M, Huber L, Parton R et al. Biogenesis of phagolysosomes proceeds through a sequential series of interactions with the endocytic apparatus. J Cell Biol 1994; 124:677-688.
12. Carruthers VB. Host cell invasion by the opportunistic pathogen Toxoplasma gondii. Acta Trop 2002; 81(2):111-122.

13. Carruthers VB, Giddings OK, Sibley LD. Secretion of micronemal proteins is associated with toxoplasma invasion of host cells. Cell Microbiol 1999; 1(3):225-235.
14. Nichols BA, Chiappino ML, O'Connor GR. Secretion from the rhoptries of Toxoplasma gondii during host-cell invasion. J Ultrastruct Res 1983; 83:85-98.
15. Coppens I, Joiner KA. Host but not parasite cholesterol controls Toxoplasma cell entry by modulating organelle discharge. Mol Biol Cell 2003; 14(9):3804-3820.
16. Porchet-Hennere E, Vivier E, Torpier G. Origine des membranes de la paroi chez Toxoplasma. Annales de parasitologie humaine et compar 1985; 60:101-110.
17. Porchet-Hennere EGT. Relations entre Toxoplasma et sa cellule-hote. Protistologica 1983; 19:357-370.
18. Mordue DG, Desai N, Dustin M et al. Invasion by Toxoplasma gondii establishes a moving junction that selectively excludes host cell plasma membrane proteins on the basis of their membrane anchoring. J Exp Med 1999; 190(12):1783-1792.
19. Charron AJ, Sibley LD. Molecular partitioning during host cell penetration by Toxoplasma gondii. Traffic 2004; 5(11):855-867.
20. Aikawa M, Miller LH, Johnson J et al. Erythrocyte entry by malaria parasites: A moving junction between erythrocyte and parasite. J Cell Biol 1978; 77:77-82.
21. Michel R, Schupp S, Raether W et al. Formation of a close junction during invasion of eythrocytes by Toxoplasma gondii in vitro. Int J Parasitol 1980; 10:309-313.
22. Aikawa M, Komata Y, Asai T et al. Transmission and scanning electron microscopy of host cell entry by Toxoplasma. Am J Pathol 1977; 87:285-296.
23. Nichols BA, O'Connor GR. Penetration of mouse peritoneal macrophages by the protozoon Toxoplasma gondii. Lab Invest 1981; 44:324-334.
24. de Souza W. Microscopy and cytochemistry of the biogenesis of the parasitophorous vacuole. Histochem Cell Biol 2005; 123(1):1-18.
25. Suss-Toby E, Zimmerberg J, Ward GE. Toxoplasma invasion: The parasitophorous vacuole is formed from host cell plasma membrane and pinches off via a fission pore. Proc Natl Acad Sci 1996; 93:8413-8418.
26. Dubremetz JF, CR, Ferreira E. Toxoplasma gondii: Redistribution of monoclonal antibodies on tachyzoites during host cell invasion. Exp Parasitol 1985; 59:24-32.
27. Alexander DL, Mital J, Ward GE et al. Identification of the moving junction complex of Toxoplasma gondii: A collaboration between distinct secretory organelles. PLoS Pathog 2005; 1(2):e17.
28. Lebrun M, Michelin A, El Hajj H et al. The rhoptry neck protein RON4 relocalizes at the moving junction during Toxoplasma gondii invasion. Cell Microbiol 2005; 7(12):1823-1833.
29. Mital J, Meissner M, Soldati D et al. Conditional expression of Toxoplasma gondii apical membrane antigen-1 (TgAMA1) demonstrates that TgAMA1 plays a critical role in host cell invasion. Mol Biol Cell 2005; 16(9):4341-4349.
30. Coppens I, Vielemeyer O. Insights into unique physiological features of neutral lipids in Apicomplexa: From storage to potential mediation in parasite metabolic activities. Int J Parasitol 2005; 35(6):597-615.
31. Foussard F, Leriche MA, Dubremetz JF. Characterization of the lipid content of Toxoplasma gondii rhoptries. Parasitology 1991; 102:367-370.
32. Carruthers VB. Armed and dangerous: Toxoplasma gondii uses an arsenal of secretory proteins to infect host cells. Parasitol Int 1999; 48(1):1-10.
33. Carruthers V, Sibley L. Sequential protein secretion from three distinct organelles of Toxoplasma gondii accompanies invasion of human fibroblasts. Eur J Cell Biol 1997; 73:114-123.
34. Dubremetz JF, Achbarou A, Bermudes D et al. Kinetics of apical organelle exocytosis during Toxoplasma gondii host cell interaction. Parasitol Res 1993; 79:402-408.
35. Ossorio PN, Schwartzman JD, Boothroyd JC. A Toxoplasma gondii rhoptry protein associated with host cell penetration has unusual charge assymetry. Mol Biochem Parasitol 1992; 50:1-16.
36. Beckers CJM, Dubremetz JF, Mercereau-Puijalon O et al. The Toxoplasma gondii rhoptry protein ROP2 is inserted into the parasitophorous vacuole membrane, surrounding the intracellular parasite, and is exposed to the host cell cytoplasm. J Cell Biol 1994; 127:947-961.
37. Sibley LD, Krahenbuhl JL, Adams GMW et al. Toxoplasma modifies macrophage phagosomes by secretion of a vesicular network rich in surface proteins. J Cell Biol 1986; 103:867-874.
38. Bermudes D, Dubremetz JF, Joiner KA. Molecular characterization of the dense granule protein GRA3 from Toxoplasma gondii. Mol Biochem Parasit 1994; 68:247-257.
39. Henriquez FL, Nickdel MB, McLeod R et al. Toxoplasma gondii dense granule protein 3 (GRA3) is a type I transmembrane protein that possesses a cytoplasmic dilysine (KKXX) endoplasmic reticulum (ER) retrieval motif. Parasitology 2005; 131(Pt 2):169-179.
40. Ahn HJ, Kim S, Nam HW. Host cell binding of GRA10, a novel, constitutively secreted dense granular protein from Toxoplasma gondii. Biochem Biophys Res Commun 2005; 331(2):614-620.
41. Lecordier L, Mercier C, Sibley LD et al. Transmembrane insertion of the Toxoplasma gondii GRA5 protein occurs after soluble secretion into the host cell. Mol Biol Cell 1999; 10(4):1277-1287.

42. Jacobs D, Dubremetz JF, Loyens A et al. Identification and heterologous expression of a new dense granule protein (GRA7) from Toxoplasma gondii. Molec Biochem Parasitol 1998; 91:237-249.
43. Carey KL, Donahue CG, Ward GE. Identification and molecular characterization of GRA8, a novel, proline-rich, dense granule protein of Toxoplasma gondii. Mol Biochem Parasitol 2000; 105(1):25-37.
44. Adjogble KD, Mercier C, Dubremetz JF et al. GRA9, a new Toxoplasma gondii dense granule protein associated with the intravacuolar network of tubular membranes. Int J Parasitol 2004; 34(11):1255-1264.
45. Mercier C, Adjogble KD, Daubener W et al. Dense granules: Are they key organelles to help understand the parasitophorous vacuole of all apicomplexa parasites? Int J Parasitol 2005; 35(8):829-849.
46. Karsten V, Qi H, Beckers CJM et al. The protozoan parasite Toxoplasma gondii targets proteins to dense granules and vacuolar space using both conserved and unusual mechanisms. J Cell Biol 1998; 141:911-914.
47. Tomavo S, Fortier B, Soete M et al. Characterization of bradyzoite-specific antigens of Toxoplasma gondii. Infect Immun 1991; 59:3750-3753.
48. Zhang YW, Halonen SK, Ma YF et al. Initial characterization of CST1, a Toxoplasma gondii cyst wall glycoprotein. Infect Immun 2001; 69(1):501-507.
49. Magno RC, Lemgruber L, Vommaro RC et al. Intravacuolar network may act as a mechanical support for Toxoplasma gondii inside the parasitophorous vacuole. Microsc Res Tech 2005; 67(1):45-52.
50. Sibley LD, Niesman IR, Parmley SF et al. Regulated secretion of multi-lamellar vesicles leads to formation of a tubulo-vesicular network in host-cell vacuoles occupied by Toxoplasma gondii. J Cell Sci 1995; 108:1669-1677.
51. Labruyere E, Lingnau M, Mercier C et al. Differential membrane targeting of the secretory proteins GRA4 and GRA6 within the parasitophorous vacuole formed by Toxoplasma gondii. Mol Biochem Parasitol 1999; 102(2):311-324.
52. Mercier C, Cesbron-Delauw MF, Sibley LD. The amphipathic alpha helices of the toxoplasma protein GRA2 mediate post-secretory membrane association. J Cell Sci 1998; 111(Pt 15):2171-2180.
53. Lecordier L, Meleon-Borodowski I, Dubremetz JF et al. Characterization of a dense granule antigen of Toxoplasma gondii (GRA6) associated to the network of the parasitophorous vacuole. Mol Biochem Parasitol 1995; 70:85-94.
54. Mercier C, Lecordier L, Darcy F et al. Molecular characterization of a dense granule antigen (GRA2) associated with the network of the parasitophorous vacuole in Toxoplasma gondii. Mol Biochem Parasitol 1993; 58:71-82.
55. Schatten H, Ris H. Three-dimensional imaging of Toxoplasma gondii-host cell interactions within the parasitophorous vacuole. Microsc Microanal 2004; 10(5):580-585.
56. Schatten H, Ris H. Unconventional specimen preparation techniques using high resolution low voltage field emission scanning electron microscopy to study cell motility, host cell invasion, and internal cell structures in Toxoplasma gondii. Microsc Microanal 2002; 8(2):94-103.
57. Coppens I, Dunn JD, Romano JD et al. Toxoplasma gondii sequesters lysosomes from mammalian hosts in the vacuolar space. Cell 2006; 125(2):261-274.
58. Magno RC, Straker LC, de Souza W et al. Interrelations between the parasitophorous vacuole of Toxoplasma gondii and host cell organelles. Microsc Microanal 2005; 11(2):166-174.
59. Melo EJ, de Souza W. Relationship between the host cell endoplasmic reticulum and the parasitophorous vacuole containing Toxoplasma gondii. Cell Struct Funct 1997; 22(3):317-323.
60. de Melo EJ, de Carvalho TU, de Souza W. Penetration of Toxoplasma gondii into host cells induces changes in the distribution of the mitochondria and the endoplasmic reticulum. Cell Struct Funct 1992; 17(5):311-317.
61. Sehgal A, Bettiol S, Pypaert M et al. Peculiarities of host cholesterol transport to the unique intracellular vacuole containing toxoplasma. Traffic 2005; 6(12):1125-1141.
62. Andrade EF, Stumbo AC, Monteiro-Leal LH et al. Do microtubules around the Toxoplasma gondii-containing parasitophorous vacuole in skeletal muscle cells form a barrier for the phagolysosomal fusion? J Submicrosc Cytol Pathol 2001; 33(3):337-341.
63. Melo EJ, Carvalho TM, De Souza W. Behaviour of microtubules in cells infected with Toxoplasma gondii. Biocell 2001; 25(1):53-59.
64. Halonen SK, Weidner E. Overcoating of Toxoplasma parasitophorous vacuoles with host cell vimentin type intermediate filaments. J Euk Microbiol 1994; 41:65-71.
65. Cintra WM, De Souza W. Immunocytochemical localization of cytoskeletal proteins and electron microscopy of detergent extracted tachyzoites of Toxoplasma gondii. J Submicrosc Cytol 1985; 17(4):503-508.
66. Sinai AP, Joiner KA. The Toxoplasma gondii protein ROP2 mediates host organelle association with the parasitophorous vacuole membrane. J Cell Biol 2001; 154(1):95-108.
67. Sadak A, Taghy Z, Fortier B et al. Characterization of a family of rhoptry proteins of Toxoplasma gondii. Mol Biochem Parasitol 1988; 29:203-211.

68. Horwich AL, Kalouse F, Mellmann I et al. A leader peptide is sufficient to direct mitochondrial import of a chimeric protein. EMBO J 1985; 4:1129-1135.
69. von Heijne G. Protein targeting signals. Curr Opin Cell Biol 1990; 2:604-608.
70. Nakaar V, Ngo HM, Aaronson EP et al. Pleiotropic effect due to targeted depletion of secretory rhoptry protein ROP2 in Toxoplasma gondii. J Cell Sci 2003; 116(Pt 11):2311-2320.
71. Jones TC, Veh S, Hirsch JG. The interaction between Toxoplasma gondii and mammalian cells. I. Mechanism of entry and intracelluar fate of the parasite. J Exp Med 1972; 136:1157-1172.
72. Jones TC, Hirsch JG. The interaction between Toxoplasma gondii and mammalian cells. II. The absence of lysosomal fusion with phagocytic vacuoles containing living parasites. J Exp Med 1972; 136:1173.
73. De Carvalho L, deSouza W. Cytochemical localization of plasma membrane enzyme markers during interiorization of tachyzoites of Toxoplasma gondii by macrophages. J Protozol 1989; 36:164-170.
74. Joiner KA, Fuhrman SA, Mietinnen H et al. Toxoplasma gondii: Fusion competence of parasitophorous vacuoles in Fc receptor transfected fibroblasts. Science 1990; 249:641-646.
75. Mordue D, Sibley L. Intracellular fate of vacuoles containing Toxoplasma gondii is determined at the time of formation and depends on the mechanism of entry. J immunol 1997; 159:4452-4459.
76. DeMelo EJT, Souza WD. Pathway for C6-NBD-ceramide on the host cell infected with Toxoplasma gondii. Cell Struct Funct 1996; 21:47-52.
77. Sibley LD, Weidner E, Krahenbuhl JL. Phagosome acidification blocked by intracellular Toxoplasma gondii. Nature 1985; 315:416-419.
78. Hakansson S, Charron AJ, Sibley LD. Toxoplasma evacuoles: A two-step process of secretion and fusion forms the parasitophorous vacuole. EMBO J 2001; 20(12):3132-3144.
79. Coppens I, Sinai AP, Joiner KA. Toxoplasma gondii exploits host low-density lipoprotein receptor-mediated endocytosis for cholesterol acquisition. J Cell Biol 2000; 149(1):167-180.
80. Morris MT, Coppin A, Tomavo S et al. Functional analysis of Toxoplasma gondii protease inhibitor 1. J Biol Chem 2002; 277(47):45259-45266.
81. Schwab JC, Beckers CJM, Joiner KA. The parasitophorous vacuole membrane surrounding intracellular Toxoplasma gondii functions as a molecular sieve. Proc Natl Acad Sci 1994; 91:509-513.
82. Vance JE, Shiao YI. Intracellular trafficking of phospholipids: Import of phosphatidylserine into mitochondria. Anticancer Res 1996; 16:1333-1340.
83. Vance JE, Vance DE. Phospholipid biosynthesis in mammalian cells. Biochem Cell Biol 2004; 82(1):113-128.
84. Charron AJ, Sibley LD. Host cells: Mobilizable lipid resources for the intracellular parasite Toxoplasma gondii. J Cell Sci 2002; 115(Pt 15):3049-3059.
85. Blader IJ, Manger ID, Boothroyd JC. Microarray analysis reveals previously unknown changes in Toxoplasma gondii-infected human cells. J Biol Chem 2001; 276(26):24223-24231.
86. Wastling JM, Burchmore R, Nelson M. Manipualtion of the host cell proteome by Toxoplasma gondii. Porticcio, Corsica, France: Paper presented at: 8th international Congress on Toxoplasmosis, 2005.
87. Molestina RE, Payne TM, Coppens I et al. Activation of NF-{kappa}B by Toxoplasma gondii correlates with increased expression of antiapoptotic genes and localization of phosphorylated I{kappa}B to the parasitophorous vacuole membrane. J Cell Sci 2003 2003; 116:4359-4371.
88. Molestina RE, Sinai AP. Host and parasite-derived IKK activities direct distinct temporal phases of NF-{kappa}B activation and target gene expression following Toxoplasma gondii infection. J Cell Sci 2005; 118(Pt 24):5785-5796.
89. Brenier-Pinchart MP, Pelloux H, Simon J et al. Toxoplasma gondii induces the secretion of monocyte chemotactic protein-1 in human fibroblasts, in vitro. Mol Cell Biochem 2000; 209(1-2):79-87.
90. Denney CF, Eckmann L, Reed SL. Chemokine secretion of human cells in response to Toxoplasma gondii infection. Infect Immun 1999; 67(4):1547-1552.
91. Kim JM, Oh YK, Kim YJ et al. Nuclear factor-kappa B plays a major role in the regulation of chemokine expression of HeLa cells in response to Toxoplasma gondii infection. Parasitol Res 2001; 87(9):758-763.
92. Ghosh S, Karin M. Missing pieces in the NF-kappaB puzzle. Cell 2002; 109(Suppl):S81-96.
93. Molestina RE, Sinai AP. Detection of a novel parasite kinase activity at the Toxoplasma gondii parasitophorous vacuole membrane capable of phosphorylating host IkBa. Cell Microbiol 2005; 7(3):351-362.
94. Aebersold R, Mann M. Mass spectrometry-based proteomics. Nature 2003; 422(6928):198-207.

Chapter 13

The Role of Host Cell Lysosomes in *Trypanosoma cruzi* Invasion

G. Adam Mott and Barbara A. Burleigh*

Abstract

The cell-invasive, trypomastigote form of *Trypanosoma cruzi* exhibits a unique relationship with lysosomes in target host cells. In contrast to many intracellular pathogens that are adept at avoiding contact with lysosomes, *T. cruzi* requires transient residence within this acidic organelle for productive infection. The low pH environment of lysosomes facilitates parasite egress from the vacuole and delivery into the host cytosol, a critical step in the *T. cruzi* developmental program. Recent studies also suggest that early lysosome fusion with invading or recently internalized parasites is critical for cellular retention of parasites. To ensure targeting to host cell lysosomes, *T. cruzi* trypomastigotes exploit two distinct modes of invasion that rapidly converge in the cell. In this chapter, we summarize the recent progress and changing views regarding the role of host cell lysosomes in the *T. cruzi* infection process where our discussion is limited to invasion of nonprofessional phagocytic cells.

Introduction

The kinetoplastid protozoan parasite, *Trypanosoma cruzi*, is the causative agent of human Chagas' disease and is endemic to regions of South and Central America. *T. cruzi* exhibits a complex life cycle that involves both mammalian hosts as well as invertebrate vectors for transmission. In the vertebrate host, *T. cruzi* is an obligate intracellular parasite that replicates in the cytoplasm of parasitized cells. It is thought that the success of the parasite may depend on its ability to infect a wide variety of host cell types including professional and nonprofessional phagocytes, however, the main pathologies associated with chronic Chagas' disease arise following infection of cardiomyocytes and smooth muscle cells. Symptomatic illness in the chronic phase of the infection is characterized by extensive focal inflammation, fibrosis and hypertrophy particularly in the heart and/or digestive tract, resulting in cardiomyopathy and gastrointestinal disease that arise in one-third of infected individuals.[1] Clearly the ability of *T. cruzi* to invade and replicate within nonprofessional phagocytic cells is a key feature of the infection process and for disease progression in mammalian hosts. A thorough understanding of the molecular basis for *T. cruzi* invasion of nonprofessional phagocytes is a critical component of our ability to comprehend the infectious process as a whole.

*Corresponding Author: Barbara A. Burleigh—Department of Immunology and Infectious Diseases, Harvard School of Public Health, Boston, Massachusetts 02115, USA.
Email: bburleig@hsph.harvard.edu

Molecular Mechanisms of Parasite Invasion, edited by Barbara A. Burleigh and Dominique Soldati-Favre.

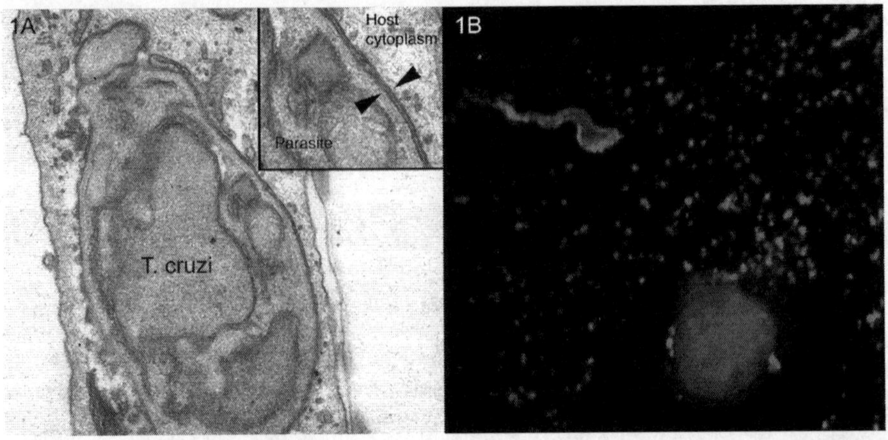

Figure 1. *T. cruzi* trypomastigotes reside within a LAMP-1 positive vacuole. Upon entry into host cells, the motile trypomastigote form of the parasite resides in a vacuole in which the parasite and vacuolar membranes are in tight apposition (1A; arrows highlight juxtaposition of parasite plasma membrane and parasitophorous vacuole (PV) membrane). These vacuoles rapidly acquire the lysosomal marker protein LAMP-1, visualized with a specific antibody shown in red (1B) at 30 minutes post-infection. Host cell and parasite DNA are stained with DAPI (blue). A color version of this figure is available online at www.eurekah.com.

General Features of *T. cruzi* Invasion of Nonphagocytic Cells

Mammalian host cells are initially penetrated by the nondividing, motile trypomastigote form of the parasite, which becomes immediately engulfed in a tight, membrane-delimited parasitophorous vacuole (PV) (Fig. 1A). As discussed below, the *T. cruzi* vacuole quickly acquires markers of host cell lysosomes. The best studied of these markers include the lysosome-associated membrane protein (LAMP)-1, a resident integral membrane protein (Fig. 1B), as well as endocytosed soluble markers that have been chased into lysosomes and are located in the luminal space of the parasite-containing vacuole. Following host cell entry, trypomastigotes reside in individual vacuoles (i.e., 1 parasite/vacuole) for several hours where they initiate a developmental program toward differentiation to amastigotes, the nonmotile, intracellular replicative form of *T. cruzi*. Parasites then escape the vacuole (beginning at ~8 hours post-infection) and are delivered to the host cytosol where amastigote replication begins at ~24 hours post-invasion. Acidification of the PV appears to be essential for parasite egress, as demonstrated by the inhibition of the process by lysomotropic agents or inhibitors of vacuole acidification.[2] While the precise mechanism used by the parasite to escape the PV has yet to be determined, it is thought to require the action of both a secreted hemolysin, referred to as TcTox,[3,4] and trans-sialidase (TS), a cell surface enzyme that catalyses the transfer of sialic acid residues from host glycoconjugates[5] including LAMP-1.[6] Parasites that invade cells lacking sialic acid residues, or those expressing higher levels of TS, are more competent for vacuole egress and for intracellular replication.[7,8] While not yet demonstrated experimentally, the current model proposes that via the concerted actions of TS, Tc-Tox and perhaps other parasite and/or host factors, the parasitophorous vacuole membrane is gradually remodeled, rendering it susceptible to lysis which is evident by ~8-10 hours post-infection.[9] The fate of the parasitophorous vacuole membrane following lysis is currently unknown; however it may fragment to form small vesicles that have been observed in the vicinity of parasites free in the cytoplasm.[2]

Pathways for Trypomastigote Invasion of Nonprofessional Phagocytes

It is clear that the interaction of internalized *T. cruzi* with the host cell lysosomal compartment is a critical event leading to cytosolic localization and survival of this pathogen within mammalian cells. The current state of research in this area indicates that two distinct pathways are used for trypomastigote entry and trafficking to lysosomes in nonprofessional phagocytic cells. These models are detailed below.

Lysosome-Dependent Entry

Given that *T. cruzi* trypomastigotes are relatively large organisms (15-20 μm in length) it was originally predicted that host cell entry by this pathogen would require the driving force of host cell actin polymerization, which is used by both phagocytic and nonphagocytic cells to engulf pathogens (reviewed in ref. 10). In contrast to this hypothesis, infection of several different cell types with *T. cruzi* trypomastigotes was shown to proceed in the absence of actin polymerization where invasion was significantly enhanced in some cases following treatment with cytochalasin D.[11,12] These observations were the first to indicate that the process of host cell entry by *T. cruzi* trypomastigotes was quite distinct from phagocytic or micropinocytic pathways exploited by many microbial pathogens.[11,12] This was soon confirmed, when it was noted that host cell lysosomes were rapidly recruited to the site of *T. cruzi* entry and that their fusion contributed to the formation of the nascent PV.[11,13] In addition, the relative proximity of lysosomes to the plasma membrane in host cells prior to *T. cruzi* invasion was shown to impact the efficiency of parasite entry. For example, pretreatment of fibroblasts with agents that cause anterograde movement of lysosomes (i.e., toward the plasma membrane) resulted in increased *T. cruzi* invasion, whereas retrograde movement of lysosomes in cells decreased the overall trypomastigote invasion capacity.[11] As might be expected from this finding, the targeting of lysosomes to the site of parasite entry depends on kinesin-based, anterograde movement along the microtubular network within the cell.[13] In addition, there is evidence that the site of interaction between the parasite and the host cell may act as a distinct site for microtubule reorganization perhaps facilitating, perhaps facilitating the targeting of lysosomes to the site.[14] Overall, these findings support the notion that the *T. cruzi* vacuole is generated at the time of parasite entry by the specific targeting and immediate fusion of host cell lysosomes with the host cell plasma membrane. This represents a highly unusual strategy for pathogen entry of mammalian cells, (reviewed in refs. 11,15,16), and suggests that signals triggered in host cells by extracellular or invading *T. cruzi* trypomastigotes are required to initiate the early events of actin rearrangements, lysosome recruitment and plasma membrane fusion.

Lysosome-Independent Entry

A key feature of lysosome-dependent model for *T. cruzi* entry is that host cell lysosomes provide the membrane needed to elaborate the parasitophorous vacuole and that lysosome fusion is required for parasite invasion. Earlier morphological studies of trypomastigote entry into HeLa cells, however, demonstrated cytochalasin D-resistant plasma membrane projections encircling membrane-associated parasites.[17] More recently, an alternate host cell entry pathway for *T. cruzi* trypomastigotes has been described that is initiated by actin-independent plasma membrane invagination.[18] In this study, the use of chimeric GFP markers which associate with phosphoinositol phosphates (PI(4,5)P_2 or PI(3,4,5)P_3) at the cytosolic face of the host cell plasma membrane, revealed that ~50% of invading trypomastigotes associated with the plasma membrane during, and immediately following, invasion (10 minutes post-infection).[18] Immunolabeling of fixed cells demonstrated that LAMP-1 or lysosomally-targeted TR-dextran did not colocalize with parasites that were positive for plasma membrane-GFP markers. In fact, only ~20% of recently internalized parasites were shown to be lysosome-associated at 10 minutes post-infection in several different cell types, and intracellular parasites were clearly observed in early endosomes.[18] These findings provided the first clear indication that *T. cruzi* trypomastigotes can invade nonprofessional phagocytic cells in a lysosome-independent fashion[18] in addition to the lysosome-dependent entry pathway (Fig. 3).[11] Other studies have shown that over-expression of dominant negative mutants of Rab5 or Rab7, GTPases involved in discrete stages of endocytosis and vesicular transport along the endocytic and phagosome maturation pathways, results in a marked reduction of *T. cruzi* invasion.[19] Given that lysosome biogenesis and maintenance is regulated by Rab7,[20] lysosome availability in these cells is expected to be dramatically reduced, which would compromise the lysosome-dependent entry pathway. Thus, it would be interesting to examine the relative effects of these dominant-interfering mutants on lysosome-dependent versus lysosome-independent *T. cruzi* invasion pathways and to determine whether Rab 5 and Rab7 act in similar or disparate pathways.

Calcium Signaling Is Required for *T. cruzi* Invasion and for Lysosome-Plasma Membrane Fusion

Given the parallels between targeted fusion of lysosomes with the plasma membrane during *T. cruzi* invasion and the more general mechanism of regulated exocytosis, investigators began to explore two questions: (1) Can lysosomes be induced to fuse with the plasma membrane in the absence of *T. cruzi*? and (2) Can *T. cruzi* activate appropriate signals in mammalian cells to promote lysosome-plasma membrane fusion? Rodriguez et al[21] were the first to demonstrate that lysosomes in fibroblasts and epithelial cells could be induced to fuse with the plasma membrane and release lumenal contents simply by elevating the intracellular calcium concentration ([Ca^{2+}]i). Strikingly, infective *T. cruzi* trypomastigotes can trigger rapid and transient increases in the intracellular free calcium concentration in mammalian cells.[22] While the ability of *T. cruzi* trypomastigotes to activate host cell [Ca^{2+}]i-signaling pathways appears to be multifactorial (refs. 22-25; discussed in Chapters 7 and 12), it is clearly required for efficient invasion.[22] The observation that increases in [Ca^{2+}]i triggered by a soluble *T. cruzi* activity could promote transient rearrangements in actin microfilaments[26] and trigger lysosome exocytosis[21] in cells, suggested a link between parasite-triggered Ca^{2+}-signaling and lysosome-plasma membrane fusion involved in facilitating invasion. In addition, cyclic AMP (c-AMP), which enhances Ca^{2+}-regulated exocytosis in many cell types, was found to regulate Ca^{2+}-regulated lysosome exocytosis as well as *T. cruzi* invasion.[27] Together, these findings provided the framework for a compelling model of *T. cruzi* invasion in which trypomastigotes trigger local actin remodeling and targeted exocytosis of lysosomes in a Ca^{2+}-dependent manner (Fig. 3). Because these signaling studies were conducted prior to the recognition of the lysosome-independent pathway for *T. cruzi* invasion, the results were interpreted solely within the context of lysosome-dependent *T. cruzi* entry. Given that Ca^{2+} is a critical regulator of a broad range of enzymatic and cellular processes, it is likely that trypomastigote-triggered [Ca^{2+}]i-transients play an important role in both pathways of host cell entry by this pathogen.

Ca^{2+}-Regulated Lysosome Exocytosis Repairs Plasma Membrane Damage

While the bigger question of why *T. cruzi* exploits lysosome exocytosis for host cell invasion remains to be answered, a possibility rests on the ubiquitous nature of lysosome exocytosis and it's critical role in Ca^{2+}-dependent plasma membrane repair.[28] This process involves lysosome exocytosis at the site of the injury, where the membranes fuse to repair damage and prevent cell death.[28] As this process is fairly universal, it is likely that the parasite, which requires transient residence in a lysosome, has evolved a strategy to enter lysosomes directly, by mimicking aspects of the plasma membrane wounding response. This provides an attractive explanation for the promiscuous nature of *T. cruzi* and its ability to invade a wide variety of mammalian host cells.

That events regulating lysosome-plasma membrane fusion and *T. cruzi* trypomastigote invasion of nonprofessional phagocytes revealed similarities to the more general process of Ca^{2+}-regulated exocytosis in cells, prompted investigation into the role of synaptotagmins, a family of proteins which are involved in regulated exocytosis.[30] Unlike other members of the

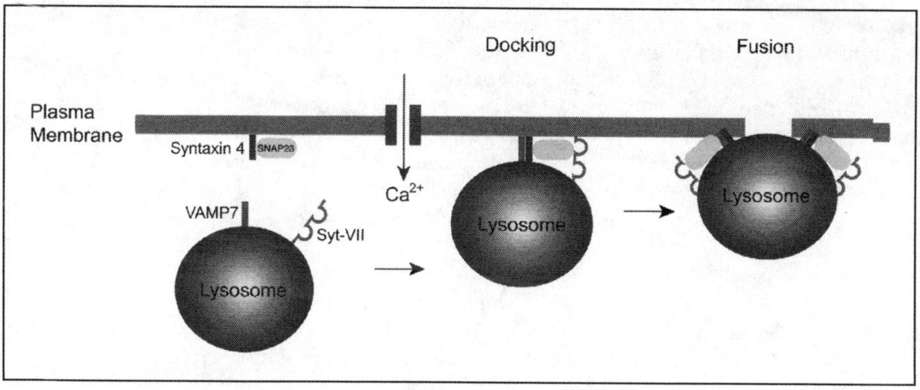

Figure 2. General model of plasma membrane repair by lysosomes. An increase in intracellular calcium levels results in binding to the lysosomal membrane Ca^{2+} sensor Synaptotagmin VII(Syt VII). Syt VII then binds to the t-SNARES SNAP-23 and syntaxin 4 in a Ca^{2+} dependant manner, which results in association with the v-SNARE VAMP7. This complex then mediates fusion of the lysosome with the plasma membrane. (Adapted from ref. 35).

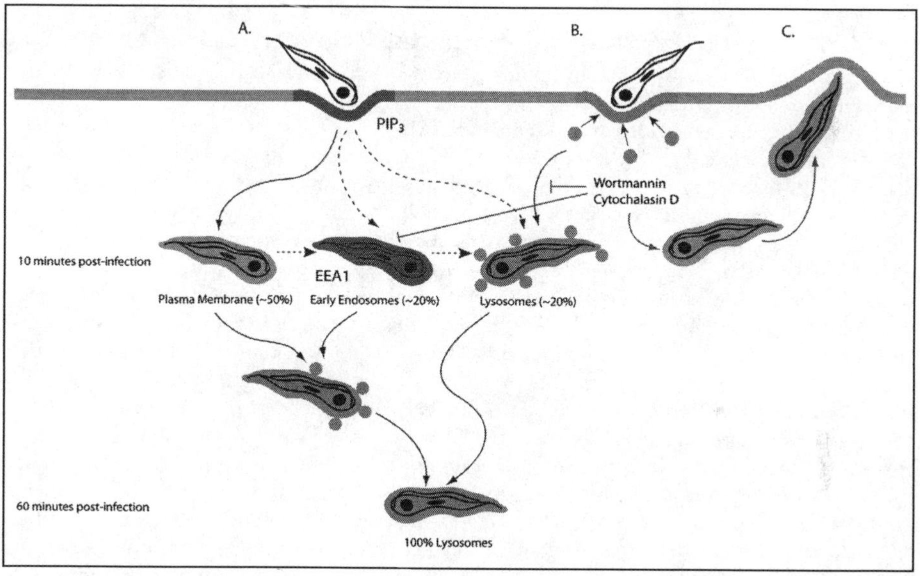

Figure 3. T. cruzi (A) invade cells via two distinct pathways. Parasites were once thought to invade the host cell through a process involving recruitment of lysosomes to the sight of entry (B). These parasites immediately entered a lysosmally-derived vacuole (B). Recent evidence suggests that the parasite can also enter a plasma membrane-derived vacuole, which can be visualized by the lipid product PIP3 (A). These vacuoles rapidly acquire markers of the endocytic pathway, including EEA1, and eventually fuse with lysosomes to form a mature parasitophorous vacuole. Addition of wortmannin or cytochalasin D results in abolishment of the lysosomal pathway, allowing the study of the plasma membrane pathway in isolation. These drugs also disrupt the cytoskeletal matrix or inhibit fusion of the parasitophorous vacuole with lysosomes, both conditions are necessary for the establishment of a successful infection. The use of these chemicals therefore results in a lower level of infection by promoting parasite exit from the cells (C). (Adapted from ref. 38).

synaptotagmin family that exhibit cell type-specific expression, synaptotagmin VII (Syt VII) was shown to be expressed in many cell types,[30] and localized to dense lysosomes.[31] Subsequent studies revealed that Syt VII is essential for efficient membrane resealing following to plasma membrane damage.[28,32,33,34] Additional components of the Syt VII-dependent lysosome-fusion machinery were identified[35] (Fig. 2) to be SNAP-23, which interacts with Syt VII via syntaxin 4 in a calcium dependent manner and the lysosomal v-SNARE TI-VAMP/VAMP7. Together, the coordinate action of these molecules, and likely others that have yet to be identified, are thought to provide the specificity for lysosome targeting and fusion with the plasma membrane.

Given the critical role for Syt VII in Ca^{2+}-regulated lysosome exocytosis and membrane resealing, it was anticipated that Syt VII would play a central role in the *T. cruzi* invasion process. As predicted, Syt VII was demonstrated on the membranes of recently formed *T. cruzi* vacuoles,[29] suggestive of their involvement in vacuole formation. However, interference with Syt VII-mediated fusion of lysosomes with the plasma membrane, either by expression of a dominant interfering region of Syt VII[29] or in fibroblasts that are Syt VII-deficient,[36] resulted in partial inhibition of *T. cruzi* invasion. While these data reinforce the role of the lysosome recruitment pathway in *T. cruzi* infection, they also demonstrate that the parasite is able to gain entry to host cells even in the absence of efficient plasma membrane repair. In other words, *T. cruzi* can readily enter cells in a lysosome-independent fashion, where they traffic to the lysosomal compartment with delayed kinetics, and go on to establish productive infection.[29,32,36]

Host Cell Phosphatidylinositol-3-Kinases Regulate *T. cruzi* Invasion

Inhibition of host cell phosphatidylinositol (PI)-3-kinases following pretreatment with wortmannin or LY294002 reduces *T. cruzi* invasion of cells by ~50% and completely abolishes early lysosome association with internalized parasites.[18,37] The fact that *T. cruzi* invasion of Syt VII-deficient fibroblasts is refractory to wortmannin pretreatment indicates that host cell PI-3 kinases differentially regulate the lysosome-dependent entry process.[36] Interestingly, cells lacking the main regulatory subunits for class I PI-3 kinases (p85α and p85β), while greatly impaired with respect to *T. cruzi* invasion, still support lysosome-dependent entry to the same relative level as WT fibroblasts. The interpretation of these experimental findings is that class I PI-3 kinases do not act in the lysosome-dependent *T. cruzi* entry pathway and further implicate the other wortmannin-sensitive enzyme, class III PI-3 kinase in this process.[38] Further studies are needed to verify these predictions. It is of note that other parasites are also known to alter PI-3 kinase signaling pathways in order to prime the host cell for infection or to promote the survival of infected cells. *Cryptosporidium parvum* initiates host cell PI-3 kinase signaling during its infective process that is required for the formation of the *Cryptosporidium* PV.[39,40] *Toxoplasma gondii* and *Theileria sp.* parasites cause constitutive activation of the PI-3 kinase signaling pathways in their host cells in order to interfere with pro-apoptotic signals (reviewed in refs. 41,42). These examples demonstrate that induction of PI-3 kinase signaling is a common attribute of parasites, which allows them to modulate the host cell cytoskeleton to promote invasion and avoid pro-apoptotic signals while in residence within the host cell by *T. cruzi*. Similarly, activation of host PI-3 kinases is an important signal to facilitate parasite entry[18,37] and can initiate anti-apoptotic signaling cascades in certain cell types.[43-45]

The Actin Cytoskeleton and Fusion with Lysosomes Are Critical for Retention of *T. cruzi* in the Host Cell

Recently, two independent groups reported that interference with the ability of invading or recently internalized trypomastigotes to fuse with components of the endocytic pathway (early endosomes, lysosomes) resulted in loss of parasites from the cells.[9,46] Woolsey et al[46] reported that treatment of cells with cytochalasin D, a compound that causes a destabilization and breakdown of the actin cytoskeleton, results in the decoupling of the process of cell penetration from subsequent fusion with endosomes and lysosomes. The treated cells showed an initial increase in total parasite invasion events, but a marked decrease in the number of internalized parasites that colocalize with the marker early endosome antigen-1 (EEA1) or with LAMP-1 at early time-points.[46] Removal of cytochalasin D in drug washout experiments released the block and permitted internalized parasites to fuse with endosomes and lysosomes. Thus, contrary to the original hypothesis, that depolymerization of host actin microfilaments increases *T. cruzi* invasion due to its facilitating effects on lysosome-plasma membrane fusion, actin depolymerization appears to facilitate invagination of the plasma membrane and parasite entry via the lysosome-independent pathway. However, it is clear that actin dynamics are required following parasite entry to retain internalized parasites.[46] Since actin polymerization faciliates fusion of endosomes with phagosomes,[47] it is possible that actin is playing a similar role during *T. cruzi* invasion and vacuole biogenesis. At this point however, the possibility that the actin cytoskeleton simply acts as a barrier to impede parasite exit from the host cell, cannot be ruled out.

Similar observations regarding *T. cruzi* retention in cells were reported in a study by Andrade and Andrews[9] where the fate of internalized parasites was examined in cells treated with wortmannin to inhibit PI-3 kinases. In these cells, early lysosome fusion with invading parasites is severely inhibited, as previously demonstrated,[18] and parasites rely upon plasma membrane invagination for entry into host cells. Similar to the effects observed following actin depolymerization,[46] parasites entering wortmannin-treated cells were less competent for lysosome fusion and did not remain cell-associated.[9] The conclusions reached in this study were that the lysosome-independent pathway of *T. cruzi* invasion is nonproductive; it was argued that parasite-induced invagination of the plasma membrane leads to the formation of a "reversible" vacuole from which parasites rapidly exit.[9] Following extrusion, it was suggested that these parasites would reinvade cells via the lysosome fusion route and only then be retained. Data from both studies supports the notion that early fusion with components of the endocytic pathway appears to be required for retention of parasites within the host cell. However, it is likely that this can happen in two (or perhaps more) ways that are outlined in the model presented in Figure 3: (1) immediate fusion of lysosomes as parasites enter cells via a Syt VII-dependent, Ca2+-regulated lysosome exocytosis pathway, and (2) delayed fusion of lysosomes, where a plasma membrane-derived vacuole subsequently fuses with early endosomes and lysosomes. Certainly, the most compelling evidence for the existence of a lysosome-independent entry mechanism is the ability of *T. cruzi* to invade and successfully infect Syt VII-deficient cells.[36] Regardless of the mechanism of host cell entry, all of the internalized parasites are housed within host cell lysosomes within the first 60 minutes of invasion.[18] It is within this lysosomal compartment that critical developmental processes are initiated which allow the parasite to eventually shed the vacuole and replicate in the host cell cytoplasm.

Concluding Remarks

Over the last decade, the view regarding the mechanism of *T. cruzi* invasion of nonphagocytic cells has evolved from a single or dominant pathway involving early recruitment and fusion of host cell lysosomes with the plasma membrane, to the understanding that this process is much more complex and that lysosome fusion is not a requirement for entry. Data from two independent groups has provided insight into the requirement for fusion of nascent *T. cruzi*-containing vacuoles with components of the endocytic pathway (early endosomes/lysosomes) as a critical step in cellular retention of parasites. While the relevance of the plasma membrane invagination pathway for the establishment of a productive intracellular infection has been challenged,[9] the ability of *T. cruzi* to establish infection in Syt VII-deficient (i.e., impaired lysosome exocytosis) fibroblasts argues that the lysosome-independent entry pathway is indeed a viable route of infection. As we currently lack data regarding relevant cell invasion mechanisms in the context of in vivo infection, it is currently unclear what roles the two pathways play in natural infection outside of in vitro systems.

The recent work in this area has highlighted a unique example of a parasite coopting a host cell pathway in order to promote its own invasion and survival. The use of the host cell plasma membrane wound repair mechanism is a fascinating example of this ability. Given the universal nature of wound repair within eukaryotic cells it may represent a highly adapted method for *T. cruzi* entry into a wide variety of host cell types. Furthermore, the study of *T. cruzi* trypomastigote invasion of nonprofessional phagocytes provided the catalyst for the discovery of Ca^{2+}-regulated lysosome exocytosis and its relationship to plasma membrane repair.[21] These studies emphasize the utility of intracellular pathogens as probes to study basic cellular processes.

References

1. Prata A. Clinical and epidemiological aspects of Chagas' disease. Lancet Infectious Diseases 2001; 1:92-100.
2. Ley V, Robbins E, Nussenzweig V et al. The exit of Trypanosoma cruzi from the phagosome is inhibited by raising the pH of acidic compartments. J Exp Med 1990; 171(2):401-413.
3. Andrews NW, Whitlow MB. Secretion by Trypanosoma cruzi of a hemolysin active at low pH. Molecular and Biochemical Parasitology 1989; 33(3):249-256.
4. Andrews NW, Abrams CK, Slatin SL et al. A T. cruzi-secreted protein immunologically related to the complement component C9: Evidence for membrane pore-forming activity at low pH. Cell 1990; 61(7):1277-1287.
5. Vandekerckhove F, Schenkman S, Carvalho LPD et al. Substrate specificity of the Trypanosoma cruzi trans-sialidase. Glycobiology 1992; 2(6):541-548.
6. Hall B, Webster P, Ma A et al. Desialylation of lysosomal membrane glycoproteins by Trypanosoma cruzi: A role for the surface neuraminidase in facilitating parasite entry into the host cell cytoplasm. J Exp Med 1992; 176(2):313-325.
7. Rubin-de-Celis SSC, Uemura H, Yoshida N et al. Expression of trypomastigote trans-sialidase in metacyclic forms of Trypanosoma cruzi increases parasite escape from its parasitophorous vacuole. Cellular Microbiology 2006; 8(12):1888-1898.
8. Lopez M, Huynh C, Andrade LO et al. Role for sialic acid in the formation of tight lysosome-derived vacuoles during Trypanosoma cruzi invasion. Molecular and Biochemical Parasitology 2002; 119(1):141-145.
9. Andrade LO, Andrews NW. Lysosomal fusion is essential for the retention of Trypanosoma cruzi inside host cells. J Exp Med 2004; 200(9):1135-1143.
10. Cossart P, Sansonetti PJ. Bacterial invasion: The paradigms of enteroinvasive pathogens. Science 2004; 304(5668):242-248.
11. Tardieux I, Webster P, Ravesloot J et al. Lysosome recruitment and fusion are early events required for trypanosome invasion of mammalian cells. Cell 1992; 71(7):1117-1130.
12. Schenkman S, Robbins ES, Nussenzweig V. Attachment of Trypanosoma cruzi to mammalian cells requires parasite energy, and invasion can be independent of the target cell cytoskeleton. Infect Immun 1991; 59(2):645-654.
13. Rodriguez A, Samoff E, Rioult M et al. Host cell invasion by trypanosomes requires lysosomes and microtubule/kinesin-mediated transport. J Cell Biol 1996; 134(2):349-362.
14. Tyler KM, Luxton GWG, Applewhite DA et al. Responsive microtubule dynamics promote cell invasion by Trypanosoma cruzi. Cellular Microbiology 2005; 7(11):1579-1591.
15. Andrade LO, Andrews NW. The trypanosoma cruzi-host-cell interplay: Location, invasion, retention. Nature Reviews Microbiology 2005; 3(10):819-823.
16. Andrews NW. Lysosomes and the plasma membrane: Trypanosomes reveal a secret relationship. J Cell Biol 2002; 158(3):389-394.
17. Schenkman S, Mortara R. HeLa cells extend and internalize pseudopodia during active invasion by Trypanosoma cruzi trypomastigotes. J Cell Sci 1992; 101(4):895-905.
18. Woolsey AM, Sunwoo L, Petersen CA et al. Novel PI 3-kinase-dependent mechanisms of trypanosome invasion and vacuole maturation. J Cell Sci 2003; 116(17):3611-3622.
19. Wilkowsky SE, Barbieri MA, Stahl PD et al. Regulation of trypanosoma cruzi invasion of nonphagocytic cells by the endocytically active GTPases dynamin, Rab5, and Rab7. Biochemical and Biophysical Research Communications 2002; 291(3):516-521.
20. Bucci C, Thomsen P, Nicoziani P et al. Rab7: A Key to lysosome biogenesis. Mol Biol Cell 2000; 11(2):467-480.
21. Rodriguez A, Webster P, Ortego J et al. Lysosomes behave as Ca2+-regulated exocytic vesicles in fibroblasts and epithelial cells. J Cell Biol 1997; 137(1):93-104.
22. Tardieux I, Nathanson M, Andrews N. Role in host cell invasion of Trypanosoma cruzi-induced cytosolic-free Ca2+ transients. J Exp Med 1994; 179(3):1017-1022.

23. Burleigh BA, Andrews NW. Signaling and host cell invasion by Trypanosoma cruzi. Current Opinion in Microbiology 1998; 1(4):461-465.
24. Scharfstein J, Schmitz V, Morandi V et al. Host cell invasion by Trypanosoma cruzi is potentiated by activation of bradykinin B2 receptors. J Exp Med 2000; 192(9):1289-1300.
25. Dorta ML, Ferreira AT, Oshiro MEM et al. Ca2+ signal induced by Trypanosoma cruzi metacyclic trypomastigote surface molecules implicated in mammalian cell invasion. Molecular and Biochemical Parasitology 1995; 73(1-2):285-289.
26. Rodriguez A, Rioult M, Ora A et al. A trypanosome-soluble factor induces IP3 formation, intracellular Ca2+ mobilization and microfilament rearrangement in host cells. J Cell Biol 1995; 129(5):1263-1273.
27. Rodriguez A, Martinez I, Chung A et al. cAMP Regulates Ca2+-dependent exocytosis of lysosomes and lysosome-mediated cell invasion by trypanosomes. J Biol Chem 1999; 274(24):16754-16759.
28. Reddy A, Caler EV, Andrews NW. Plasma membrane repair is mediated by Ca2+-regulated exocytosis of lysosomes. Cell 2001; 106(2):157-169.
29. Caler EV, Chakrabarti S, Fowler KT et al. The Exocytosis-regulatory protein synaptotagmin VII mediates cell invasion by Trypanosoma cruzi. J Exp Med 2001; 193(9):1097-1104.
30. Li C, Ullrich B, Zhang JZ et al. Ca2+ -dependent and independent activities of neural and nonneural synaptotagmins. Nature 1995; 375(6532):594-599.
31. Martinez I, Chakrabarti S, Hellevik T et al. Synaptotagmin VII regulates Ca2+-dependent exocytosis of lysosomes in Fibroblasts. J Cell Biol 2000; 148(6):1141-1150.
32. Chakrabarti S, Kobayashi KS, Flavell RA et al. Impaired membrane resealing and autoimmune myositis in synaptotagmin VII-deficient mice. J Cell Biol 2003; 162(4):543-549.
33. Roy D, Liston DR, Idone VJ et al. A process for controlling intracellular bacterial infections induced by membrane injury. Science 2004; 304(5676):1515-1518.
34. Andrews NW. Membrane resealing: Synaptotagmin VII keeps running the show. Sci STKE 2005; 2005(282):19.
35. Rao SK, Huynh C, Proux-Gillardeaux V et al. Identification of SNAREs involved in synaptotagmin VII-regulated lysosomal exocytosis. J Biol Chem 2004; 279(19):20471-20479.
36. Chakrabarti S, Andrade LO, Andrews NW. Trypanosoma cruzi invades synaptotagmin VII-deficient cells by a PI-3 kinase independent pathway. Molecular and Biochemical Parasitology 2005; 141(1):125-128.
37. Wilkowsky SE, Barbieri MA, Stahl P et al. Trypanosoma cruzi: Phosphatidylinositol 3-kinase and protein kinase B activation is associated with parasite invasion. Experimental Cell Research 2001; 264(2):211-218.
38. Burleigh BA. Host cell signaling and trypanosoma cruzi invasion: Do all roads lead to lysosomes? Sci STKE 2005; 2005(293):pe36-.
39. Chen XM, Splinter PL, Tietz PS et al. Phosphatidylinositol 3-kinase and frabin mediate cryptosporidium parvum cellular invasion via activation of Cdc42. J Biol Chem 2004; 279(30):31671-31678.
40. Forney JR, DeWald DB, Yang S et al. A role for host phosphoinositide 3-kinase and cytoskeletal remodeling during cryptosporidium parvum infection. Infect Immun 1999; 67(2):844-852.
41. Kim L, Denkers EY. Toxoplasma gondii triggers Gi-dependent PI 3-kinase signaling required for inhibition of host cell apoptosis. J Cell Sci 2006; 119(10):2119-2126.
42. Dobbelaere DA, Kuenzi P. The strategies of the Theileria parasite: A new twist in host-pathogen interactions. Current Opinion in Immunology 2004; 16(4):524-530.
43. Chuenkova MV, Furnari FB, Cavenee WK et al. Trypanosoma cruzi trams-sialidase: a potent and specific survival factor for human Schwann cells by means of phosphatidyl inositol 3-kinase/Akt signaling. Proc Natl Acad Sci USA 2001; 98(17):9936-9941.
44. Petersen CA, Krumholz KA, Carmen J et al. Trypanosoma cruzi Infection and nuclear factor kappa B activation prevent apoptosis in cardiac cells. Infect Immun 2006; 74(3):1580-1587.
45. Aoki MDP, Cano RC, Pellegrini AV et al. Different signaling pathways are involved in cardiomyocyte survival induced by a Trypanosoma cruzi glycoprotein. Microbes and Infection 2006; 8(7):1723-1731.
46. Woolsey AM, Burleigh BA. Host cell actin polymerization is required for cellular retention of Trypanosoma cruzi and early association with endosomal/lysosomal compartments. Cellular Microbiology 2004; 6(9):829-838.
47. Kjeken R, Egeberg M, Habermann A et al. Fusion between phagosomes, early and late endosomes: A role for actin in fusion between late, but not early endocytic organelles. Mol Biol Cell 2004; 15(1):345-358.

Chapter 14

Leishmania Invasion and Phagosome Biogenesis

Robert Lodge and Albert Descoteaux*

Abstract

Whereas bacterial pathogens take over the control of their host cell actin cytoskeleton by delivering an array of protein effectors through specialized secretion systems, promastigotes of the protozoan parasite *Leishmania donovani* rely entirely upon a cell surface glycolipid to achieve this feat. Here, we review recent evidence that *L. donovani* promastigotes subvert host macrophage actin dynamics during the establishment of infection and we discuss the potential mechanisms involved.

Leishmania

During their life cycle, protozoan parasites of the genus *Leishmania* alternate between two distinct developmental stages. In the mammalian host, the parasite proliferates intracellularly as nonmotile amastigotes, within the acidic and hydrolase-rich phagolysosomal compartment of host macrophages.[1] Transmission of the parasite is mediated by the blood-sucking sand fly, of either the genus *Phlebotomus* or the genus *Lutzomyia*. When feeding on an infected mammal, the sand fly takes up amastigote-containing macrophages/monocytes. During digestion of the bloodmeal, amastigotes differentiate into motile promastigotes, which attach to the sand fly midgut epithelium to avoid being excreted together with the digested bloodmeal. Virulence is acquired during metacyclogenesis, a process by which dividing, noninfective promastigotes (procyclic) transform into nondividing infective forms.[2] These metacyclic promastigotes detach from the gut epithelial cells and migrate towards the anterior end of the digestive tract. Upon the next bloodmeal of the infected sand fly, metacyclic promastigotes are inoculated into the mammalian host where they ultimately end up inside a macrophage, in a phagolysosome. The various *Leishmania* species are responsible for a spectrum of human diseases ranging from a relatively confined cutaneous disease to a progressive visceral disease that can be fatal. In the latter case, *L. donovani*, the causative agent of visceral leishmaniasis (Kala-azar), disseminates and infects macrophages of the liver, the spleen, and the bone marrow.

To sucessfully colonize both hosts, *Leishmania* requires the expression of stage-specific virulence determinants, including a family of related glycoconjugates known as the the phosphoglycans and which have in common repeating Galβ1,4Manα1-PO_4 units.[3] This family of glycoconjugates includes the secreted acid phosphatase and phosphoglycan (PG), the proteophosphoglycan, as well as the membrane-anchored lipophosphoglycan (LPG).

*Corresponding Author: Albert Descoteaux—INRS-Institut Armand-Frappier, 531, boul. des Prairies, Laval, Quebec H7V 1B7, Canada. Email: albert.descoteaux@iaf.inrs.ca

Molecular Mechanisms of Parasite Invasion, edited by Barbara A. Burleigh and Dominique Soldati-Favre.

Lipophosphoglycan

LPG is the most abundant promastigote surface glycolipid, with approximately 5 million copies per cell. In contrast, LPG levels are down-regulated by at least three orders of magnitude in amastigotes. The structure of this molecule, which comprises four domains, is illustrated by the prototypic *L. donovani* LPG (Fig. 1).[4] The four domains of the *L. donovani* LPG are (i) a 1-*O*-alkyl-2-*lyso*-phosphatidyl(*myo*)inositol anchor, (ii) a glycan core, (iii) repeating Galβ1,4Manα1-PO_4 units, and (iv) a small oligosaccharide cap. Species- and strain-specific polymorphisms occur in the cap and repeating units structures, whereas the lipid anchor and the glycan core are conserved.[4] In addition, during metacyclogenesis LPG undergoes a doubling in the number of repeating units, from about 15 in procyclic promastigotes to about 30 in metacyclic forms. Elongation of LPG is also accompanied by changes in cap structure and substitutions to the repeating units which allow to regulate binding and release of the parasite from the sand fly midgut epithelium.[5]

A combination of LPG- and Galβ1,4Manα1-PO_4-defective mutants for several *Leishmania* species have been used to investigate the roles of these phosphoglycans in the invertebrate and mammalian hosts. Hence, the *lpg1*-KO mutant secretes repeating Galβ1,4Manα1-PO_4 unit-containing molecules to the same extent as wild type promastigotes, but lacks the ability to assemble a functional LPG glycan core,[6] thereby precluding surface expression of full-lenght LPG. The *lpg2*-KO mutant expresses the truncated LPG Gal(α1,6)Gal(α1,3) Gal*f*(β1,3)[Glc(α1-P)]Man(α1,3)Man(α1,4)GN(α1,6)-PI, and is unable to synthesize repeating Galβ1,4Manα1-PO_4 units.[7] These mutants revealed that the requirement for phosphoglycans in the establishment of infection inside macrophages is not universal. In the case of *L. mexicana*, Galβ1,4Manα1-PO_4-containing glycoconjugates are not required for promastigote virulence,[8] whereas LPG- and Galβ1,4Manα1-PO_4-defective mutants of both *L. donovani* and *L. major* do not survive phagocytosis.[9-11]

Entry into Macrophages

Uptake of *Leishmania* promastigotes is a classical receptor-mediated phagocytic event. Several *Leishmania* and macrophage surface molecules have been implicated in the attachment of promastigotes to macrophages. On macrophages, the complement receptors (CR)1 and CR3 (Mac-1), the mannose-fucose receptor, and p150,95 mediate promastigote binding, although their respective importance and contribution remain to be firmly established.[12,13] *Leishmania* surface glycoconjugates including LPG and gp63 participate in the attachment process, either directly or indirectly, as they are acceptors for various opsonins including C3b and iC3b,[13,14] the mannan-binding protein,[15] and galectins.[16,17] However, LPG is not essential for attachment, as phagocytosis of LPG-defective mutants is similar or even superior to that of wild type promastigotes.[9]

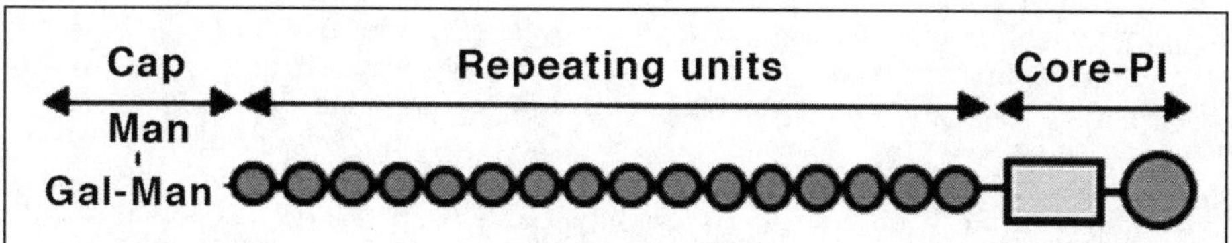

Figure 1. Structure of LPG. LPG is attached to the membrane by a phosphatidylinositol anchor (large circle), which is connected to a glycan core (rectangle). Repeating Galβ1,4Manα1-PO_4 units (small circles) are linked to the core; the number of units varies among *Leishmania* species and at different stages in the parasite's life cycle. The molecule is capped by a small oligosaccharide structure.

Subsequent to their attachment to macrophage receptors, promastigotes are internalized in a phagosome and differentiate into amastigotes. Little is known on the molecular mechanism(s) by which *Leishmania* promastigotes are engulfed by macrophages and this may be related to the fact that internalization results from the simultaneous action of several receptors. Treatment of mouse macrophages with the actin-depolymerizing agent cytochalasin inhibited promastigotes intake,[18] establishing that the actin cytoskeleton is active in the phagocytosis process. Further insight into the mechanisms by which *Leishmania* promastigotes are internalized by macrophages was provided by identifying the Rho-family guanosine triphosphatases (GTPases) involved in this process. These GTPases play a central role in phagocytosis by regulating rearrangements of the actin cytoskeleton.[19,20] In their activated GTP-bound conformation, Rho-family GTPases interact with several effectors involved in the formation of filaments of F-actin around the nascent phagosome. The best characterized cascade of events deals with Rac1/Cdc42 downstream effectors involved in Fcγ-receptor mediated intake.[21] Cdc42 is found early on the protruding pseudopods enveloping the target, whereas Rac1 is mainly recruited to the phagocytic cup.[21] Wiscott-Aldrich syndrome protein (WASP) associates to the phagosome by interacting with GTP-bound Cdc42, which in turn activates the Arp2/3 complex involved in actin nucleation and assembly. At this point, adaptors (Nck), cytoskeletal proteins (VASP) and molecules needed in phosphoinositide signaling are recruited.[22] Cytoplasmic extensions, pseudopods and the heavy involvement of the actin cytoskeleton underline the very dynamic nature of Fcγ-receptor driven entry. The more passive CR-mediated phagocytosis involves RhoA, ROK (Rho kinase), and myosin II, leading to Arp2/3-mediated actin polymerization.[23] In this case, the target 'sinks' into the forming vacuole, which contrasts greatly with Fcγ-receptor entry. Of note, CR3 can perform phagocytosis of nonopsonized particles, which requires Rac1 and Cdc42 instead of RhoA.[24] Even though promastigotes use a combination of receptors, it is likely that rapid opsonisation by complement in the host favors a RhoA, CR-mediated internalisation by circulating macrophages. Hence, uptake of unopsonized *L. donovani* promastigotes by murine macrophages is mainly dependent on the action of Rac1 and to a lesser extent of Cdc42, whereas entry of serum-opsonized *L. donovani* promastigotes is mediated by RhoA.[25] In addition, the small GTPase Arf6, the only Arf-family GTPase implicated in phagocytosis, is also involved in the Rac1/Cdc42 pathway (Lodge and Descoteaux, unpublished). Of note, presence of LPG and other Galβ1,4ManαPO$_4$ units-containing molecules has no influence on the phagocytic pathway, regardless of the opsonization status of the parasite. This observation suggests that although LPG can act as a C3 acceptor, other promastigote surface molecules play this role in the absence of LPG.

The mechanisms by which amastigotes enter macrophages are less understood. In one study, it was reported that uptake of *L. amazonenesis* amastigotes by macrophages requires actin polymerization, and involves the colocalization of F-actin, paxillin, and talin to phagocytic cups that are formed around amastigotes during internalization.[26] Similar to *L. donovani* promastigotes, the opsonization status of *L. amazonensis* amastigotes determines the mode of entry. Hence, amastigote internalization in CHO cells involves the GTPases Rho and Cdc42. When uptake is mediated by fibronectin or when amastigotes are opsonized with immunoglobulin G and internalized by Fc receptor-expressing CHO cells, Rac1 is activated and required for parasite internalization.[27] Collectively, these observations are consistent with the notion that internalization of both *Leishmania* promastigotes and amastigotes is a classical receptor-mediated phagocytic process requiring cytoskeletal reorganization in which the parasite plays no or little active role.

Phagosome Remodelling

Within minutes after internalization of a particle, the F-actin associated to the phagocytic cup dissociates, and the newly formed phagosome interacts with early endosomes. These interactions may be facilitated by the greater access of the phagosomal membrane as the F-actin-rich phagocytic cup is disassembled.[28] The small GTPase Rab5 is necessary for this interaction.[29] In its GTP-bound active form, this small GTPase recognizes either type I (p85/p110) or type III (hVps34) PI 3-kinases. The interaction with hVps34 is involved in Rab5's role in phagosome maturation. The combined presence of Rab5, PI(3)P and the subsequent recruitment of EEA1 through its FYVE domain, participate in the tethering of endosomes to the phagosome, since disrupting the function of either Rab5 or EEA1 affects phagosome biogenesis[30,31] and impairs the later recruitment of the LAMPs and other lysosomal proteins. Rab7 is involved in phagolysosome biogenesis by regulating interactions between the phagosome and late endosomes.[31] Unlike early endosomes, interactions between late endosomes and *L. donovani* and *L. major* promastigote-containing phagosomes are inhibited,[32,33] and recruitment of proteins such as Rab7 and the LAMPs is delayed[34,35] (Fig. 2). This inhibition of phagososomal biogenesis and maturation is strictly dependent on the LPG repeating unit domain,[32,34] since phagosomes harboring LPG-defective promastigotes quickly mature into functional phagolysosomes.[32,34] Insertion of LPG in lipid bilayers alters their fusogenic properties,[36] and this may be responsible for inefficient interactions between endosomes and the phagosome.[37] Furthermore, LPG has been shown to disrupt lipid rafts.[38,39] These cholesterol-rich specialized microdomains are enriched in flotillin-1, α and β subunits of heterotrimeric G proteins,[38] and other proteins such as the NADPH oxidase.[40] Therefore, by inserting itself into the phagosomal lipid raft, LPG enables *L. donovani* promastigotes to target phagosomal raft integrity and hinder important signaling pathways and phagosomal functions.[38,39]

Whereas *L. donovani* and *L. major* promastigotes inhibit phagosome maturation, this is not necessarily the case for all *Leishmania* species. Hence, *L. amazonensis* are internalized in phagosomes which acquire lysosomal features.[41] In addition, LPG-deficient *L. mexicana* promastigotes differentiate into amastigotes and proliferate as efficiently as their wild-type

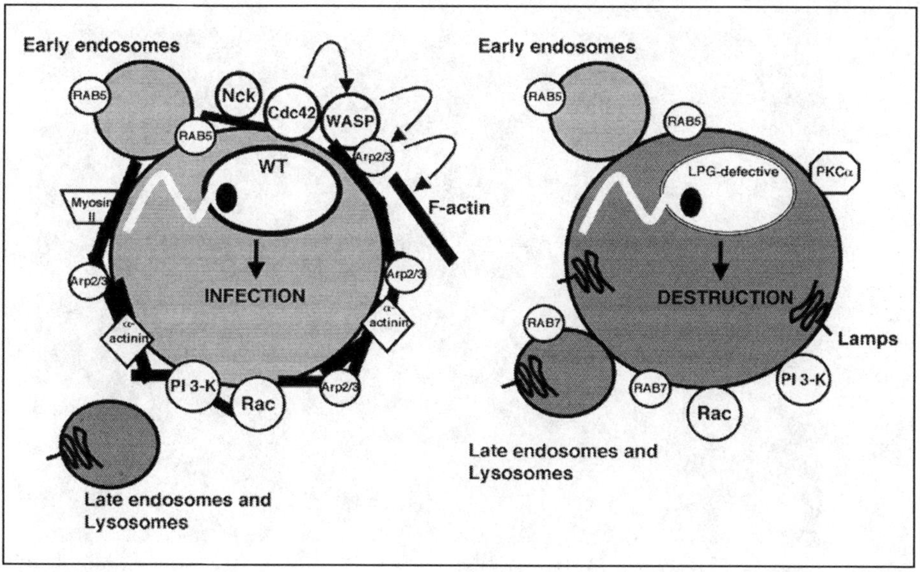

Figure 2. Phagosomes harboring either WT or LPG-defective promastigotes. WT promastigotes remodel the phagosome through the action of LPG: F-actin and proteins involved in F-actin filament assembly aberrantly accumulate on the WT promastigote-containing phagosomes. Interaction of these phagosomes with late endosomes and lysosomes is greatly delayed. This is not the case for phagosomes harboring LPG-deficient *L. donovani*, which mature normally, resulting in the destruction of the mutant parasites.

counterparts in mouse peritoneal macrophages.[8] Of interest, both *L. amazonensis* and *L. mexicana* amastigotes reside in large communal parasitophorous vacuoles, compared to the tight single-parasite harboring phagosomes found in *L. donovani* or *L. major* infected macrophages.[42]

F-Actin Accumulation on the Phagosome

LPG-mediated phagosome remodelling is characterized by a progressive accumulation of periphagosomal F-actin, which might interfere with the recruitment of signal transducers and vesicle trafficking to *L. donovani* promastigote-containing phagosomes.[35] Insights into the underlying mechanism was provided by investigating the impact of LPG on the phagosomal recruitment/retention of host cell molecules involved in cytoskeleton rearrangements.[25] In the case of Rho-family GTPases, Racl and RhoA are present on promastigote-containing phagosomes, regardless on the presence of LPG (Fig. 3). In contrast, Cdc42 is present only on phagosomes containing wild type promastigotes or promastigotes coated with purified LPG. Kinetic analysis revealed that Cdc42 remains associated to these phagosomes for more than two hours after phagocytosis, whereas it rapidly dissociates from phagosomes containing LPG-defective promastigotes. Other effectors, including Arp2/3, WASP, α-actinin, Myosin II, and Nck accumulate with kinetics similar to F-actin and Cdc42 on *L. donovani* promastigote-containing phagosomes, in a LPG-dependent manner. Cdc42 plays an important role in this phenomenon, since expression of a dominant-negative mutant of Cdc42 prevented LPG-mediated accumulation of periphagosomal F-actin and retention of WASP and Myosin.[25] WASP, Arp2/3 and α-actinin are involved in phagocytosis mediated by various receptors, whereas Myosin II is usually associated to CR-driven internalization and Nck is associated to Fcγ-receptor mediated phagocytosis. Therefore, the aberrant accumulation of both

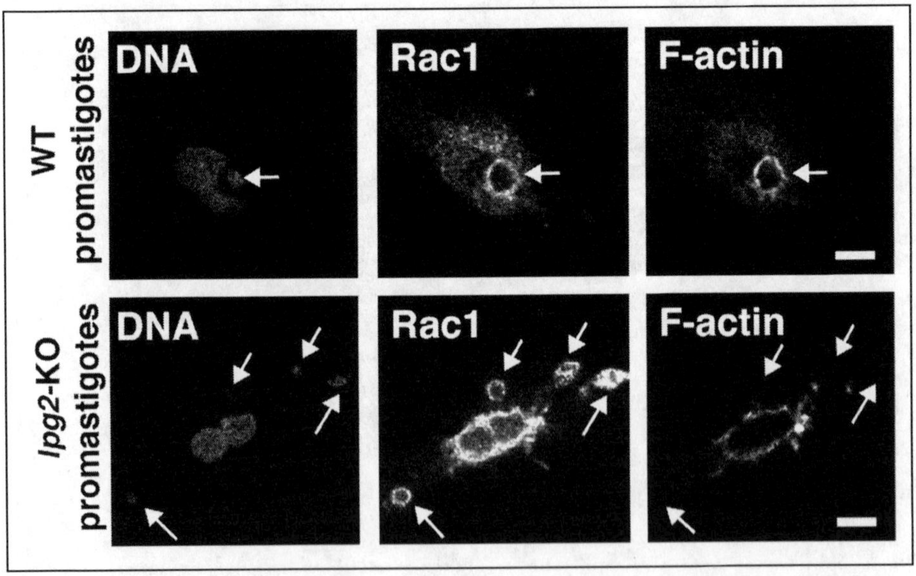

Figure 3. LPG mediates periphagosomal F-actin accumulation. Murine peritoneal exudate macrophages were infected with either WT (top) or *lpg2*-KO (bottom) *L. donovani* promastigotes and stained for DNA (first column), the GTPase Rac1 (second column), or F-actin using phalloidin (third column) and visualized by confocal microscopy. Unlike Rac1, which is found on all phagosomes, F-actin persists on phagosomes harboring WT promastigotes compared to those containing their LPG-deficient counterparts. Phagosomes are identified with arrows; bar = 3 μm.

Nck and Myosin II on promastigote-containing phagosomes suggests that LPG affects an array of pathways involved in F-actin polymerisation. Thus, in the presence of LPG, the molecular machinery required for F-actin assembly is retained on the phagosome, and active Cdc42 drives periphagosomal F-actin accumulation.

How are F-actin and its associated proteins accumulating on *L. donovani* promastigote-containing phagosomes? The fact that Cdc42 is necessary for the LPG-driven process provides some important clues. Cdc42 is recruited very early on the phagosomal membrane,[21] and therefore represents a prime target for luring other F-actin molecules onto the forming organelle. How exactly LPG gets the small GTPase to remain membrane-associated is unknown. One possibility is that LPG activates Cdc42-specific RhoGEFs or inhibits Cdc42 GTP release by inactivating RhoGAPs. Since lipid raft integrity is altered by LPG,[39] hindering these signaling platforms may affect RhoGEF function. Another protein associated to rafts is PKCα. Absence of PKCα on the phagosome, either induced by overexpression of a dominant-negative enzyme or by the presence of LPG, results in F-actin accumulation and the impaired recruitment of LAMP1.[35,43] This might be related to the fact that PKCα phosphorylates MARCKS (myristoylated alanine-rich C kinase substrate), a well-known F-actin-crosslinking protein.[44,45] Therefore, a combination of factors, starting with Cdc42 activation, followed by defective lipid rafts and impaired PKCα recruitment on phagosomes, could result in LPG-mediated F-actin accumulation. Other F-actin associated proteins, such as Coronin (or TACO, for tryptophane-aspartate-containing coat), which are usually found on the phagosome, may also act as regulators of F-actin assembly. Recent work by the group of Grinstein indicated that factors other than inactivation of the Rho-family GTPases contribute to the disassembly of F-actin during phagocytosis.[46] One such factor may be $PI(4,5)P_2$, since in contrast to Rac1 and Cdc42, its disappearance from the phagocytic cup parallels actin dissociation very closely. Based on their findings, they proposed that disappearance of $PI(4,5)P_2$ has a critical role in the termination of actin assembly and in its ensuing disassembly from the phagocytic cup. One consequence of this event would be to create access for incoming endomembrane organelles targeted to fuse with the forming phagosome. However, the exact mechanism by which $PI(4,5)P_2$ is metabolised is not known. Whether LPG causes an alteration of phosphoinositide metabolism at the phagosome is an attractive hypothesis that deserves consideration.

Concluding Remarks

Similar to several pathogenic microbes,[47] *L. donovani* promastigotes target the host cell's cytoskeleton and manipulate Rho-family GTPases to remodel their intracellular habitat and establish infection.[25] In contrast to several bacterial pathogens that use secreted protein effectors, *L. donovani* promastigotes rely predominantly, if not entirely, on a surface glycolipid to achieve these feats. Further studies on the underlying mechanisms will provide novel knowledge on the regulation of phagosome maturation, an important aspect of macrophage biology in the context of host defense against infections. In addition, the consequences of these LPG-mediated events on the ability of infected macrophages to deploy their antimicrobial arsenal remain to be addressed.

References

1. Alexander J, Russell DG. The interaction of Leishmania species with macrophages. Adv Parasitol 1992; 31:175-254.
2. Sacks DL. Metacyclogenesis in Leishmania promastigotes. Exp Parasitol 1989; 69:100-103.
3. Descoteaux A, Turco SJ. Glycoconjugates in Leishmania infectivity. Biochim Biophys Acta 1999; 1455:341-352.
4. Turco SJ, Descoteaux A. The lipophosphoglycan of Leishmania parasites. Annu Rev Microbiol 1992; 46:65-94.
5. Sacks D, Kamhawi S. Molecular aspects of parasite-vector and vector-host interactions in leishmaniasis. Annu Rev Microbiol 2001; 55:453-483.

6. Huang C, Turco SJ. Defective galactofuranose addition in lipophosphoglycan biosynthesis in a mutant of Leishmania donovani. J Biol Chem 1993; 268:24060-24066.
7. Descoteaux A, Luo Y, Turco SJ et al. A specialized pathway affecting virulence glycoconjugates of Leishmania. Science 1995; 269:1869-1872.
8. Ilg T, Demar M, Harbecke D. Phosphoglycan repeat-deficient Leishmania mexicana parasites remain infectious to macrophages and mice. J Biol Chem 2001; 276:4988-4997.
9. McNeely TB, Turco SJ. Requirement of lipophosphoglycan for intracellular survival of Leishmania donovani within human monocytes. J Immunol 1990; 144:2745-2750.
10. Späth GF, Epstein L, Leader B et al. Lipophosphoglycan is a virulence factor distinct from related glycoconjugates in the protozoan parasite Leishmania major. Proc Natl Acad Sci USA 2000; 97:9258-9263.
11. Turco SJ, Späth GF, Beverley SM. Is lipophosphoglycan a virulence factor? A surprising diversity between Leishmania species. Trends Parasitol 2001; 17(5):223-226.
12. Blackwell JM. Receptors and recognition mechanisms of Leishmania species. Trans R Soc Trop Med Hyg 1985; 79:606-612.
13. Kane MM, Mosser DM. Leishmania parasites and their ploys to disrupt macrophage activation. Curr Opin Hematol 2000; 7:26-31.
14. Joshi PB, Kelly BL, Kamhawi S et al. Targeted gene deletion in Leishmania major identifies leishmanolysin (GP63) as a virulence factor. Mol Biochem Parasitol 2002; 120:33-40.
15. Green PJ, Feizi T, Stoll MS et al. Recognition of the major cell surface glycoconjugates of Leishmania parasites by the human serum mannan-binding protein. Mol Biochem Parasitol 1994; 66:319-328.
16. Pelletier I, Sato S. Specific recognition and cleavage of galectin-3 by Leishmania major through species-specific polygalactose epitope. J Biol Chem 2002; 277:17663-17670.
17. Pelletier I, Hashidate T, Urashima T et al. Specific recognition of Leishmania major poly-beta-galactosyl epitopes by galectin-9: Possible implication of galectin-9 in interaction between L. major and host cells. J Biol Chem 2003; 278:22223-22230.
18. Privé C, Descoteaux A. Leishmania donovani promastigotes evade the activation of mitogen-activated protein kinases p38, c-Jun N-terminal kinase, and extracellular signal-regulated kinase-1/2 during infection of naive macrophages. Eur J Immunol 2000; 30:2235-2244.
19. Chimini G, Chavrier P. Function of Rho family proteins in actin dynamics during phagocytosis and engulfment. Nat Cell Biol 2000; 2:E191-E196.
20. Etienne-Manneville S, Hall A. Rho GTPases in cell biology. Nature 2002; 420:629-635.
21. Hoppe AD, Swanson JA. Cdc42, Rac1, and Rac2 display distinct patterns of activation during phagocytosis. Mol Biol Cell 2004; 15:3509-3519.
22. Coppolino MG, Krause M, Hagendorff P et al. Evidence for a molecular complex consisting of Fyb/SLAP, SLP-76, Nck, VASP and WASP that links the actin cytoskeleton to Fcγ receptor signalling during phagocytosis. J Cell Sci 2001; 114:4307-4318.
23. Olazabal IM, Caron E, May RC et al. Rho-kinase and myosin-II control phagocytic cup formation during CR, but not FcγR, phagocytosis. Curr Biol 2002; 12:1413-1418.
24. Le Cabec V, Carreno S, Moisand A et al. Complement receptor 3 (CD11b/CD18) mediates type I and type II phagocytosis during nonopsonic and opsonic phagocytosis, respectively. J Immunol 2002; 169:2003-2009.
25. Lodge R, Descoteaux A. Leishmania donovani promastigotes induce periphagosomal F-actin accumulation through retention of the GTPase Cdc42. Cell Microbiol 2005; 7:1647-1658.
26. Love DC, Mentink Kane M, Mosser DM. Leishmania amazonensis: The phagocytosis of amastigotes by macrophages. Exp Parasitol 1998; 88:161-171.
27. Morehead J, Coppens I, Andrews NW. Opsonization modulates Rac-1 activation during cell entry by Leishmania amazonensis. Infect Immun 2002; 70:4571-4580.
28. Aderem A, Underhill DM. Mechanisms of phagocytosis in macrophages. Annu Rev Immunol 1999; 17:593-623.
29. Duclos S, Diez R, Garin J et al. Rab5 regulates the kiss and run fusion between phagosomes and endosomes and the acquisition of phagosome leishmanicidal properties in RAW 264.7 macrophages. J Cell Sci 2000; 113:3531-3541.
30. Vieira OV, Botelho RJ, Rameh L et al. Distinct roles of class I and class III phosphatidylinositol 3-kinases in phagosome formation and maturation. J Cell Biol 2001; 155:19-25.
31. Vieira OV, Bucci C, Harrison RE et al. Modulation of Rab5 and Rab7 recruitment to phagosomes by phosphatidylinositol 3-kinase. Mol Cell Biol 2003; 23:2501-2514.
32. Desjardins M, Descoteaux A. Inhibition of phagolysosomal biogenesis by the Leishmania lipophosphoglycan. J Exp Med 1997; 185:2061-2068.

33. Dermine JF, Scianimanico S, Privé C et al. Leishmania promastigotes require lipophosphoglycan to actively modulate the fusion properties of phagosomes at an early step of phagocytosis. Cell Microbiol 2000; 2:115-126.
34. Scianimanico S, Desrosiers M, Dermine JF et al. Impaired recruitment of the small GTPase rab7 correlates with the inhibition of phagosome maturation by Leishmania donovani promastigotes. Cell Microbiol 1999; 1:19-32.
35. Holm Å, Tejle K, Magnusson KE et al. Leishmania donovani lipophosphoglycan causes periphagosomal actin accumulation: Correlation with impaired translocation of PKCα and defective phagosome maturation. Cell Microbiol 2001; 3:439-447.
36. Tolson DL, Turco SJ, Pearson TW. Expression of a repeating phosphorylated disaccharide lipophosphoglycan epitope on the surface of macrophages infected with Leishmania donovani. Infect Immun 1990; 58:3500-3507.
37. Miao L, Stafford A, Nir S et al. Potent inhibition of viral fusion by the lipophosphoglycan of Leishmania donovani. Biochemistry 1995; 34:4676-4683.
38. Dermine JF, Duclos S, Garin J et al. Flotillin-1-enriched lipid raft domains accumulate on maturing phagosomes. J Biol Chem 2001; 276:18507-18512.
39. Dermine JF, Goyette G, Houde M et al. Leishmania donovani lipophosphoglycan disrupts phagosome microdomains in J774 macrophages. Cell Microbiol 2005; 7:1263-1270.
40. Vilhardt F, Van Deurs B. The phagocyte NADPH oxidase depends on cholesterol-enriched membrane microdomains for assembly. EMBO J 2004; 23:739-748.
41. Courret N, Frehel C, Gouhier N et al. Biogenesis of Leishmania-harbouring parasitophorous vacuoles following phagocytosis of the metacyclic promastigote or amastigote stages of the parasites. J Cell Sci 2002; 115:2303-2316.
42. Antoine JC, Prina E, Lang T et al. The biogenesis and properties of the parasitophorous vacuoles that harbour Leishmania in murine macrophages. Trends Microbiol 1998; 6:392-401.
43. Holm Å, Tejle K, Gunnarsson T et al. Role of protein kinase Cα for uptake of unopsonized prey and phagosomal maturation in macrophages. Biochem Biophys Res Commun 2003; 302:653-658.
44. Allen LH, Aderem A. A role for MARCKS, the alpha isozyme of protein kinase C and myosin I in zymosan phagocytosis by macrophages. J Exp Med 1995; 182:829-840.
45. Allen LA, Aderem A. Protein kinase C regulates MARCKS cycling between the plasma membrane and lysosomes in fibroblasts. EMBO J 1995; 14:1109-1121.
46. Scott CC, Dobson W, Botelho RJ et al. Phosphatidylinositol-4,5-bisphosphate hydrolysis directs actin remodeling during phagocytosis. J Cell Biol 2005; 169:139-149.
47. Gruenheid S, Finlay BB. Microbial pathogenesis and cytoskeletal function. Nature 2003; 422:775-781.

Chapter 15

Plasmodium Sporozoite Passage across the Sinusoidal Cell Layer

Ute Frevert,* Ivan Usynin, Kerstin Baer and Christian Klotz

Abstract

Malaria sporozoites must cross at least two cell barriers to reach their initial site of replication in the mammalian host. After transmission into the skin by an infected mosquito, they migrate towards small dermal capillaries, traverse the vascular endothelial layer,[1,2] and rapidly home to the liver. To infect hepatocytes, the parasites must cross the sinusoidal cell layer, composed of specialized highly fenestrated sinusoidal endothelia and Kupffer cells, the resident macrophages of the liver (Fig. 1). The exact route *Plasmodium* sporozoites take to hepatocytes has been subject of controversial discussions for many years. Recent cell biological, microscopic, and genetic approaches have considerably enhanced our understanding of the initial events leading to the establishment of a malaria infection in the liver (for recent reviews see refs. 3-8).

Targeting to the Liver

The initial arrest of the parasites in the liver sinusoid is thought to be mediated by a multivalent interaction between two major sporozoite surface proteins, the circumsporozoite protein (CSP) and the thrombospondin-related adhesive protein (TRAP), and hepatic heparan sulfate proteoglycans located in the extracellular matrix (ECM) inside the space of Disse.[9-14] The degree of sulfation of heparan sulfate from liver approaches that of heparin and is therefore markedly higher than in any other tissue.[15] According to our current model (Fig. 2), large stellate cell-derived ECM proteoglycans protrude from the space of Disse through the endothelial sieve plates into sinusoidal lumen,[11] thus providing a basis for sporozoites to glide along the sinusoidal endothelium.[16] When a parasite encounters a Kupffer cell, CSP engages in a multivalent interaction with selected chondroitin sulfate and heparan sulfate proteoglycans that are expressed on the surface of these macrophages.[9] Once in the space of Disse, the sporozoite recognizes small proteoglycans on the surface of hepatocytes and enters these parenchymal cells.[14,17-22]

In addition to cell surface heparan sulfate proteoglycans, CSP also binds to the low density lipoprotein receptor-related protein 1 (LRP-1),[23] a multifunctional scavenger receptor that is predominantly found in the liver.[24-26] LRP, also known as the alpha-2-macroglobulin receptor (α2MR) or CD91, is responsible for the clearance from the blood of a large number of molecules including activated α2M, proteases and their complexes with inhibitors, matrix proteins, growth factors, and small particles such as lipoprotein remnants.[24-26] In vitro, binding of soluble recombinant CSP to LRP on hepatocytes leads to endocytic uptake followed by lysosomal

*Corresponding Author: Ute Frevert—NYU School of Medicine, Department of Medical Parasitology, 341 E. 25 Street, New York, New York 10010, USA.
Email: ute.frevert@med.nyu.edu

Molecular Mechanisms of Parasite Invasion, edited by Barbara A. Burleigh and Dominique Soldati-Favre.

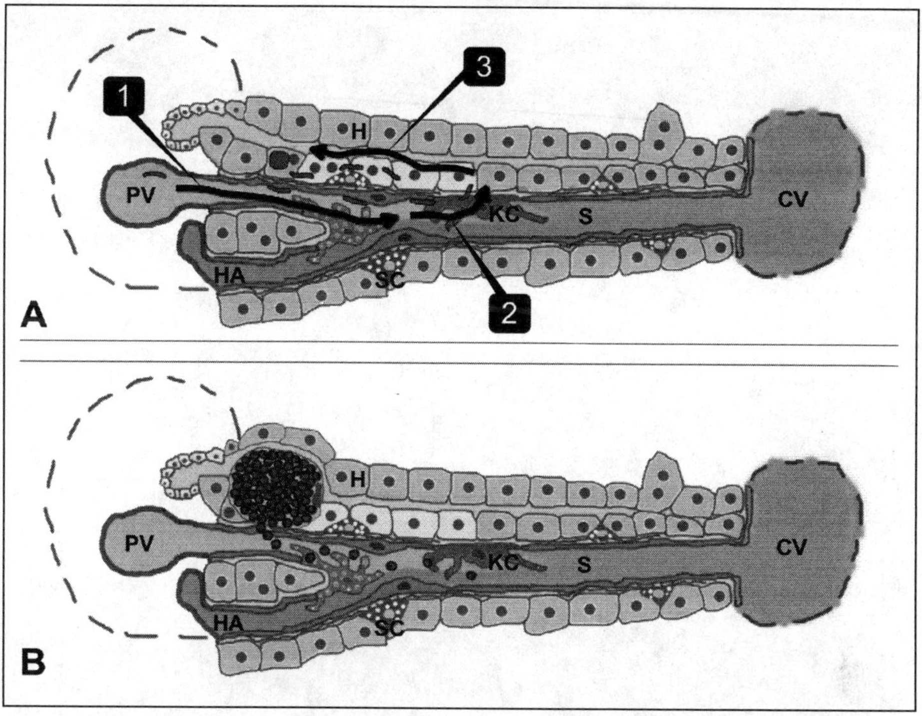

Figure 1. Model of *Plasmodium* sporozoite infection of the liver. The portal venous (PV) and hepatic arterial (HA) blood supplies merge at the portal field, flow along the liver sinusoid (S), and exit the liver lobules via the central vein (CV). Liver sinusoids are lined by a layer of fenestrated endothelia, interspersed with Kupffer cells (KC). The space of Disse (D) separates the sinusoidal cell layer from the cords of hepatocytes (H). Stellate cells (SC) are located in a niche between hepatocytes inside the space of Disse. Sporozoite infection of the liver follows a complex cascade of events. A) Sporozoite entry into the liver. Arrow 1: Abrupt arrest in the sinusoid and gliding along sinusoidal endothelia, most likely by binding to ECM proteoglycans. Arrow 2: Recognition of, deactivation, active entry into, and passage through Kupffer cells. Arrow 3: Transmigration through several hepatocytes and infection of a final hepatocyte. B) Merozoite release into the bloodstream. Mature liver schizonts release merozoites either individually or in groups, still enveloped by host cell membrane.

degradation.[23] To determine the contribution of hepatocyte LRP-1 to sporozoite infection of the liver, transgenic mice with a functional defect in LRP-1 expression were generated by conditional gene targeting.[27,28] The recombinant Ad5 virus used for this approach infects predominantly hepatocytes and to a minor degree also sinusoidal endothelia,[29] but not Kupffer cells.[30] Consequently, Cre adenovirus injection into LRP-1$^{flox/flox}$ and double mutant LDLR$^{-/-}$;LRP-1$^{flox/flox}$ mice resulted in LRP-1 gene inactivation in 100% of the hepatocytes.[28] Since infection with *P. yoelii* sporozoites was not diminished in these LRP-1$^{flox/flox}$ Cre^{+} mice,[31] it appears that LRP-1 expression on hepatocytes is not obligatory for sporozoite infection of the liver. Interestingly, however, the livers of the LDLR$^{-/-}$;LRP-1$^{flox/flox}$ mice and, in particular the LRP-1$^{flox/flox}$ mice, still contained low levels of LRP-1.[27,28] Because Kupffer cells also express a high amount of LRP on their surface,[32,33] we propose that the residual LRP-1 derived from these macrophages. CSP binding to Kupffer cells, but not hepatocytes, induces an intracellular signal transduction cascade (see below), which was most likely unaffected in the LRP-1 knockout mice.

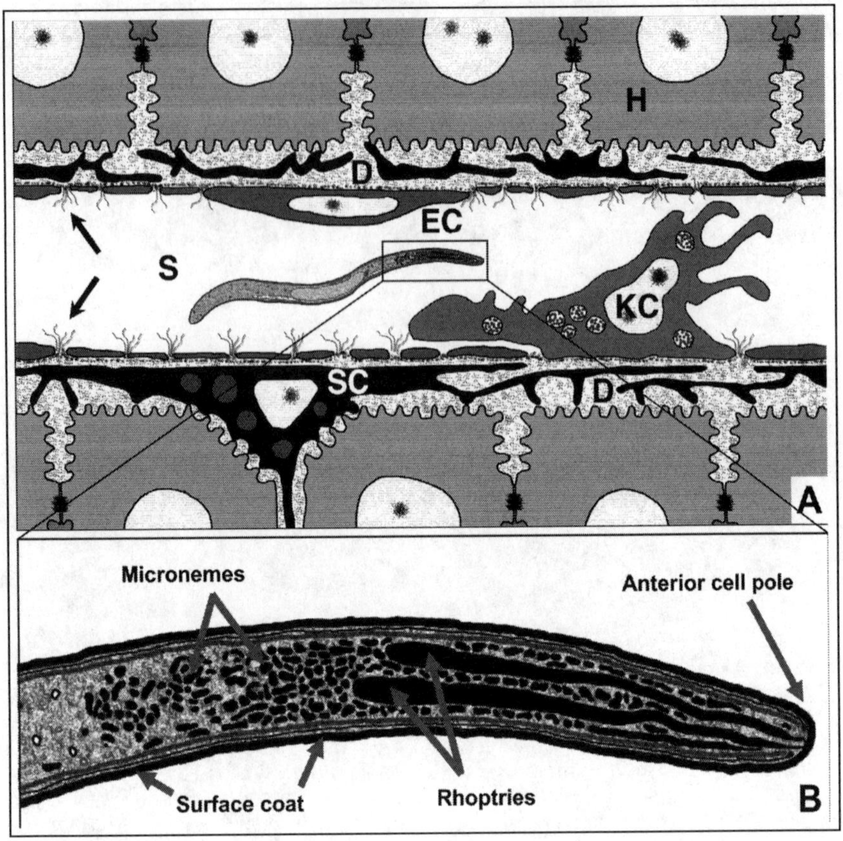

Figure 2. *Plasmodium* and host molecules involved in sporozoite interaction with the liver sinusoid. A) Invasion-related liver proteins and glycoconjugates. Stellate cells (SC) are located in a niche between adjacent hepatocytes (H) and secrete sulfated extracellular matrix proteoglycans into the space of Disse (D). The long glycosaminoglycan chains that are attached to the core proteins of these proteoglycans are thought to protrude through the endothelial fenestration into the sinusoidal lumen (arrows), where they are recognized by sporozoites traveling in a sinusoid (S). Kupffer cells (KC), but not endothelial cells (EC) express CSP-binding proteoglycans and LRP-1 on their surface. Hepatocytes carry CSP- and TRAP-binding heparan sulfate proteoglycans on their surface as well as LRP-1, CD81, and the HGF receptor MET. B) Invasion-related sporozoite proteins. CSP is a major component of the thick coat that covers the entire sporozoite surface. It can be detected in rhoptries and micronemes, from where it is transported to the cell surface at the apical cell pole. By interacting with liver-specific proteoglycans, CSP is thought to mediate sporozoite arrest in the sinusoid. By binding to proteoglycans and LRP-1 on Kupffer cells, CSP induces a signaling cascade that leads to inhibition of the respiratory burst. P36p is also expressed on the sporozoite surface and contributes to liver stage development by a yet undefined mechanism. The following micronemal proteins are secreted upon cell invasion as suggested by in vitro studies, however the precise sequence of events in the intact liver has not been established. The transmembrane protein TRAP plays a pivotal role in sporozoite motility and invasion. Like CSP, TRAP also binds to liver proteoglycans. Under "resting" conditions, only a small proportion of TRAP is expressed on the sporozoite surface; the vast majority is stored intracellularly and released upon invasion. Another micronemal protein, AMA-1, is also secreted upon hepatocyte infection. SPECT proteins are stored in micronemes; the exact time of their release is unknown, but their postulated pore-forming property suggests secretion upon hepatocyte transmigration or, alternatively, upon parasite escape from an intracellular compartment.

Crossing the Sinusoidal Cell Barrier

A plethora of studies have demonstrated that basically all mammalian *Plasmodium* species can develop in hepatocytes or hepatoma cells in vitro[34-43] indicating that Kupffer cells are not required for infection of parenchymal liver cells. In vivo, however, sporozoites must cross the sinusoidal cell layer to enter the liver parenchyma. Biochemical and physiological studies as well as various microscopic and genetic approaches have all suggested that sporozoites enter the liver by passing through Kupffer cells, not endothelia.[9,11,16,44-48] In vitro, *P. berghei* and *P. yoelii* sporozoites attach to and actively enter Kupffer cells, but not sinusoidal endothelia, and survive for several hours inside these macrophages, enclosed in a vacuole that does not acidify or undergo lysosomal fusion.[44] However, interpretation of these studies as a whole has been hampered by various factors including the scarcity of the invasion events, the complex architecture of the liver and its large size, and mismatch of parasite host combination.

To confirm the role of Kupffer cells as a gate for sporozoites to the liver, two different approaches have been used to functionally or physically eliminate these phagocytes from the liver: (1) Saturation of the phagocytic capacity with silica particles[49,50] and (2) removal from the liver by intoxication with liposome-encapsulated dichloromethylene diphosphonate (clodronate).[51] Saturation of the phagocytic capacity of the Kupffer cells by inoculation of silica particles prior to infection decreased the number of *P. berghei* liver stages.[49,50] This finding was initially interpreted as the consequence of a successful blockage of Kupffer cell phagocytosis (and thereby sporozoite passage), but later attributed to the silica-induced Kupffer cell activation resulting in an increase in interleukin-6,[52] which is known to inhibit sporozoite invasion and schizont development.[51] Clodronate, on the other hand, had an enhancing effect on *P. berghei* liver stage development.[51] In addition, clodronate treatment also restored the infectivity of SPECT-deficient *P. berghei* sporozoites, which are normally unable to pass through cells.[48] Because intravenous clodronate application also depletes the spleen of macrophages thus causing a temporary functional splenectomy, this may have increased the number of liver stages by allowing more sporozoites to repeatedly circulate through the liver. These conflicting results obscured the role of Kupffer cells in sporozoite entry into the liver.

An argument against malaria sporozoite passage through Kupffer cells has been that the parasites can cross endothelial cell barriers, since they are clearly able to penetrate the vascular capillary wall in the skin.[1,2] However, the conditions for sporozoite passage across the dermal capillary wall and the sinusoidal barrier differ markedly. While sporozoites approach sinusoidal endothelia from their apical (vascular) surface, they first encounter the adventitial surface of dermal blood capillaries. Vascular endothelia are polarized cells,[53] which secrete heparan sulfate proteoglycans into the basement membrane that typically surrounds blood capillaries.[54-56] Because sporozoites recognize vascular basement membrane proteoglycans in various tissues,[9,11,14] it is reasonable to assume that the parasites recognize these glycoconjugates when they encounter a dermal blood capillary. The route the parasites take across this barrier, whether trans- or para-endothelial, remains to be established.

Sporozoites also enter dermal lymphatic vessels from where they are transported to the regional lymph node,[2] but it should be noted that terminal lymph vessels are quite distinct from blood capillaries. For example, while vascular capillaries possess a clear basement membrane and are surrounded by a layer of pericytes,[53,55,57-59] lymphatic vessels lack pericytes and their basement membrane is sparse and lacks proteoglycans.[56,60,61] Further, vascular and lymphatic endothelia exhibit distinct molecular surface markers.[56] Most importantly and in contrast to the vascular endothelium, the lining of lymphatic vessels is discontinuous. Lymphatic endothelia form adjustable inlet valves for the collection of interstitial fluid, whose diameter can range from 1-6 μm depending on the demand.[61,62] The large size of these openings (up to several times the diameter of a sporozoite) and the continuous lymph flow towards and through these valves render the possibility likely that sporozoites enter lymphatic vessels passively by passing through endothelial inlet valves. The observed lateral drifting of sporozoites along lymph vessels[2] supports this interpretation. SPECT-deficient *P. berghei* parasites, which lack cell passage ability (see below), are ideal tools to address the question of sporozoite entry into dermal lymph vessels. It is known that intravenously inoculated SPECT mutants are unable to infect the intact liver[48] because of the continuity of the sinusoidal cell layer.[63] Similarly, intradermally inoculated SPECT mutants are unable to infect mice (Photini Sinnis, personal communication) suggesting that infection is also blocked after transmission by mosquito bite. However, taking into account the existence of lymphatic inlet valves, SPECT-deficient sporozoites should be able to enter dermal lymph vessels normally. Experiments testing this hypothesis will provide valuable new insights into the mechanisms *Plasmodium* sporozoites use to cross host cell barriers and are therefore eagerly awaited.

In conclusion, each of the cell barriers malaria sporozoites have to overcome during their journey to hepatocytes has its own characteristics, composition and orientation and must therefore be viewed individually. With regard to the passage across the sinusoidal barrier, all available evidence points to Kupffer cells being the gate to the liver.

Kupffer Cell Location and Motility

For clearance and defense function, Kupffer cells are located in strategic positions in the sinusoid, frequently at bifurcations of the highly anastomozed sinusoidal capillary bed of the liver.[64-66] When labeled with tissue macrophage-specific antibodies such as F4/80 or anti-CD68, Kupffer cells exhibit an irregular elongated or star-like shape[67-70] and a predominantly periportal distribution within the liver lobule.[71,72] Scanning electron microscopy demonstrates that, while most of the Kupffer cell body is located inside the lumen of the sinusoid, it is anchored to the sinusoidal cell wall with numerous microvillous projections, lamellipodia, blebs, and protrusions.[73,74] Due to this largely intra-sinusoidal location, transmission electron micrographs often give the impression that Kupffer cells are attached to the luminal surface of the endothelium, rather than being part of the lining.[75,76] However, ultrathin sections reveal that Kupffer cell microvilli intermingle over large distances with those from hepatocytes thus demonstrating that Kupffer cells are, in fact, focally embedded in the endothelial cell layer.[66,73,77-80] Thus, Kupffer cells are true inhabitants of the sinusoidal wall that can provide sporozoites with direct access to the space of Disse and hepatocytes.

Traditionally, Kupffer cells have been defined as stationary tissue macrophages of a characteristic star-like shape.[65,66] However, some recently published liver pathology textbooks and hepato-immunology reviews neglect the majority of the primary observations and describe Kupffer cells as motile macrophages that patrol the liver sinusoids.[81-84] This conclusion is based on a single reflection microscopy study purporting to show that Kupffer cells have a rounded shape and crawl along liver sinusoids at a speed of several micrometers per minute.[85] Interestingly, the migratory cells shown in this report were rounded and exhibited generally low phagocytic activity, suggesting that they were in fact monocytes, while other phagocytically highly active cells were irregularly shaped and nonmigratory, which is typical of Kupffer cells. The generation of lys-EGFP-ki mice that express GFP in various types of macrophages,[86] including Kupffer cells,[16] helped shed light on this issue. Kupffer cells of lys-EGFP-ki mouse livers can be easily differentiated from blood macrophages by their distinct shape and much lower fluorescence intensity.[16,86] During several hours of intravital examination, bright rounded blood macrophages were observed flowing with blood and sometimes moving along the sinusoidal surface, but Kupffer cells did not change their shape or position in the sinusoid[87] thus confirming the fixed nature of these tissue macrophages (Robert McCuskey, personal communication).[88,89] On a long-term basis, however, Kupffer cells are capable of migration. For example, the Kupffer cell population is generally thought to be self-renewing under physiologic conditions, but under certain circumstances it can be replenished by influx and differentiation of bone marrow-derived monocytes.[66,77,90-93] Also, Kupffer cells have been found in the portal tract and in hepatic lymph nodes a few days after selective loading with phagocytic markers.[94] Marginated leukocytes are dissimilar to Kupffer cells; they have a spherical or oval shape and much less surface contact with the sinusoidal cell layer.[66,77,92] Intravital analysis of lys-EGFP-ki mice confirmed this and also demonstrated that blood-derived macrophages crawl along sinusoids, gradually infiltrating the liver and eventually accumulating in large numbers under its capsule.[87] We attribute this infiltration to surgical exposure of the liver for microscopic examination, because transient influx of large numbers of mononuclear phagocytes and neutrophils is a characteristic response of the liver to inflammatory stimuli.[95,96] In conclusion, Kupffer cells can undoubtedly exit liver lobules and their precursors can be recruited from extrahepatic sites, but once positioned in the liver sinusoid, Kupffer cells remain immobile for hours if not days.

Are Kupffer Cells Essential for Sporozoite Infection of the Liver?

To verify that Kupffer cells are required for sporozoite infection of the liver, we used op/op mice with a genetic defect in the production of macrophage colony-stimulating factor 1 (CSF-1).[87] Homozygous op/op mice exhibit a severe deficiency in macrophage maturation in contrast to their heterozygous phenotypically normal littermates.[97] Op/op mice possess only a small fraction of the normal number of Kupffer cells, and these are in addition morphologically and functionally immature. As a consequence, such mice exhibit a markedly reduced clearance capacity of intravenously inoculated colloidal carbon.[87] Quantitative real-time PCR revealed that op/op mice are significantly less susceptible to *P. yoelii* infection than their phenotypically normal heterozygous siblings with normal Kupffer cell numbers.[87] Op/op mice exhibit lower than normal levels of pro-inflammatory cytokines[97-99] and are therefore classified as immunosuppressed.[100] Because there is no indication of increased IFN-γ, IL-6 or other cytokines known to suppress *Plasmodium* liver stage growth, it appears unlikely that factors other than the reduced number of Kupffer cells are responsible for the refractoriness of the op/op mice to *P. yoelii* infection.

To understand why sporozoite invasion is increased in clodronate-treated mice and decreased in op/op mice, we analyzed the ultrastructure of clodronate-depleted and op/op mouse livers. Because Kupffer cells represent an integral component of the lining of the liver capillaries (see above),[78,79] they leave behind small gaps in the normally continuous sinusoidal cell layer when synchronously expelled from the liver tissue by clodronate intoxication.[87] The data indicate that these gaps are large enough to allow sporozoite passage, but too small to generate hemorrhages, which generally require more extensive damage to the microcirculation such as toxic injury to endothelial cells.[101-103] In op/op mice, by contrast, the sinusoidal cell barrier is intact.[104] It appears, therefore, that the openings created by clodronate treatment in the sinusoidal cell layer provide sporozoites with direct access to hepatocytes and ECM proteoglycans in the space of Disse, thus explaining why liver infection by wild-type sporozoites is enhanced and by SPECT-deficient mutants is restored. In conclusion, Kupffer cells are obligatory for the establishment of a *Plasmodium* infection in the mammalian host.

Mode of Sporozoite Entry into Mammalian Cells: Membrane Wounding vs. Invagination

After entering the liver parenchyma, *Plasmodium* sporozoites migrate through several hepatocytes before settling down in a final one for development to a liver stage. Sporozoites are able to enter cells in at least two different ways. Transmigration through hepatocytes involves focal membrane wounding,[16,105] while invasion of the final hepatocyte, similar to entry into Kupffer cells,[44] is thought to occur by membrane invagination.[105-107] Entry by membrane wounding, which provides the parasites with direct access to the cytoplasm, has been observed in vitro for hepatoma cells and a variety of other nonphagocytic cells of epithelial origin, but not for Kupffer cells.[44,47,48,105,108] Disruption of the expression of either SPECT1, a micronemal protein of unknown function, or SPECT2, a sporozoite protein with a putative membrane attack complex domain, causes *P. berghei* sporozoites to lose the ability for transmigration, but does not affect sporozoite infectivity for hepatoma cells in vitro.[48] SPECT knockout mutants are unable to cross the sinusoidal cell layer after intravenous inoculation into mice, suggesting that the Kupffer cell passage is blocked, but details on the mechanism of action of these proteins are not available to date.[4,47,48,109]

Another interesting molecule in this context is a sporozoite surface phospholipase with membrane lytic activity.[110] *P. berghei* sporozoites deficient in this enzyme are impaired in crossing epithelial cell barriers in vitro. When transmitted by mosquito bite, phospholipase-deficient parasites are drastically reduced in their ability to infect mice suggesting that they have lost the ability to enter dermal capillaries. This finding suggests that passage through dermal vascular endothelia requires membrane wounding. When inoculated intravenously, however, phospholipase-deficient sporozoites infect the liver at the same rate as wild-type parasites. Thus, sporozoite entry into the bloodstream in the skin[1] is distinct from exit from the bloodstream in the liver, further supporting the notion that liver infection occurs via Kupffer cells rather than sinusoidal endothelia and by membrane invagination rather than membrane wounding.[16,44]

The exact time of formation of the parasitophorous vacuole has long been unclear. Individual *P. yoelii* sporozoites were found to be surrounded by a membrane early after invasion of murine hepatocytes in vitro[107] and in vivo,[16] but the scarcity of the invasion event made systematic analysis of the parasitophorous vacuole in statu nascendi difficult. Sporozoites may theoretically wound the hepatocyte membrane upon entry or exit, and appropriate live cell studies that distinguish between these two possibilities have not been performed to date. The possibility that the vacuole is generated only after sporozoite entry by membrane wounding[111] has therefore not been formally excluded. A recent study has shed light onto this issue.[106] *P. falciparum* and *P. yoelii* sporozoites depend on cholesterol and the tetraspanin CD81 for hepatocyte infection and parasitophorous vacuole formation.[106,107,112] Interestingly, a certain proportion of *P. yoelii* and *P. falciparum* sporozoites develop to early liver stages inside hepatoma cell nuclei. These Intranuclear parasites had all entered the hepatoma cells by membrane wounding, i.e., independently of CD81, and consequently lacked a parasitophorous vacuole. In contrast, productive infection, which requires vacuole formation, occurred exclusively in the cytoplasm of nonwounded cells. These findings support the concept that the parasitophorous vacuole is formed by membrane invagination at the time of sporozoite entry and also confirm earlier work documenting that vacuole formation is not required for sporozoite-to-liver stage transformation.[113]

Plasmodium Proteins Involved in Crossing the Blood Liver Barrier

Transgenic parasites have greatly advanced our understanding of the liver phase of a malaria infection. For example, *P. berghei* engineered to express CSPs with modified amino acid sequences have provided information on the motifs involved in sporozoite binding to the sinusoid. The CSPs from all mammalian *Plasmodium* species contain two heparin binding motifs: a cluster of basic amino acids upstream from region I and the conserved TSR motif at the C terminus.[14,18,19,114,115] Disruption of the conserved thrombospondin repeat type I domain (TSR or region II-plus) abolished parasite motility and infectivity,[116] supporting the notion that this cell-adhesive motif, which is also present in many mammalian proteins,[13] is crucial for sporozoite arrest in the liver. Upon target cell contact, the entire N-terminus of surface-expressed CSP is proteolytically removed by a papain family cysteine protease of parasite origin.[117] The processing, which is crucial for sporozoite infectivity in rodents, is thought to occur within the short conserved region I (KLKQP). Further experimental confirmation of this hypothesis is needed, since elimination of region I from *P. falciparum* CSP expressed by transgenic *P. berghei* parasites had no effect on parasite motility or infectivity.[116] The central repeats appear not to have a role in parasite recognition of the liver, because this region is interchangeable between different mammalian *Plasmodium* species: *P. berghei* sporozoites, which express *P. falciparum* or *P. yoelii* CSP instead of the native protein, or hybrid *P. berghei* parasites, in which the repeats were replaced by those from *P. falciparum* CSP, develop normally in rodents.[116,118] These findings suggest that exposure of the TSR motif of CSP is required for liver infection, while the repeat region does not play a direct role. Interestingly, replacement of *P. berghei* CSP with the *P. gallinaceum* homolog had no effect on sporozoite motility, but abolished infectivity for mice.[119] Notably, compared to the selective development of mammalian *Plasmodium* species in the liver, *P. gallinaceum* exhibits a much more promiscuous cell and tissue tropism: cells supporting the growth of exoerythrocytic stages include capillary endothelia in the brain and macrophages in all organs of the reticulo-endothelial system.[120,121] *P. gallinaceum* CSP lacks the two heparin binding motifs[122] so that the most likely explanation for the observed attenuation of these hybrid parasites is a failure to home to the liver, the only organ that supports merozoite formation in mammalian *Plasmodium* species. Taken together, the data are consistent with a scenario in which the two heparin-binding motifs of CSP mediate the gliding of the parasites along the liver sinusoid and the subsequent recognition of Kupffer cell surface proteoglycans. When the transmigration phase has ended and the sporozoite has found its final nurse hepatocyte, CSP is proteolytically processed and the parasite commences to the next stage of its life cycle. Once confirmed with a suitable host parasite combination in vivo, these genetic approaches will help understand the molecular steps the complex liver infection cascade.

The micronemal thrombospondin-related adhesive protein (TRAP) is involved in gliding motility and also in sporozoite infection of the liver.[123,124] Intravenously inoculated TRAP knockout parasites or mutants with modifications in the adhesive domains are markedly less infectious to rats compared to wild-type parasites,[123,125] but no information is available as to which step in the liver infection cascade is impaired in the absence of TRAP. Together with aldolase, actin, and myosin A, the cytoplasmic tail of TRAP forms an intracellular motor complex that is essential for sporozoite motility[126,127] so that sinusoidal gliding could be blocked. Alternatively, sporozoite passage through Kupffer cells could be impaired, because gliding motility is generally thought to propel apicomplexan parasites through the moving junction formed during cell entry by membrane invagination.[128] Finally, the thrombospondin type I repeat and the integrin-like A domain of TRAP are involved in host cell invasion in vitro and in vivo,[10,125,129] raising the possibility that TRAP is required for hepatocyte transmigration and/or invasion. Insight into the possible mechanism was provided by the finding that complex formation between the A domain and the sialylated glycoprotein fetuin-A mediates binding of TRAP to hepatoma cells and liver infection is delayed in fetuin-A deficient mice.[129] Similar to TRAP, the micronemal protein apical membrane antigen 1 (AMA-1) is relocated to the parasite surface upon microneme exocytosis and proteolytically processed during *P. falciparum* sporozoite invasion of human hepatocytes in vitro.[130] The finding that antibodies against AMA-1 inhibit hepatocyte invasion in vitro emphasizes the complex nature of sporozoite entry into the liver and makes the involvement of additional parasite proteins likely.

Tachyzoites of a distant apicomplexan relative of *Plasmodium*, *Toxoplasma gondii,* use the interaction between the surface protein MIC2 and ICAM-1 for crossing epithelial cell barriers at tight junctions, where cellular and junctional adhesion molecules are concentrated.[131] This mechanism, however, may not be directly applicable to *Plasmodium* sporozoites for the following reasons. (1) The paracellular passage of *Toxoplasma* tachyzoites across epithelial barriers occurs in the absence of any host cell damage,[131] while *Plasmodium* sporozoites transmigrate through hepatocytes (cells of epithelial origin) by wounding their membrane and causing them to die by necrosis.[16,105] (2) Tight junctions restrict the solute flow across epithelial cell barriers. Not surprisingly, therefore, they are absent from sinusoidal endothelia, which are designed to facilitate the transport of blood plasma to towards the underlying hepatocytes.[63,132,133] (3) In contrast to epithelial cells, ICAM-1 expression is not restricted to cellular junctions, but distributed homogeneously on the surface of sinusoidal endothelia.[134] (4) ICAM-1 is absent from lymphatic endothelia,[56] excluding the involvement of this adhesion molecule in sporozoite recognition of dermal lymph vessels.[2] Taken together, it appears that the para-epithelial path of *Toxoplasma* tachyzoites is distinct from the route *Plasmodium* sporozoites take across the sinusoidal cell barrier.

Further insight into the mechanisms malaria sporozoites possess to traverse the sinusoidal cell barrier may come from genetically attenuated *P. berghei* mutants. For example, *P. berghei* mutants deficient in P36p, a member of the P48/45 family of sporozoite surface proteins, are able to transmigrate through and infect HepG2 cells in vitro, but are arrested at a very early stage of development in these cells, probably related to their inability to maintain a parasitophorous vacuole.[135] In vivo, Ishino and coworkers found similar numbers of

P36p-deficient and wild-type sporozoites associated with rat livers shortly after parasite inoculation.[109] Immunofluorescence analysis of relatively thick frozen sections, in which sinusoidal cells are difficult to identify, led the authors to propose that the parasites had crossed the sinusoidal cell layer and entered the liver parenchyma normally, but then remained outside of hepatocytes suggesting a defect in transmigration and invasion. On the other hand, van Dijk and coworkers showed that immunization with P36p-deficient parasites protects mice against challenge with wild-type *P. berghei* sporozoites.[135] Because hepatocyte infection is required for the generation of a protective immune response against malaria liver stages,[136] these data suggest that at least some of the P36p knockout parasites had successfully entered parenchymal cells. Future analysis of the function of P36p, the exact localization of P36p-deficient parasites and their long-term fate in the liver should provide further information as to whether parasite persistence is required for long-term immunity or whether cross-presentation of *Plasmodium* antigens, for example by dendritic cells, is adequate to maintain protection.

Genetically attenuated *P. berghei* strains, in which expression of the small liver stage-specific proteins UIS3 and UIS4 was disrupted, are clearly impaired in events occurring after hepatocyte invasion.[137,138] UIS3 and UIS4 knockout mutants fail to develop to mature liver stages and therefore do not proceed to clinical malaria. UIS4 is normally exported into the parasitophorous vacuole that surrounds developing liver stages inside hepatocytes, suggesting the possibility that vacuole maintenance is impaired in UIS4-deficient parasites. It remains to be explored if UIS3 and UIS4 play a general role in sporozoite invasion by membrane invagination, in which case these proteins might also be involved in the passage through Kupffer cells. Alternatively, the proteins may be required solely for maintenance of the parasitophorous vacuole in hepatocytes so that their role would be limited to liver stage development.

How are parasite and hepatocyte primed for the days of intimate relationship culminating in the release of thousands of erythrocyte-infective merozoites from the remains of an exhausted nurse cell? An interesting concept is that sporozoites must be activated for productive liver infection by transmigrating through hepatocytes.[108] However, sporozoites must cross at least two vascular barriers, the capillary wall in the skin and the sinusoidal cell layer of the liver, before reaching any hepatocyte. Because passage through Kupffer cells occurs by membrane invagination,[9,44,46] sporozoites must already be activated when they arrive in the liver. Further, disruption of SPECT, which abolishes the capability for sporozoite transmigration, has no effect on liver stage development[48] suggesting that sporozoites are constitutively "activated". The interpretation of the activation hypothesis has therefore been challenged.[6] Another study demonstrated that sporozoite transmigration caused secretion of hepatocyte growth factor (HGF) from wounded hepatoma cells in vitro.[139] This finding led the authors to propose that in the liver, release of HGF from wounded hepatocytes enhances expression of the HGF receptor MET in neighboring parenchymal cells thus rendering them susceptible for productive *Plasmodium* infection. In vivo, however, the predominant source of HGF is nonparenchymal cells, in particular Kupffer cells.[140-142] An alternative explanation for the observed increased MET expression in the liver[139] is therefore that hepatocyte death, a consequence of sporozoite transmigration,[16] induced the normal cascade of liver regeneration.[143] In response to liver injury, Kupffer cells typically proliferate and produce HGF, which primes hepatocytes in a paracrine or endocrine manner for replication via the receptor tyrosine kinase MET.[140-142] Future work will hopefully unravel the details of this complex molecular interplay between host and parasite.

Malaria Sporozoites Modulate Kupffer Cell Function

Kupffer cells are professional macrophages and antigen-presenting cells.[144] Strategically positioned in the liver sinusoids, they efficiently clear foreign and altered-self substances from the blood.[64,65] Despite their ability to phagocytoze microorganisms, Kupffer cells surprisingly do not kill malaria sporozoites. For example, sporozoites can pass through peritoneal macrophages rapidly and actively in vitro and morphologically alter and eventually destroy the phagocytes.[145-148] These findings have suggested that *Plasmodium* sporozoites are able to either evade or actively suppress the respiratory burst in macrophages. To test this hypothesis, we analyzed intracellular signaling events known to interfere with the production of reactive oxygen species. We found that *P. berghei* and *P. yoelii* sporozoites as well as recombinant *P. falciparum* CSP increase, in a dose- and time-dependent fashion, the intracellular concentration of cAMP and inositol 1,4,5-triphosphate (IP_3) in Kupffer cells, but not in other liver cell types such as hepatocytes and sinusoidal endothelia.[32] The CSP-mediated cAMP elevation suppresses the zymosan- and phorbolester-induced production of reactive oxygen species (respiratory burst) in Kupffer cells. Similar effects were obtained when Kupffer cells were preincubated with cAMP analogs or when cAMP degradation was prevented with specific phosphodiesterase inhibitors. Blockage of the adenylyl cyclase abrogated the suppressive effect of sporozoites on the respiratory burst. Taken together, these results suggest that the sporozoite-induced elevation of the intracellular cAMP concentration inhibits the assembly of the NADPH oxidase. Because the archetypal second messenger cAMP is responsible for the conversion of many intercellular signals to intracellular events, it must be expected that malaria sporozoites are able to manipulate other Kupffer cell functions as well. Indeed, *P. berghei* sporozoites were shown to up-regulate the anti-inflammatory cytokine IL-10 and down-modulate the pro-inflammatory IL-12 as well as MHC class I expression in vivo.[149]

Merozoite Release across the Sinusoidal Cell Barrier

The generation of a transgenic *P. yoelii* parasite strain that fluoresces throughout the life cycle has enabled us to observe by intravital microscopy the development of exoerythrocytic parasites in the mouse liver and to document the release of merozoites from mature schizonts.[150] Interestingly, liver stages exhibited an irregular, highly flexible outline and released membrane-bound "extrusomes" or "merosomes", packets of parasite material of variable size and shape, across the sinusoidal cell barrier into the sinusoidal lumen.[151] Studies are in progress to analyze these events in greater detail.

Evolutionary Aspects

The obligatory role of Kupffer cells for sporozoite infection of the liver and the ability of the parasites to control macrophage function does not come as a surprise. Avian and reptilian malaria species such as *P. gallinaceum* do not develop in hepatocytes, but initially infect dermal macrophages at the mosquito bite site and later tissue macrophages in all organs of the reticulo-endothelial system as well as vascular endothelia of the brain.[120] In the liver, tissue stages of *P. gallinaceum* appear to develop exclusively in Kupffer cells[120,152] and remain considerably smaller than their mammalian counterparts maturing in hepatocytes (Frevert, unpublished data).[121] It seems therefore plausible that the evolutionarily younger mammalian *Plasmodium* species have retained the ability for recognition of and stealth entry into Kupffer cells, but have acquired new mechanisms to continue their journey across the space of Disse and into parenchymal cells. With their enormous metabolic activity, scarce expression of histocompatibility antigens, and relatively concealed location, hepatocytes conceivably provide a nutritionally richer and immunologically safer environment, thus better supporting the rapid growth of the liver stages to a size much larger than their original host cell.

Conclusion

Sporozoite infection of the liver follows a complex cascade of events, involving arrest by binding to liver-specific proteoglycans, gliding along sinusoidal endothelia, deactivation of and passage through Kupffer cells, transmigration through several hepatocytes, and finally invasion of a suitable hepatocyte for schizogony and merozoite formation. Kupffer cells do not only play a key role in the onset of a malaria infection, but also in a unique property of the liver: the induction of a tolerance rather than inflammation and immunity against incoming antigens.[144,153] Thus, by choosing the liver as a port of entry into the mammalian host, *Plasmodium* is able to exploit the tolerogenic properties of this exceptional immune organ.

References

1. Vanderberg JP, Frevert U. Intravital microscopy demonstrating antibody-mediated immobilization of Plasmodium berghei sporozoites injected into skin by mosquitoes. Int J Parasitol 2004; 34:991-996.
2. Amino R, Thiberge S, Martin B et al. Quantitative imaging of Plasmodium transmission from mosquito to mammal. Nat Med 2006; 12:220-224.
3. Frevert U. Sneaking in through the back entrance: The biology of malaria liver stages. Trends Parasitol 2004; 20:417-424.
4. Yuda M, Ishino T. Liver invasion by malarial parasites—how do malarial parasites break through the host barrier? Cell Microbiol 2004; 6:1119-1125.
5. Amino R, Menard R, Frischknecht F. In vivo imaging of malaria parasites—recent advances and future directions. Curr Opin Microbiol 2005; 8:407-414.
6. Silvie O, Franetich JF, Renia L et al. Malaria sporozoite: Migrating for a living. Trends Mol Med 2004; 10:97-100.
7. Mota MM, Rodriguez A. Migration through host cells: The first steps of Plasmodium sporozoites in the mammalian host. Cell Microbiol 2004; 6:1113-1118.
8. Frevert U, Usynin I, Baer K et al. Nomadic or sessile: Can Kupffer cells function as portals for malaria sporozoites to the liver? Cell Microbiol 2006; 8:1537-1546.
9. Pradel G, Garapaty S, Frevert U. Proteoglycans mediate malaria sporozoite targeting to the liver. Mol Microbiol 2002; 45:637-651.
10. Robson KJH, Frevert U, Reckmann I et al. Thrombospondin related adhesive protein (TRAP) of Plasmodium falciparum: Expression during sporozoite ontogeny and binding to human hepatocytes. EMBO J 1995; 14:3883-3894.
11. Pradel G, Garapaty S, Frevert U. Kupffer and stellate cell proteoglycans mediate malaria sporozoite targeting to the liver. Comp Hepatol 2004; 3(Suppl 1):S47.
12. Gressner AM, Schäfer S. Comparison of sulphated glycosaminoglycan and hyaluronate synthesis and secretion in cultured hepatocytes, fat storing cells, and Kupffer cells. J Clin Chem Clin Biochem 1989; 27:141-149.
13. Cerami C, Frevert U, Sinnis P et al. The basolateral domain of the hepatocyte plasma membrane bears receptors for the circumsporozoite protein of Plasmodium falciparum sporozoites. Cell 1992; 70:1021-1033.
14. Frevert U, Sinnis P, Cerami C et al. Malaria circumsporozoite protein binds to heparan sulfate proteoglycans associated with the surface membrane of hepatocytes. J Exp Med 1993; 177:1287-1298.
15. Lyon M, Denkin JA, Gallagher JT. Liver heparan sulfate structure. A novel molecular design. J Biol Chem 1994; 269:11208-11215.
16. Frevert U, Engelmann S, Zougbédé S et al. Intravital observation of Plasmodium berghei sporozoite infection of the liver. PLoS Biol 2005; 3:e192.
17. Pinzon-Ortiz C, Friedman J, Esko J et al. The binding of the circumsporozoite protein to cell surface heparan sulfate proteoglycans is required for Plasmodium sporozoite attachment to target cells. J Biol Chem 2001; 276:26784-26791.
18. Ying P, Shakibaei M, Patankar MS et al. The malaria circumsporozoite protein: Interaction of the conserved regions I and II-plus with heparin-like oligosaccharides in heparan sulfate. Exp Parasitol 1997; 85:168-182.
19. Ancsin JB, Kisilevsky R. A binding site for highly sulfated heparan sulfate is identified in the amino-terminus of the circumsporozoite protein: Significance for malarial sporozoite attachment to hepatocytes. J Biol Chem 2004; 279:21824-21832.
20. Rathore D, Kumar S, Lanar DE et al. Disruption of disulfide linkages of the Plasmodium falciparum circumsporozoite protein: Effects on cytotoxic and antibody responses in mice. Mol Biochem Parasitol 2001; 118:75-82.
21. Rathore D, McCutchan TF. Heparin can regulate the binding of Plasmodium falciparum circumsporozoite protein. Mol Biochem Parasitol 2000; 108:253-256.
22. Rathore D, McCutchan TF, Garboczi DN et al. Direct measurement of the interactions of glycosaminoglycans and a heparin decasaccharide with the malaria circumsporozoite protein. Biochemistry 2001; 40:11518-11524.
23. Shakibaei M, Frevert U. Dual interaction of the malaria circumsporozoite protein with the low density lipoprotein receptor-related protein (LRP) and cell surface heparan sulfate. J Exp Med 1996; 184:1699-1711.
24. Strickland DK, Kounnas MZ, Argraves WS. LDL receptor-related protein: A multiligand receptor for lipoprotein and proteinase catabolism. FASEB J 1995; 9:890-897.

25. Strickland DK, Kounnas MZ, Williams SE et al. LDL receptor-related protein (LRP): A multiligand receptor. Fibrinolysis 1994; 8(Suppl):204-215.
26. Herz J. The LDL-receptor-related protein—portrait of a multifunctional receptor. Curr Opin Lipidol 1993; 4:107-113.
27. Rohlmann A, Gotthardt M, Hammer RE et al. Inducible inactivation of hepatic LRP gene by cre-mediated recombination confirms role of LRP in clearance of chylomicron remnants. J Clin Invest 1998; 101:689-695.
28. Rohlmann A, Gotthardt M, Willnow TE et al. Sustained somatic gene inactivation by viral transfer of Cre recombinase. Nat Biotechnol 1996; 14:1562-1565.
29. Herz J, Gerard RD. Adenovirus-mediated transfer of low density lipoprotein receptor gene acutely accelerates cholesterol clearance in normal mice. Proc Natl Acad Sci USA 1993; 90:2812-2816.
30. Shayakhmetov DM, Li ZY, Ni S et al. Analysis of adenovirus sequestration in the liver, transduction of hepatic cells, and innate toxicity after injection of fiber-modified vectors. J Virol 2004; 78:5368-5381.
31. Marshall P, Rohlmann A, Nussenzweig V et al. Plasmodium sporozoites invade cells with targeted deletions in the LDL receptor related protein. Mol Biochem Parasitol 2000; 106:293-298.
32. Usynin I, Klotz C, Frevert U. Malaria circumsporozoite protein inhibits the respiratory burst in Kupffer cells. Submitted.
33. Moestrup SK, Gliemann J, Pallesen G. Distribution of the α_2-macroglobulin receptor / low density lipoprotein receptor-related protein in human tissues. Cell Tissue Res 1992; 269:375-382.
34. Mazier D, Collins WE, Mellouk S et al. Plasmodium ovale: In vitro development of hepatic stages. Exp Parasitol 1987; 64:393-400.
35. Millet P, Collins WE, Fisk TL et al. In vitro cultivation of exoerythrocytic stages of the human malaria parasite Plasmodium malariae. Am J Trop Med Hyg 1988; 38:470-473.
36. Millet P, Fisk TL, Collins WE et al. Cultivation of exoerythrocytic stages of Plasmodium cynomolgi, P. knowlesi, P. coatneyi, and P. inui in Macaca mulatta hepatocytes. Am J Trop Med Hyg 1988; 39:529-534.
37. Karnasuta C, Pavanand K, Chantakulkij S et al. Complete development of the liver stage of Plasmodium falciparum in a human hepatoma cell line. Am J Trop Med Hyg 1995; 53:607-611.
38. Karnasuta C, Watt G. Enhanced detection of Plasmodium vivax liver stages by cytocentrifugation. Parasitol Today 1996; 12:451-453.
39. Sinden RE, Suhrbier A, Davies CS et al. The development and routine application of high-density exoerythrocytic-stage culture of Plasmodium berghei. WHO Bulletin 1990; 68(Suppl):115-125.
40. Calvo-Calle JM, Moreno A, Eling WMC et al. In vitro development of infectious liver stages of P. yoelii and P. berghei malaria in human cell lines. Exp Parasitol 1994; 79:362-373.
41. Mota MM, Rodríguez A. Invasion of mammalian host cells by Plasmodium sporozoites. BioEssays 2002; 24:149-156.
42. Hollingdale MR. Biology and immunology of sporozoite invasion of liver cells and exoerythrocytic development of malaria parasites. Prog Allergy 1988; 41:15-48.
43. Hollingdale MR. Malaria and the liver. Hepatology 1985; 5:327-335.
44. Pradel G, Frevert U. Plasmodium sporozoites actively enter and pass through Kupffer cells prior to hepatocyte invasion. Hepatology 2001; 33:1154-1165.
45. Meis JFGM, Verhave JP, Brouwer A et al. Electron microscopic studies on the interaction of rat Kupffer cells and Plasmodium berghei sporozoites. Z Parasitenkd 1985; 71:473-483.
46. Meis JFGM, Verhave JP, Jap PHK et al. An ultrastructural study on the role of Kupffer cells in the process of infection by Plasmodium berghei sporozoites in rats. Parasitology 1983; 86:231-242.
47. Ishino T, Chinzei H, Yuda M. A Plasmodium sporozoite protein with a membrane attack complex domain is required for breaching the liver sinusoidal cell layer prior to hepatocyte infection. Cell Microbiol 2005; 7:199-208.
48. Ishino T, Yano K, Chinzei Y et al. Cell-passage activity is required for the malarial parasite to cross the liver sinusoidal cell layer. PLoS Biol 2004; 2:E4.
49. Sinden RE, Smith JE. The role of the Kupffer cell in the infection of rodents by sporozoites of Plasmodium: Uptake of sporozoites by perfused liver and the establishment of infection in vivo. Acta Trop 1982; 39:11-27.
50. Verhave JP, Meuwissen JHET, Golenser J. The dual role of macrophages in the sporozoite-induced malaria infection. A hypothesis. Int J nucl Med Biol 1980; 7:149-156.
51. Vreden SGS, Sauerwein RW, Verhave JP et al. Kupffer cell elimination enhances development of liver schizonts of Plasmodium berghei in rats. Infect Immun 1993; 61:1936-1939.
52. Gosset P, Lassalle P, Vanhée D et al. Production of tumor necrosis factor-α and interleukin-6 by human alveolar macrophages exposed in vitro to coal mine dust. Am J Resp Cell Mol Biol 1991; 5:431-436.

53. Braverman IM. Ultrastructure and organization of the cutaneous microvasculature in normal and pathologic states. J Invest Dermatol 1989; 93:2S-9S.
54. Perlmutter LS, Chui HC. Microangiopathy, the vascular basement membrane and Alzheimer's disease: A review. Brain Res Bull 1990; 24:677-686.
55. Nerlich AG, Schleicher E. Identification of lymph and blood capillaries by immunohistochemical staining for various basement membrane components. Histochemistry 1991; 96:449-453.
56. Erhard H, Rietveld FJ, Brocker EB et al. Phenotype of normal cutaneous microvasculature. Immunoelectron microscopic observations with emphasis on the differences between blood vessels and lymphatics. J Invest Dermatol 1996; 106:135-140.
57. Tilton RG. Capillary pericytes: Perspectives and future trends. J Electron Microsc Tech 1991; 19:327-344.
58. Sims DE. Diversity within pericytes. Clin Exp Pharmacol Physiol 2000; 27:842-846.
59. Gerhardt H, Betsholtz C. Endothelial-pericyte interactions in angiogenesis. Cell Tissue Res 2003; 314:15-23.
60. Schmid-Schönbein GW. Microlymphatics and lymph flow. Physiol Rev 1990; 70:987-1028.
61. Jeltsch M, Tammela T, Alitalo K et al. Genesis and pathogenesis of lymphatic vessels. Cell Tissue Res 2003; 314:69-84.
62. Zöltzer H. Initial lymphatics—morphology and function of the endothelial cells. Lymphology 2003; 36:7-25.
63. Wisse E, Zanger RB, Charels K et al. The liver sieve: Considerations concerning the structure and function of endothelial fenestrae, the sinusoidal wall and the space of Disse. Hepatology 1985; 5:683-692.
64. Gumucio JJ, Bilir BM, Moseley RH et al. The biology of the liver cell plate. In: Arias IM, Boyer JL, Fausto N, Jakoby WB, Schachter DA, Shafritz DA, eds. The Liver: Biology and Pathobiology. New York: Raven Press, 1994:1143-1163.
65. Kuiper J, Brouwer A, Knook DL et al. Kupffer and sinusoidal endothelial cells. In: Arias IM, Boyer JL, Fausto N, Jakoby WB, Schachter DA, Shafritz DA, eds. The Liver: Biology and Pathobiology. New York: Raven Press, Ltd.,1994:791-818.
66. Wake K, Decker K, Kirn A et al. Cell biology and kinetics of Kupffer cells in the liver. Int Rev Cytol 1989; 118:173-229.
67. Lee SH, Starkey PM, Gordon S. Quantitative analysis of total macrophage content in adult mouse tissues. Immunochemical studies with monoclonal antibody F4/80. J Exp Med 1985; 161:475-489.
68. Naito M, Hayashi S, Yoshida H et al. Abnormal differentiation of tissue macrophage populations in 'osteopetrosis' (op) mice defective in the production of macrophage colony-stimulating factor. Am J Pathol 1991; 139:657-667.
69. Witmer-Pack MD, Hughes DA, Schuler G et al. Identification of macrophages and dendritic cells in hte osteopetrotic (op/op) mouse. J Cell Sci 1993; 104:1021-1029.
70. Cecchini MG, Dominguez MG, Mocci S et al. Role of colony stimulating factor-1 in the establishment and regulation of tissue macrophages during postnatal development of the mouse. Development 1994; 120:1357-1372.
71. Verhave JP, Meis JFGM, Boo TMd et al. The delivery of exoerythrocytic parasites of Plasmodium berghei: A hormone controlled process. Ann Soc Belge Med Trop 1985; 65:35-44.
72. Sleyster EC, Knook DL. Relation between localization and function of rat liver Kupffer cells. Lab Invest 1982; 47:484-490.
73. Motta PM. The three-dimensional microanatomy of the liver. Arch Histol Jpn 1984; 47:1-30.
74. Motta PM, Muto M, Fujita T. The vascular system. In: Motta PM, Muto M, Fujita T, eds. The Liver: An Atlas of Electron Microscopy. Tokyo: Igaku-Shoin, 1978:83-118.
75. Arii S, Imamura M. Physiological role of sinusoidal endothelial cells and Kupffer cells and their implication in the pathogenesis of liver injury. J Hepatobiliary Pancreat Surg 2000; 7:40-48.
76. McCuskey RS, Reilly FD. Hepatic microvasculature: Dynamic structure and its regulation. Semin Liver Dis 1993; 13:1-12.
77. Bouwens L, Wisse E. The origin of Kupffer cells and their relationship to hepatocytes. In: Billiar TR, Curran RD, eds. Hepatocyte and Kupffer Cell Interactions. Boca Raton: CRC Press, 1992:3-21.
78. Wisse E. Observations on the fine structure and peroxidase cytochemistry of normal rat liver Kupffer cells. J Ultrastruct Res 1974; 46:393-426.
79. Wisse E. Ultrastructure and function of Kupffer cells and other sinusoidal cells in the liver. In: Wisse E, Knook DL, eds. Kupffer Cells and other Liver Sinusoidal Cells. Amsterdam: Elsevier/North Holland Biomedical Press, 1977:33-60.
80. Wisse E, Braet F, Luo D et al. Structure and function of sinusoidal lining cells in the liver. Toxicol Pathol 1996; 24:100-111.
81. Mackay IR. Hepatoimmunology: A perspective. Immunol Cell Biol 2002; 80:36-44.

82. Bertolino P, McCaughan GW, Bowen DG. Role of primary intrahepatic T-cell activation in the 'liver tolerance effect'. Immunol Cell Biol 2002; 80:84-92.
83. MacSween RNM, Desmet VJ, Roskams T et al. Developmental anatomy and normal structure. In: MacSween RNM, Burt AD, Portmann BC, Ishak K, Scheuer GPJ, Anthony PP, eds. Pathology of the Liver. New York: Churchill Livingstone, 2002:1-66.
84. Markiewski MM, DeAngelis RA, Lambris JD. Liver inflammation and regeneration: Two distinct biological phenomena or parallel pathophysiologic processes? Mol Immunol 2006; 43:45-56.
85. MacPhee PJ, Schmidt EE, Groom AC. Evidence for Kupffer cell migration along liver sinusoids, from high-resolution in vivo microscopy. Am J Physiol 1992; 263:G17-23.
86. Faust N, Varas F, Kelly LM et al. Insertion of green fluorescent protein into the lysozyme gene creates mice with green fluorescent granulocytes and macrophages. Blood 2000; 96:719-726.
87. Baer K, Roosevelt M, Van Rooijen N et al. Kupffer cells are obligatory for Plasmodium sporozoite infection of the liver. Cell Microbiol 2006; 8:online early.
88. McCuskey RS. Morphological mechanism for regulating blood flow through hepatic sinusoids. Liver 2000; 20:3-7.
89. McCuskey RS. The hepatic microvascular system. In: Arias IM, Boyer JL, Fausto N, Jakoby WB, Schachter DA, Shafritz DA, eds. The Liver: Biology and Pathobiology. New York: Raven Press, 1994:1089-1106.
90. Bouwens L, Baekeland M, De Zanger R et al. Quantitation, tissue distribution and proliferation kinetics of Kupffer cells in normal rat liver. Hepatology 1986; 6:718-722.
91. Crofton RW, Diesselhoff-den Dulk MM, van Furth R. The origin, kinetics, and characteristics of the Kupffer cells in the normal steady state. J Exp Med 1978; 148:1-17.
92. Diesselhoff-den Dulk MM, Crofton RW, van Furth R. Origin and kinetics of Kupffer cells during an acute inflammatory response. Immunology 1979; 37:7-14.
93. van Furth R. [Characteristics of mononuclear phagocytes in their normal state and in inflammation]. Folia Haematol Int Mag Klin Morphol Blutforsch 1984; 111:131-140.
94. Hardonk MJ, Dijkhuis FW, Grond J et al. Evidence for a migratory capability of rat Kupffer cells to portal tracts and hepatic lymph nodes. Virchows Arch B Cell Pathol Incl Mol Pathol 1986; 51:429-442.
95. McCuskey RS. Hepatic and splanchnic microvascular responses to inflammation and shock. Hepatogastroenterology 1999; 46(Suppl 2):1464-1467.
96. Lawson JA, Farhood A, Hopper RD et al. The hepatic inflammatory response after acetaminophen overdose: Role of neutrophils. Toxicol Sci 2000; 54:509-516.
97. Wiktor-Jedrzejczak W, Gordon S. Cytokine regulation of the macrophage (Mτ) system studied using the colony stimulating factor-1-deficient op/op mouse. Physiol Rev 1996; 76:927-947.
98. Nishioji K, Okanoue T, Mori T et al. Experimental liver injury induced by Propionibacterium acnes and lipopolysaccharide in macrophage colony stimulating factor-deficient osteopetrotic (op/op) mice. Dig Dis Sci 1999; 44:1975-1984.
99. Szperl M, Ansari AA, Urbanowska E et al. Increased resistance of CSF-1-deficient, macrophage-deficient, TNF alpha-deficient, and IL-1 alpha-deficient op/op mice to endotoxin. Ann NY Acad Sci 1995; 762:499-501.
100. Shultz LD, Sidman CL. Genetically determined murine models of immunodeficiency. Annu Rev Immunol 1987; 5:367-403.
101. Knight TR, Jaeschke H. Peroxynitrite formation and sinusoidal endothelial cell injury during acetaminophen-induced hepatotoxicity in mice. Comp Hepatol 2004; 3(Suppl 1):S46.
102. Yee SB, Hanumegowda UM, Copple M BL et al. Endothelial cell injury and coagulation system activation during synergistic hepatotoxicity from monocrotaline and bacterial lipopolysaccharide coexposure. Toxicol Sci 2003; 74:203-214.
103. DeLeve LD, Ito Y, Bethea NW et al. Embolization by sinusoidal lining cells obstructs the microcirculation in rat sinusoidal obstruction syndrome. Am J Physiol Gastrointest Liver Physiol 2003; 284:G1045-1052.
104. Naito M, Hasegawa G, Takahashi K. Development, differentiation, and maturation of Kupffer cells. Microsc Res Tech 1997; 39:350-364.
105. Mota MM, Pradel G, Vanderberg JP et al. Migration of Plasmodium sporozoites through cells before infection. Science 2001; 291:141-144.
106. Silvie O, Greco C, Franetich JF et al. Expression of human CD81 differently affects host cell susceptibility to malaria sporozoites depending on the Plasmodium species. Cell Microbiol 2006; 8:1134-1146.
107. Silvie O, Rubinstein E, Franetich JF et al. Hepatocyte CD81 is required for Plasmodium falciparum and Plasmodium yoelii sporozoite infectivity. Nature Med 2002; 9:93-96.

108. Mota MM, Hafalla JC, Rodriguez A. Migration through host cells activates Plasmodium sporozoites for infection. Nat Med 2002; 8:1318-1322.
109. Ishino T, Chinzei Y, Yuda M. Two proteins with 6-cys motifs are required for malarial parasites to commit to infection of the hepatocyte. Mol Microbiol 2005; 58:1264-1275.
110. Bhanot P, Schauer K, Coppens I et al. A surface phospholipase is involved in the migration of Plasmodium sporozoites through cells. J Biol Chem 2005; 280:6752-6760.
111. Stewart MJ, Schulman S, Vanderberg JP. Rhoptry secretion of membrane whorls by Plasmodium berghei sporozoites. J Protozool 1985; 32:280-283.
112. Silvie O, Charrin S, Billard M et al. Cholesterol contributes to the organization of tetraspanin-enriched microdomains and to CD81-dependent infection by malaria sporozoites. J Cell Sci 2006; 119:1992-2002.
113. Kaiser K, Camargo N, Kappe SH. Transformation of sporozoites into early exoerythrocytic malaria parasites does not require host cells. J Exp Med 2003; 197:1045-1050.
114. Rathore D, Sacci JB, de la Vega P et al. Binding and invasion of liver cells by Plasmodium falciparum sporozoites. J Biol Chem 2002; 277:7092-7098.
115. Sinnis P, Clavijo P, Fenyö D et al. Structural and functional properties of region-II plus of the malaria circumsporozoite protein. J Exp Med 1994; 180:297-306.
116. Tewari R, Spaccapelo R, Bistoni F et al. Function of region I and II adhesive motifs of Plasmodium falciparum circumsporozoite protein in sporozoite motility and infectivity. J Biol Chem 2002; 277:47613-47618.
117. Coppi A, Pinzon-Ortiz C, Hutter C et al. The Plasmodium circumsporozoite protein is proteolytically processed during cell invasion. J Exp Med 2005; 201:27-33.
118. Persson C, Oliveira GA, Sultan AA et al. Cutting edge: A new tool to evaluate human preerythrocytic malaria vaccines: rodent parasites bearing a hybrid Plasmodium falciparum circumsporozoite protein. J Immunol 2002; 169:6681-6685.
119. Tewari R, Rathore D, Crisanti A. Motility and infectivity of Plasmodium berghei sporozoites expressing avian Plasmodium gallinaceum circumsporozoite protein. Cell Microbiol 2005; 7:699-707.
120. Huff CG. Exoerythrocytic stages of avian and reptilian malarial parasites. Exp Parasitol 1969; 24:383-421.
121. Frevert U, Bennett B, Spaeth G et al. Exoerythrocytic development of Plasmodium gallinaceum in the domestic chicken. In preparation.
122. McCutchan TF, Kissinger JC, Touray MG et al. Comparison of circumsporozoite proteins from avian and mammalian malarias: Biological and phylogenetic implications. Proc Natl Acad Sci USA 1996; 93:11889-11894.
123. Sultan AA, Thathy V, Frevert U et al. TRAP is necessary for gliding motility and infectivity of Plasmodium sporozoites. Cell 1997; 90:511-522.
124. Gantt SM, Clavijo P, Bai X et al. Cell adhesion to a motif shared by the malaria circumsporozoite protein and thrombospondin is mediated by its glycosaminoglycan-binding region and not by CSVTCG. J Biol Chem 1997; 272:19205-19213.
125. Matuschewski K, Nunes AC, Nussenzweig V et al. Plasmodium sporozoite invasion into insect and mammalian cells is directed by the same dual binding system. EMBO J. 2002; 21:1-10.
126. Sultan AA, Briones MRS, Gerwin N et al. Sporozoites of Plasmodium yoelii infect mice with targeted deletions in ICAM-1 and ICAM-2 or complement components C3 and C4. Mol Biochem Parasitol 1997; 88:263-266.
127. Wengelnik K, Spaccapelo R, Robson KJH et al. The A-domain and the thrombospondin-related motif of Plasmodium falciparum TRAP are implicated in the invasion process of mosquito salivary glands. EMBO J 1999; 18:5195-5204.
128. Sibley LD. Intracellular parasite invasion strategies. Science 2004; 304:248-253.
129. JethwaneyD, Lepore T, Hassan S et al. Fetuin-A, a hepatocyte-specific protein that binds Plasmodium berghei thrombospondin-related adhesive protein: A potential role in infectivity. Infect Immun 2005; 73:5883-5891.
130. Silvie O, Franetich JF, Charrin S et al. A role for apical membrane antigen 1 during invasion of hepatocytes by Plasmodium falciparum sporozoites. J Biol Chem 2004; 279:9490-9496.
131. Barragan A, Brossier F, Sibley LD. Transepithelial migration of Toxoplasma gondii involves an interaction of intercellular adhesion molecule 1 (ICAM-1) with the parasite adhesin MIC2. Cell Microbiol 2005; 7:561-568.
132. Lalor PF, Shields P, Grant A et al. Recruitment of lymphocytes to the human liver. Immunol Cell Biol 2002; 80:52-64.
133. Smedsrod B, De Bleser PJ, Braet F et al. Cell biology of liver endothelial and Kupffer cells. Gut 1994; 35:1509-1516.

134. Wang HH, Nance DM, Orr FW. Murine hepatic microvascular adhesion molecule expression is inducible and has a zonal distribution. Clin Exp Metastasis 1999; 17:149-155.
135. van Dijk MR, Douradinha B, Franke-Fayard B et al. Genetically attenuated, P36p-deficient malarial sporozoites induce protective immunity and apoptosis of infected liver cells. Proc Natl Acad Sci USA 2005; 102:12194-12199.
136. Nussenzweig V, Nussenzweig RS. Rationale for the development of an engineered sporozoite malaria vaccine. Adv Immunol 1989; 45:283-334.
137. Mueller AK, Labaied M, Kappe SH et al. Genetically modified Plasmodium parasites as a protective experimental malaria vaccine. Nature 2005; 433:164-167.
138. Mueller AK, Camargo N, Kaiser K et al. Plasmodium liver stage developmental arrest by depletion of a protein at the parasite-host interface. Proc Natl Acad Sci USA 2005; 102:3022-3027.
139. Carrolo M, Giordano S, Cabrita-Santos L et al. Hepatocyte growth factor and its receptor are required for malaria infection. Nature Med 2003; 9:1363-1369.
140. Fausto N. Liver regeneration. J Hepatol 2000; 32:19-31.
141. Rosario M, Birchmeier W. How to make tubes: Signaling by the Met receptor tyrosine kinase. Trends Cell Biol 2003; 13:328-335.
142. Zarnegar R, DeFrances MC, Michalopoulos GK. Hepatocyte growth factor. In: Arias IM, Boyer JL, Fausto N, Jakoby WB, Schachter DA, Shafritz DA, eds. The Liver: Biology and Pathobiology. New York: Raven Press, 1994:1047-1057.
143. Fausto N, Campbell JS. The role of hepatocytes and oval cells in liver regeneration and repopulation. Mech Dev 2003; 120:117-130.
144. Knolle PA, Gerken G. Local control of the immune response in the liver. Immunol Rev 2000; 174:21-34.
145. Danforth HD, Aikawa M, Cochrane AH et al. Sporozoites of mammalian malaria: Attachment to, interiorization and fate within macrophages. J Protozool 1980; 27:193-202.
146. Vanderberg JP, Chew S, Stewart MJ. Plasmodium sporozoite interactions with macrophages in vitro: A videomicroscopic analysis. J Protozool 1990; 37:528-536.
147. Vanderberg JP, Stewart MJ. Plasmodium-sporozoite-host cell interactions during sporozoite invasion. WHO Bulletin 1990; 68(Suppl.):74-79.
148. Smith JE, Alexander J. Evasion of macrophage microbicidal mechanisms by mature sporozoites of Plasmodium yoelii yoelii. Parasitology 1986; 93:33-38.
149. Steers N, Schwenk R, Bacon DJ et al. The immune status of Kupffer cells profoundly influences their responses to infectious Plasmodium berghei sporozoites. Eur J Immunol 2005; 35:2335-2346.
150. Tarun AS, Baer K, Dumpit RF et al. Quantitative isolation and in vivo imaging of malaria parasite liver stages. Int J Parasitol 2006; 36:1283-1293.
151. Sturm A, Amino R, van de Sand C et al. Manipulaation of host hepatocytes by the malaria parasite for delivery into liver sinusoids. Science 2006; 313:1287-1290.
152. Bray RS. Studies on the exo-erythrocytic cycle in the genus Plasmodium. Mem London School Hyg trop Med 1957; 12:1-92.
153. Cantor HM, Dumont AE. Hepatic suppression of sensitization to antigen absorbed into the portal system. Nature 1967; 215:744-745.

Chapter 16

Transepithelial Migration by *Toxoplasma*

Antonio Barragan* and Niclas Hitziger

Abstract

A hallmark of *T. gondii* infections is passage of parasites across restrictive biological barriers—intestine, blood-brain barrier, blood-retina barrier and placenta—during primary infection or reactivation of chronic disease. Traversal of cellular barriers permits rapid dissemination of parasites to gain access to biologically restricted organs. This process involves active parasite motility and tightly regulated interactions between host cell receptors and parasite adhesins that facilitate paracellular transfer. Mounting evidence also suggests that parasites use migrating leukocytes as Trojan horses to disseminate in the organism while avoiding immune attack. Thus, the interaction of Toxoplasma with biological barriers is a determinant factor of human toxoplasmosis. The elucidation of determinants involved in the process of migration may reveal virulence factors and novel therapeutic targets to combat disease.

Gate-Keeping Functions of Biological Barriers

One of many important functions of biological barriers is to regulate permeability of fluids to maintain a homeostatic balance by sealing the paracellular space and thereby allowing separation of apical and basal fluid compartments. Biological barriers also represent a first shield against microbial infection. A limited number of pathogenic microbes, including parasites, bacteria, viruses and fungi, are capable of crossing physiologically nonpermissive biological barriers such as the blood-brain barrier, the placenta and the intestine. Thus, while biological barriers provide a first line of defense against most pathogens, those microbes that have successfully thwarted this defense are among the most severe causes of human disease. The strategies these pathogens use to breach barriers include trafficking within leukocytes (i.e., Trojan horse), transcytosis, and paracellular migration.[1,2] The precise mechanisms by which bacterial and viral pathogens gain entry into the target tissue are partly understood and include the induction of endocytosis / phagocytosis[3,4] or host actin-based translocation.[5] In contrast, Toxoplasma, and other apicomplexans, actively penetrate their host cells in vitro[6] and this process is also instrumental in vivo. Thus, whereas the mechanisms leading to bacterial and viral penetration of biological barriers have been partly elucidated, very little is known to date about mechanisms used by parasites.

Crossing Barriers IIs a Requisite for the Establishment of Toxoplasma Infection

Following oral infection, Toxoplasma initially crosses the intestinal epithelium, disseminates into the deep tissues and traverses biological barriers such as the placenta and the

*Corresponding Author: Antonio Barragan—Swedish Institute for Infectious Disease Control and Center for Infectious Medicine, Karolinska Institutet, SE-141 86 Stockholm, Sweden. Email: antonio.barragan@ki.se

Molecular Mechanisms of Parasite Invasion, edited by Barbara A. Burleigh and Dominique Soldati-Favre.

blood-brain barrier to reach immunologically privileged sites. Here, the parasite causes the most severe pathology, i.e., disseminated congenital infections in the developing fetus,[7] severe neurological complications in immunocompromised individuals[8] and ocular pathology in otherwise healthy individuals.[9] Within the vertebrate host, Toxoplasma is capable of infecting all types of nucleated cells and of crossing several important cellular barriers. During the initial phases of infection, the parasite penetrates intestinal epithelial cells, and either develops within them, or exits through the basolateral side and burrows across the basement membrane.[10] Parasites rapidly disseminate following oral infection in mice and rats and are found in distant sites, such as the lung and heart, within several hours of ingestion.[11] Importantly, the migration of the parasite to such distant sites as the central nervous system (CNS), retina and placenta are situations most often associated with severe toxoplasmosis.

Animal Models Provide Important Clues for Understanding Dissemination

Animal models are indispensable to reveal the intricate complexity of parasite-host interactions in vivo. The murine model of Toxoplasmosis is well established for oral and intraperitoneal infections by tissue cysts or tachyzoites respectively. The generation of genetically modified parasite and mouse mutants has provided important tools to analyze host-parasite interactions.[12,13] Studies have concentrated on the disease progression determined by the appearance of the infected animal or lethality.[14,15] Other investigations e.g., assessment of tissue destruction or parasite tissue burden are limited by the need to sacrifice animals for further histological examination or parasite quantification from organ homogenates by plaquing assays or real-time PCR. Clearly, advancing our understanding of the role that cellular and molecular determinants play in the pathobiology of infection requires an integrative approach that permits evaluation of determinants of infection in the context of intact organs using noninvasive assays with fine temporal resolution.

In bioluminescence imaging (BLI), the use of internal biological sources of light, luciferases, to tag cells constitutes in vivo indicators that can be detected externally.[16] The bioluminescent photons emitted by the reporter expressed in these cells pass through the host tissue and are detectable at the surface where they provide an indication of cell localization and biological function.[17] BLI can be successfully applied to the Toxoplasma mouse model.[18] Luciferase expressing parasites injected into mice were easily detected in different organs such as spleen and liver (Fig. 1). Furthermore, dissemination of parasites to immunoprivileged organs (brain, eyes) could be monitored and the testes were identified as a site of immune privilege with significant parasite dissemination and replication[18] (Fig. 1). Interestingly, photon emission analyses revealed that dissemination of parasites in the organism occurred very rapidly and in a strain dependent fashion.[18] Importantly, dissemination to immunoprivileged organs, e.g., brain, eyes and testes, was followed by a dramatic expansion of parasite biomass and often associated with lethal outcome. BLI was also proven useful in studying host determinants of infection in vivo, e.g., Toll-like receptor signaling[18] or the implication of leukocyte trafficking in parasite dissemination.[19] Thus, BLI provides a noninvasive tool to visualize Toxoplasma infections over time in different organs and allows the quantification of parasite burden in vivo. Hence, experimental animals can be individually assessed during the course of infection, thereby eliminating inter-animal differences and substantially reducing the number of experimental animals used.

The Migratory Capacity of Toxoplasma Is a Virulence Factor during Infection

T. gondii has a highly unusual population structure comprised of three clonal lineages (I, II and III) and the parasite genotype influences the pathogenicity of the strain in mice.[20,21] Type I strains have been associated to acute virulence in the murine model and a relative over-representation of type I strains in congenital toxoplasmosis and severe cases has been

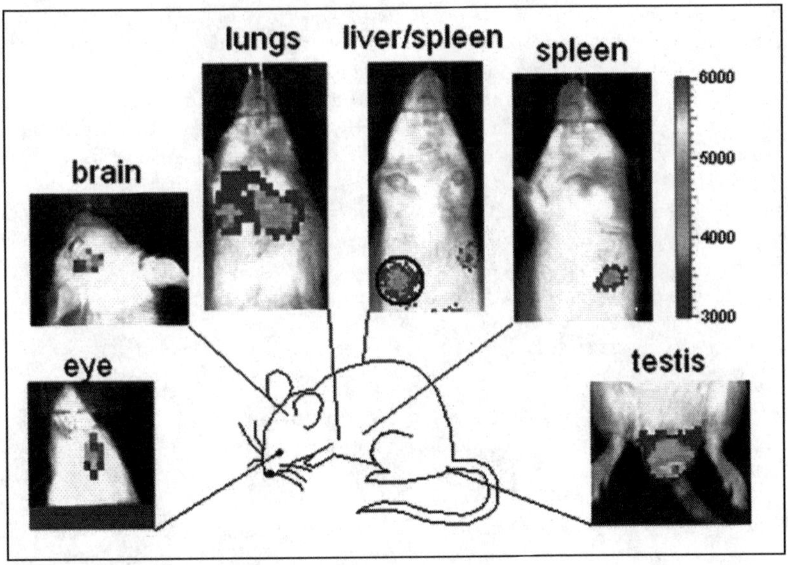

Figure 1. Dissemination of *Toxoplasma gondii* in the mouse model visualized by in vivo bioluminescence imaging. Male Balb/c mice were challenged with *T. gondii* tachyzoites that constitutively express firefly luciferase and parasite dissemination was assessed by bioluminescence imaging 3 to 7 days post infection. After injection of D-luciferin (luciferase substrate), the emitted light from the enzymatic reaction is detected externally in anesthetized mice using a highly sensitive charge-coupled device (CCD) camera. Photonic emissions (photons/sec/cm^2) are measured at each locus and are depicted using a color scale that allows organ or tissue-specific quantification of parasites. Distinct signals, i.e., parasitic foci, can be detected from eyes, CNS, liver, spleen and testes after intraperitoneal injection of *T. gondii*. Detection of parasites in the lungs was obtained after intravenous injection.

described.[21-24] One key feature of infections by type I strains in mice, is that they rapidly disseminate and reach high tissue burdens even from a very low initial inoculum.[18,25,26] This is in line with the observation that type I strains consistently exhibit a superior migratory capacity than the nonvirulent type II and type III strains.[27] Type I strain parasites also demonstrate a greater capacity for transmigration across mouse intestine ex vivo, and direct penetration into the lamina propria and vascular endothelium.[27] A subpopulation of virulent type I parasites exhibit a long distance migration (LDM) phenotype in vitro, that is not expressed by nonvirulent type II and type III strains. Also, parasites expressing the LDM phenotype present a substantial increase of migratory capacity in vivo. These studies showed that Toxoplasma is able to actively migrate across polarized cellular monolayers and extracellular matrix. Furthermore, the inherent property of Toxoplasma parasites to up-regulate their migratory capacity likely plays an important role in establishing new infections and in dissemination upon reactivation of chronic infections.

The genetic determinants of the migratory phenotype remain to be elucidated. In clones generated by a genetic cross between a virulent type I strain and a nonvirulent type III strain, an association has been found between migratory capacity and acute virulence that maps to a quantitative trait locus on chromosome VIIa (Barragan, Su, Wootton, Sibley, unpublished and ref. 28). Although advances have been made in the understanding of the underlying genetic basis for parasitic virulence,[28] the contribution of host-related factors still remains poorly understood.

Gliding Motility Powers Migration in the Microenvironment

Toxoplasma lacks cilia or flagella and the mode of locomotion, termed gliding motility, is used to enter host cells by active penetration, a rapid process that does not rely on the host cell machinery for uptake.[6,29] Gliding motility also plays a determinant role in infections by other apicomplexan parasites[30] including Plasmodium[31] and cryptosporidium.[32] Motility depends on polymerization of new actin filaments[33] and is powered by a small myosin called TgMyoA that is localized beneath the plasma membrane.[13,34] Thus, cytoskeletal components play a determinant role in parasite motility[35] and include transmembrane parasite adhesins and recently identified bridging molecules between adhesins and motor complexes.[36,37] Passage of Toxoplasma across polarized epithelium also requires active motility and occurs by direct penetration[27] but does not result in damage to the monolayer or to individual cells.[38] Thus, the use of a paracellular route for transmigration by Toxoplasma bypasses the onset of intracellular replication and also avoids damage to the host cell that would initiate an inflammatory response.[38] In contrast, transcellular migration of plasmodium sporozoites into the liver occurs by a process that may result in damage to some host cells[39] and parasite activation.[40,41] Interestingly, extravasation of neutrophils involves both paracellular and transcellular pathways,[42] and it is also possible that parasites are able to cross polarized monolayers using several different mechanisms.

Thus, mounting evidence indicates that active motility is not only used by Toxoplasma for cell invasion but also provides the parasite with an effective mechanism of dissemination in its microenvironment in tissues.[27,43] The ability to rapidly cross epithelial barriers and reach the circulation within hours after infection may be an important component of dissemination in vivo, particularly to sites of immune privilege, e.g., the CNS and the developing fetus. Following oral infection, it is possible that the ability for deep tissue migration is upregulated, for example by selection for those parasites that successfully cross the intestinal epithelial barrier. Establishment of this early wave of rapidly migrating parasites may be crucially important to assure dissemination prior to the onset of an effective immune response.

Parasitizing the Host's Trafficking Machinery

In higher organisms, the restrictive and regulative role of cellular barriers must be reconciliated with the body's need for cellular traffic, e.g., the process by which leukocytes extravasate. Leukocyte trafficking is facilitated by specialized sites at intercellular junctions.[44] The paracellular transmigration process of leukocytes involves a multistep cascade, including rolling on endothelial cells mediated by selectins, rapid activation of integrins, tight adhesion to members of the Immunoglobulin superfamily and finally, diapedesis.[1] Recent evidence suggests that Toxoplasma makes use of these systems to cross biological barriers.

Members of the apicomplexan group of parasites share an evolutionary conserved family of transmembrane proteins that belong to a family known as TRAP (thrombospondin-related anonymous protein).[45,46] These transmembrane proteins are stored inside secretory organelles and are briefly exposed to the parasite cell surface. In their ectodomains, TRAP proteins may contain adhesive modules composed of combinations of an integrin-like I/A domain and various type 1 repeats of thrombospondin (TSR). Integrin-like I/A domains are found in a variety of taxa, including prokaryotes, protozoa, plants and metazoan animals.[47] Notably, TRAP has been shown to be essential for motility and invasion by Plasmodium sporozoites and the I/A domain and TSR domain are important for productive infection in both vertebrate and invertebrate hosts.[48-50] In Toxoplasma, the TRAP-homologue MIC2 is secreted apically and translocated rapidly in a process that depends on the actin-based cytoskeleton before being released by proteolytic processing.[51,52] It has recently been shown that TRAP homologues are linked to the cytoskeleton via an interaction between the cytoplasmic C-domain and the actin-binding protein aldolase.[36] Thus, TRAP homologues function as motility receptors in a similar fashion as integrins, which connect the extracellular matrix to the actin cytoskeleton.

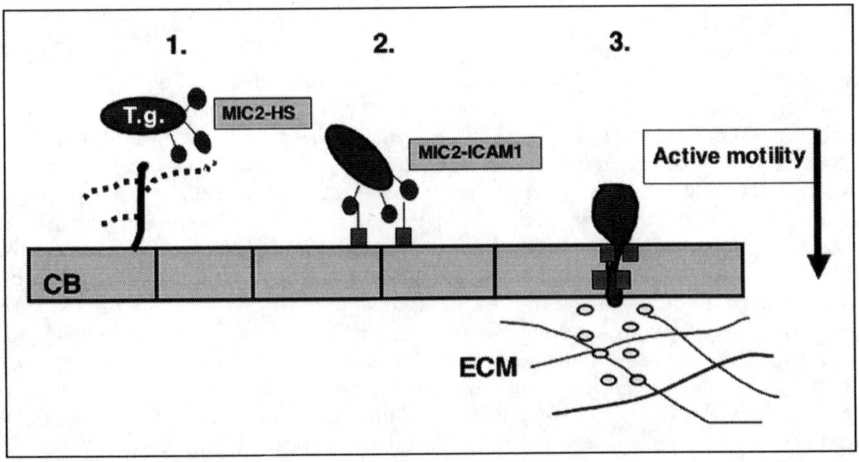

Figure 2. Transmigration of Toxoplasma across biological barriers shares features of leukocyte extravasation. Hypothetical model for transmigration of Toxoplasma gondii (T.g.) across biological barriers, e.g., the intestine, the placenta or the blood-brain barrier. Rectangles represent a cellular barrier (CB) with underlying extracellular matrix/ basal lamina (ECM). The process of transmigration shares similarities with leukocyte extravasation and implicate interactions of parasite adhesins with host cell receptors. 1) Interactions with low affinity high capacity glycan moieties, e.g., heparan sulfate (HS) with parasite-derived adhesin MIC2. 2) Firm adhesion involves ligand-receptor interactions of MIC2 with the host-cell receptor ICAM-1. 3) Transmigration occurs by a paracellular route and is powered by active parasite motility. The process of transmigration likely involves additional receptor-ligand interactions between the parasite and host cells, and secretion of proteolytic enzymes that digest the extracellular matrix to facilitate passage.

An interaction between human ICAM-1 and the parasite adhesin MIC2 was recently implicated in the process of parasite transmigration.[38] The parasite surface form of MIC2, but not the fully processed secreted form of MIC2, bound to ICAM-1 indicating that binding of this adhesin to ICAM-1 is a regulated process in transmigration. Also, soluble human ICAM-1 and ICAM-1 antibodies inhibited transmigration of parasites across cellular barriers. The interaction of mature MIC2 with ICAM-1 is highly reminiscent of the role of integrin A domains in the β2 integrins LFA-1 and Mac-1.[53] It is likely this represents an example of recent functional convergence for an otherwise ancient protein domain. Also, interactions with complex glycan moieties, glycosaminoglycans, e.g., heparan sulfate, have been shown to be important for parasite invasion[54] and these interactions are mediated in part by MIC2.[55] Consistent with this, soluble glycosaminoglycans partly inhibit transmigration of Toxoplasma across polarized cell monolayers (Barragan, unpublished).

Altogether, these findings reveal that Toxoplasma exploits the natural cell trafficking pathways in the host to cross cellular barriers and disseminate to deep tissues, likely using strategies reminiscent of leukocyte extravasation (Fig. 2). Interactions with the host's natural pathways for traffic of cells mediating immunoregulation could be determinant for the establishment of infections and for reactivation of chronic infections.

Parasite Dissemination and Host Immunity: A Race for Balance?

Early during infection, ingested parasites (oocysts and bradyzoites in tissue cysts) invade the intestine and differentiate into tachyzoites followed by spread of the organism hematogenously and via lymphatics.[56] Tachyzoites, the rapidly replicating stage, play an important role in pathogenesis during acute toxoplasmosis and upon reactivation of chronic infections in humans.[57] Natural immunity against *T. gondii* is dependent on the induction of strong parasite-specific immunity in the host.[58] Recent work indicates that apicomplexan parasites can potently modulate immune responses during the initial phases of infection.[59,60] As unlimited parasite spreading and replication would kill the host and limit parasite transmission it is necessary for *T. gondii* to alert the immune system. Hence, a strong IL-12 dependent parasite-specific interferon-γ (IFN-γ) response is triggered in the host.[58,61] By raising IFN-γ release Toxoplasma inhibits its own replication[62] and induces the transition from tachyzoite to bradyzoite stage.[63]

Because neutrophils rapidly migrate to a site of infection, this initial response may be important for parasite control before adaptive immunity is established.[14,64] These cells likely represent a first source of secreted IL-12 at the infection site.[65] Furthermore they activate dendritic cells (DC) for further IL-12 production and induction of a Th1-response to the parasite.[66] Activation of DC by Toxoplasma has been shown to involve ligation of the chemokine receptor CCR5 by a *T. gondii* protein, cyclophilin-18 (C-18).[60,67] Also, Toll-like receptors (TLR), a broad family of receptors for pathogen associated molecular patterns that signal via the adaptor molecule MyD88, have been implicated in this process and neutrophils and DC derived from MyD88$^{-/-}$-mice exhibit defective IL-12 secretion.[68]

Toll/Interleukin-1 receptor (TIR) signaling plays an important role in host resistance to Toxoplasma infection but the precise contribution of Toll-like receptor(s) remains unclear.[68,69] Importantly, assessment by BLI of the Toll/interleukin-1 receptor signaling pathway during Toxoplasma infection revealed that signal transduction to the adaptor protein MyD88 is mediated by Toll-like receptor(s) rather than by IL-1R or IL-18R signaling.[18] An increased susceptibility of TLR2 deficient mice to a high infection dose applied orally has been reported.[70] However, after intraperitoneal challenge with Toxoplasma TLR1-, TLR2-, TLR4-, TLR6-, and TLR9-deficient mice did not exhibit enhanced susceptibility to infection.[18] Noteworthy, a Toxoplasma profilin-like protein has been described to be a natural ligand for TLR11 in vitro but TLR11 deficient mice did not show an increased lethality upon Toxoplasma infections.[71] Hence, individual TLRs may be redundant and signaling via multiple TLR or yet unknown receptors may utilize MyD88 as adaptor molecule for signaling.

Thus, intricate immune evasion and immune stimulatory mechanisms assure wide dissemination and survival of parasites during the acute phase of infection prior to the onset of protective host responses.

Subversion of the Host Cell Trafficking Machinery

Infection with *T. gondii* results in recruitment of intraepithelial lymphocytes to the gut,[72,73] and inflammatory monocytes,[15] macrophages[74] and neutrophils[75] to the peritoneal cavity. Toxoplasma actively infects leukocytes in vitro[76] and tachyzoites have been identified within leukocytes in the murine intestine.[73,77] Given their migratory properties, DC constitute an especially appealing cell type. As sentinels of the innate immune system, DC recognize pathogens with subsequent maturation and migration to lymph nodes where they normally alert the adaptive immune system.[78] Upon infection by Toxoplasma, DC exhibit dramatically enhanced motility in vitro and enhanced migration in vivo.[19] In vitro studies on human and murine DC revealed that active invasion of DC by Toxoplasma induces a state of hypermotility in DC, enabling transmigration of infected DC across endothelial cell monolayers in the absence of chemotactic stimuli. Furthermore, adoptive transfer of infected DC in mice suggested that induced migration of infected DC may assure wide dissemination of parasites early during the acute phase of infection prior to the onset of protective host responses. Thus, the extravasation properties of DC in conjunction with the induction of migration of DC by *T. gondii*, present a possible strategy for the parasite to disseminate and traverse biological barriers. This is consistent with findings of DC in immunologically privileged sites in association with Toxoplasma encephalitis.[79,80] Moreover, in line with this is also the observation that high numbers of infected CD11c^{+} DC were found in circulation and in peripheral organs (Lambert, Barragan, unpublished and[81]) supporting the notion that the "Trojan horses" protect the parasite from immune attack. Manipulating DC migration could benefit the parasite by assuring its dissemination and survival, especially as infected DC appear to be compromised in their ability to activate T cells.[82,83] In this regard, it is conceivable that the ability of Toxoplasma to survive in leukocytes while modulating host cell signaling and apoptosis[84-86] may also contribute to dissemination. Thus, trafficking of leukocytes could contribute to dissemination of intracellular parasites via a Trojan horse type of mechanism. Whether *T. gondii* modulates adhesive and homing properties of infected DC, thereby facilitating penetration of parasites through biological barriers such as the blood-brain barrier, blood-testis barrier and placenta, remains to be elucidated.

Acknowledgements

We are grateful to Daria Donati for help with figures. Supported by grants from the Swedish Society of Medicine, the Swedish Research Council and the Swedish Foundation for Strategic Research. NH is the recipient of a postdoctoral fellowship from the KI infection network.

References

1. Kerr JR. Cell adhesion molecules in the pathogenesis of and host defence against microbial infection. Mol Pathol 1999; 52:220-230.
2. Huang SH, Jong AY. Cellular mechanisms of microbial proteins contributing to invasion of the blood-brain barrier. Cell Microbiol 2001; 3:277-287.
3. Meresse S et al. Controlling the maturation of pathogen-containing vacuoles: A matter of life and death. Nat Cell Biol 1999; 1:E183-188.
4. Koi H et al. The mechanisms of placental viral infection. Ann NY Acad Sci 2001; 943:148-156.
5. Cossart P, Bierne H. The use of host cell machinery in the pathogenesis of Listeria monocytogenes. Curr Opin Immunol 2001; 13:96-103.
6. Dobrowolski JM, Sibley LD. Toxoplasma invasion of mammalian cells is powered by the actin cytoskeleton of the parasite. Cell 1996; 84:933-939.
7. Desmonts G, Couvreur J. Congenital toxoplasmosis: A prospective study of 378 pregnancies. N Engl J Med 1974; 290:1110-1116.
8. Luft BJ et al. Toxoplasmic encephalitis in patients with the acquired immunodeficiency syndrome. N Engl J Med 1993; 329:995-1000.
9. Roberts F, McLeod R. Pathogenesis of toxoplasmic retinochoroiditis. Parasitology Today 1999; 15:51-57.
10. Dubey JP. Advances in the life cycle of Toxoplasma gondii. Int J Parasitol 1998; 28:1019-1024.
11. Dubey JP. Comparative infectivity of Toxoplasma gondii bradyzoites in rats and mice. J Parasitol 1998; 84:1279-1282.
12. Yap GS, Sher A. The use of germ line-mutated mice in understanding host-pathogen interactions. Cell Microbiol 2002; 4:627-634.
13. Meissner M et al. Role of Toxoplasma gondii myosin A in powering parasite gliding and host cell invasion. Science 2002; 298:837-840.
14. Khan IA et al. Mice lacking the chemokine receptor CCR1 show increased susceptibility to Toxoplasma gondii infection. J Immunol 2001; 166:1930-1937.
15. Robben PM et al. Recruitment of Gr-1+ monocytes is essential for control of acute toxoplasmosis. J Exp Med 2005; 201:1761-1769.
16. Contag CH, Bachmann MH. Advances in in vivo bioluminescence imaging of gene expression. Annu Rev Biomed Eng 2002; 4:235-260.
17. Doyle TC et al. In vivo bioluminescence imaging for integrated studies of infection. Cell Microbiol 2004; 6:303-317.
18. Hitziger N et al. Dissemination of Toxoplasma gondii to immunoprivileged organs and role of Toll/interleukin-1 receptor signaling for host resistance assessed by in vivo bioluminescence imaging. Cell Microbiol 2005; 7:837-848.
19. Lambert H et al. Induction of dendritic cell migration upon Toxoplasma gondii infection potentiates parasite dissemination. Cell Microbiol 2006; 8(10):1611-1623.
20. Sibley LD, Boothroyd JC. Virulent strains of Toxoplasma gondii comprise a single clonal lineage. Nature (Lond.) 1992; 359:82-85.
21. Howe DK, Sibley LD. Toxoplasma gondii comprises three clonal lineages: Correlation of parasite genotype with human disease. Journal Infectious Diseases 1995; 172:1561-1566.

22. Fuentes I et al. Genotypic characterization of Toxoplasma gondii strains associated with human toxoplasmosis in Spain: Direct analysis from clinical samples. J Clin Microbiol 2001; 39:1566-1570.
23. Khan A et al. Genotyping of toxoplasma gondii strains from immunocompromised patients reveals high prevalence of Type I strains. J Clin Microbiol 2005; 43:5881-5887.
24. Gallego C et al. Direct genotyping of animal and human isolates of Toxoplasma gondii from Colombia (South America). Acta Trop 2005; 97:161-162.
25. Mordue DG et al. Acute toxoplasmosis leads to lethal overproduction of Th1 cytokines. J Immunol 2001; 167:4574-4584.
26. Gavrilescu LC, Denkers EY. IFN-γ overproduction and high level apoptosis are associated with high but not low virulence Toxoplasma gondii infection. J Immunol 2001; 167:902-909.
27. Barragan A, Sibley LD. Transepithelial migration of Toxoplasma gondii is linked to parasite motility and virulence. J Exp Med 2002; 195:1625-1633.
28. Su C et al. Identification of quantitative trait loci controlling acute virulence in Toxoplasma gondii. Proc Natl Acad Sci USA 2002; 99:10753-10758.
29. Morisaki JH et al. Invasion of Toxoplasma gondii occurs by active penetration of the host cell. J Cell Sci 1995; 108:2457-2464.
30. Russell DG, Sinden RE. The role of the cytoskeleton in the motility of coccidian sporozoites. J Cell Sci 1981; 50:345-359.
31. Vanderberg JP. Studies on the motility of Plasmodium sporozoites. Journal of Protozoology 1974; 21:527-537.
32. Wetzel DM et al. Gliding motility leads to active cellular invasion by Cryptosporidium parvum sporozoites. Infect Immun 2005; 73:5379-5387.
33. Wetzel DM et al. Actin filament polymerization regulates gliding motility by apicomplexan parasites. Mol Biol Cell 2003; 14:396-406.
34. Herm-Gotz A et al. Toxoplasma gondii myosin A and its light chain: A fast, single-headed, plus-end-directed motor. EMBO J 2002; 21:2149-2158.
35. Morrissette NS, Sibley LD. Cytoskeleton of apicomplexan parasites. Microbiol Mol Biol Rev 2002; 66:21-38.
36. Jewett TJ, Sibley LD. Aldolase forms a bridge between cell surface adhesins and the actin cytoskeleton in apicomplexan parasites. Mol Cell 2003; 11:885-894.
37. Opitz C, Soldati D. 'The glideosome': A dynamic complex powering gliding motion and host cell invasion by Toxoplasma gondii. Mol Microbiol 2002; 45:597-604
38. Barragan A, Brossier F, Sibley LD. Transepithelial migration of Toxoplasma gondii involves an interaction of intercellular adhesion molecule 1 (ICAM-1) with the parasite adhesin MIC2. Cell Microbiol 2005; 7:561-568.
39. Mota MM et al. Migration of Plasmodium sporozoites through cells before infection. Science 2001; 291:141-144.
40. Mota MM et al. Migration through host cells activates Plasmodium sporozoites for infection. Nat Med 2002; 8:1318-1322.
41. Carrolo M et al. Hepatocyte growth factor and its receptor are required for malaria infection. Nat Med 2003; 9:1363-1369.
42. Feng D et al. Neutrophils emigrate from venules by a transendothelial cell pathway in response to FMLP. J Exp Med 1998; 187:903-915.
43. Barragan A, Sibley LD. Migration of Toxoplasma gondii across biological barriers. Trends Microbiol 2003; 11:426-430.
44. Tsukita S et al. Multifunctional strands in tight junctions. Nat Rev Mol Cell Biol 2001; 2:285-293.
45. Sibley LD. Intracellular parasite invasion strategies. Science 2004; 304:248-253.
46. Baum J et al. A conserved molecular motor drives cell invasion and gliding motility across malaria lifecycle stages and other apicomplexan parasites. J Biol Chem 2006; 281:5197-5208.
47. Whittaker CA, Hynes RO. Distribution and evolution of von willebrand/integrin a domains: Widely dispersed domains with roles in cell adhesion and elsewhere. Mol Biol Cell 2002; 13:3369-3387.
48. Kappe S et al. Conservation of a gliding motility and cell invasion machinery in Apicomplexan parasites. J Cell Biol 1999; 147:937-944.
49. Sultan AA et al. TRAP is necessary for gliding motility and infectivity of Plasmodium sporozoites. Cell 1997; 90:511-522.
50. Matuschewski K et al. Plasmodium sporozoite invasion into insect and mammalian cells is directed by the same dual binding system. EMBO J 2002; 21:1597-1606.
51. Carruthers VB et al. The Toxoplasma adhesive protein MIC2 is proteolytically processed at multiple sites by two parasite-derived proteases. J Biol Chem 2000; 275:14346-14353.
52. Brossier F et al. C-terminal processing of the toxoplasma protein MIC2 is essential for invasion into host cells. J Biol Chem 2003; 278:6229-6234.

53. Hynes RO. Integrins: Bidirectional, allosteric signaling machines. Cell 2002; 110:673-687.
54. Carruthers VB et al. Toxoplasma gondii uses sulfated proteoglycans for substrate and host cell attachment. Infect Immun 2000; 68:4005-4011.
55. Harper JM et al. Multimerization of the Toxoplasma gondii MIC2 integrin-like A-domain is required for binding to heparin and human cells. Mol Biochem Parasitol 2004; 134:201-212.
56. Dubey JP. Bradyzoite-induced murine toxoplasmosis: Stage conversion pathogenesis, and tissue cyst formation in mice fed bradyzoites of different strains of Toxoplasma gondii. J Eukaryot Microbiol 1997; 44:592-602.
57. Reiter-Owona I et al. Is stage conversion the initiating event for reactivation of Toxoplasma gondii in brain tissue of AIDS patients? J Parasitol 2000; 86:531-536.
58. Sher A et al. Induction and regulation of host cell-mediated immunity by Toxoplasma gondii. CIBA Found Symp 1995; 195:95-104, (discussion 104-109).
59. Ocana-Morgner C et al. Malaria blood stage suppression of liver stage immunity by dendritic cells. J Exp Med 2003; 197:143-151.
60. Aliberti J et al. Molecular mimicry of a CCR5 binding-domain in the microbial activation of dendritic cells. Nat Immunol 2003; 4:485-490.
61. Yap G et al. Cutting edge: IL-12 is required for the maintenance of IFN-γ production in T cells mediating chronic resistance to the intracellular pathogen, Toxoplasma gondii. J Immunol 2000; 165:628-631.
62. Pfefferkorn ER. Interferon-γ blocks the growth of Toxoplasma gondii in human fibroblasts by inducing the host to degrade tryptophan. Proc Natl Acad Sci USA 1984; 81:908-912.
63. Bohne W et al. Induction of bradyzoite-specific Toxoplasma gondii antigens in gamma interferon-treated mouse macrophages. Infect Immun 1993; 61:1141-1145.
64. Del Rio L et al. CXCR2 deficiency confers impaired neutrophil recruitment and increased susceptibility during Toxoplasma gondii infection. J Immunol 2001; 167:6503-6509.
65. Bliss SK et al. Rapid recruitment of neutrophils containing prestored IL-12 during microbial infection. J Immunol 2000; 165:4515-4521.
66. Bennouna S et al. Cross-talk in the innate immune system: Neutrophils instruct recruitment and activation of dendritic cells during microbial infection. J Immunol 2003; 171:6052-6058.
67. Aliberti J et al. CCR5 provides a signal for microbial induced production of IL-12 by CD8-α dendritic cells. Nature Immunology 2000; 1:83-87.
68. Scanga CA et al. Cutting edge: MyD88 is required for resistance to Toxoplasma gondii infection and regulates parasite-induced IL-12 production by dendritic cells. J Immunol 2002; 168:5997-6001.
69. Gazzinelli RT et al. Role of the Toll/interleukin-1 receptor signaling pathway in host resistance and pathogenesis during infection with protozoan parasites. Immunol Rev 2004; 201:9-25.
70. Mun HS et al. TLR2 as an essential molecule for protective immunity against Toxoplasma gondii infection. Int Immunol 2003; 15:1081-1087.
71. Yarovinsky F et al. TLR11 activation of dendritic cells by a protozoan profilin-like protein. Science 2005; 308:1626-1629.
72. Luangsay S et al. CCR5 mediates specific migration of Toxoplasma gondii-primed CD8 lymphocytes to inflammatory intestinal epithelial cells. Gastroenterology 2003; 125:491-500.
73. Egan CE et al. A Requirement for the V{gamma}1+ subset of peripheral {gamma}{delta} T cells in the control of the systemic growth of toxoplasma gondii and infection-induced pathology. J Immunol 2005; 175:8191-8199.
74. Suzuki Y et al. Interferon-γ: The major mediator of resistance against Toxoplasma gondii. Science 1988; 240:516-518.
75. Bliss SK et al. Neutrophil depletion during Toxoplasma gondii infection leads to impaired immunity and lethal systemic pathology. Infect Immun 2001; 69:4898-4905.
76. Channon JY et al. Differential infectivity and division of Toxoplasma gondii in human peripheral blood leukocytes. Infect Immun 2000; 68:4822-4826.
77. Dubey JP et al. Oocyst-induced murine toxoplasmosis: Life cycle, pathogenicity, and stage conversion in mice fed Toxoplasma gondii oocysts. J Parasitol 1997; 83:870-882.
78. Steinman RM. Some interfaces of dendritic cell biology. APMIS 2003; 111:675-697.
79. Fischer HG et al. Phenotype and functions of brain dendritic cells emerging during chronic infection of mice with Toxoplasma gondii. J Immunol 2000; 164:4826-4834.
80. Suzuki Y et al. Microglia and macrophages as innate producers of interferon-gamma in the brain following infection with Toxoplasma gondii. Int J Parasitol 2005; 35:83-90.
81. Courret N et al. CD11c- and CD11b-expressing mouse leukocytes transport single Toxoplasma gondii tachyzoites to the brain. Blood 2006; 107:309-316.
82. Wei S et al. Toxoplasma gondii-infected human myeloid dendritic cells induce T-lymphocyte dysfunction and contact-dependent apoptosis. Infect Immun 2002; 70:1750-1760.

83. McKee AS et al. Functional inactivation of immature dendritic cells by the intracellular parasite Toxoplasma gondii. J Immunol 2004; 173:2632-2640.
84. Butcher BA et al. Toxoplasma gondii tachyzoites inhibit proinflammatory cytokine induction in infected macrophages by preventing nuclear translocation of the transcription factor NF-κB. J Immunol 2001; 167:2193-2201.
85. Shapira S et al. Suppression of NF-κB activation by infection with Toxoplasma gondii. J Infect Dis 2002; 185(Suppl 1):S66-72.
86. Luder CG, Gross U. Apoptosis and its modulation during infection with Toxoplasma gondii: Molecular mechanisms and role in pathogenesis. Curr Top Microbiol Immunol 2005; 289:219-237.

CHAPTER 17

The Microsporidian Polar Tube and Its Role in Invasion

Frédéric Delbac* and Valérie Polonais

Introduction

The Microsporidia are a phylum of small unicellular eukaryotes comprising more than 150 genera and 1200 species. They are obligate intracellular parasites which are able to form environmentally resistant spores. Historically, *Nosema bombycis*, was the first described species in this phylum and is the etiological agent of "pébrine" disease that nearly destroyed the silk-worm industry in the nineteenth century. Although the majority of microsporidia that have been described are found in arthropods and fishes, being responsible of important economic losses in the apiculture and fish farms, there are several species of medical and veterinary significance which infect animals and humans.[1,2] Cerebral microsporidian infections attributed to *Encephalitozoon cuniculi* were initially described in 1922 in rabbits with granulomatous encephalitis and several infections were then reported in most vertebrate groups. The first case of microsporidiosis in a human was identified in 1959 in a nine-year-old boy suffering from neurological disorders.[3] While reports of humans infected with microsporidia were extremely rare before the AIDS epidemic, these organisms are now recognized as significant emerging pathogens in immunocompromised hosts (HIV-infected patients with AIDS and organ transplant recipients) and is a cause of intestinal, ocular, muscular and systemic diseases.[2] Some clinical manifestations have been also reported in immunocompetent hosts. Serological studies with blood donors and pregnant women revealed a prevalence of about 8%,[4] suggesting that infections by microsporidia may be common in humans. So far, species belonging to seven different genera *Brachiola* (recently renamed *Anncaliia*), *Encephalitozoon*, *Enterocytozoon*, *Nosema*, *Pleistophora*, *Trachipleistophora* and *Vittaforma* have been found in human infections. Although the origin of infection and epidemiology still remain to be documented for human microsporidiosis, horizontal transmission of most microsporidia occurs by oral ingestion of spores, with the site of initial infection being the gastrointestinal tract.[5]

In addition to their medical importance, microsporidia have attracted attention as amitochondriate eukaryotes containing prokaryote-like ribosomes and were assumed to be very ancient protozoans. However, data from molecular phylogenies based on various gene sequences argued against an early origin of microsporidia and supported a placement of these organisms among the Fungi.[6,7,8] Furthermore, mitochondria-like organelles were identified in these peculiar fungi,[9] consistent with the finding in the *E. cuniculi* genome of several genes coding for proteins of mitochondrial origin.[10]

The growing interest in the study of microsporidia is also prompted by two further characteristics: (i) some microsporidian species harbor a remarkable degree of nuclear genome

*Corresponding Author: Frédéric Delbac—Equipe Parasitologie Moléculaire et Cellulaire, LBP, UMR CNRS 6023, Université Blaise Pascal, 24 Avenue des Landais 63177 Aubière Cedex, France. Email address: frederic.delbac@univ-bpclermont.fr

Molecular Mechanisms of Parasite Invasion, edited by Barbara A. Burleigh and Dominique Soldati-Favre. ©2008 Landes Bioscience and Springer Science+Business Media.

compaction including *E. cuniculi* (2.9 Mbp, ~2000 potential protein-coding genes) whose genome has been fully sequenced,[10] and (ii) all of the Microsporidia possess an invasive apparatus composed of a highly specialised structure, the polar tube, that is unique in the eukaryotic world and represents one of the most sophisticated infection mechanisms among all of the eukaryotics. The invasion process involves the sudden extrusion of the sporal polar tube (also called spore germination) for initiating the entry of the parasite into a new host cell.

The purpose of this chapter is to review the available data on the composition and structure of the microsporidian polar tube and the role of this unique organelle in host cell invasion. Some alternative mechanisms used by microsporidia to invade cells are also discussed.

The Microsporidian Spore

The extracellular infective stage of microsporidia is a small spore from 1 to 12 μm in size which is protected by a thick wall (Fig. 1). The content of the spore can be viewed as consisting of two functionally distinct parts, namely the sporoplasm and the extrusion apparatus. The main identifiable intrasporoplasmic structures are the nuclear apparatus (one or two nuclei, according to genus), ribosomes and endoplasmic reticulum membranes. No dictyosomes are seen in microsporidian cells; however, a vesicular and tubular complex exhibiting thiamine

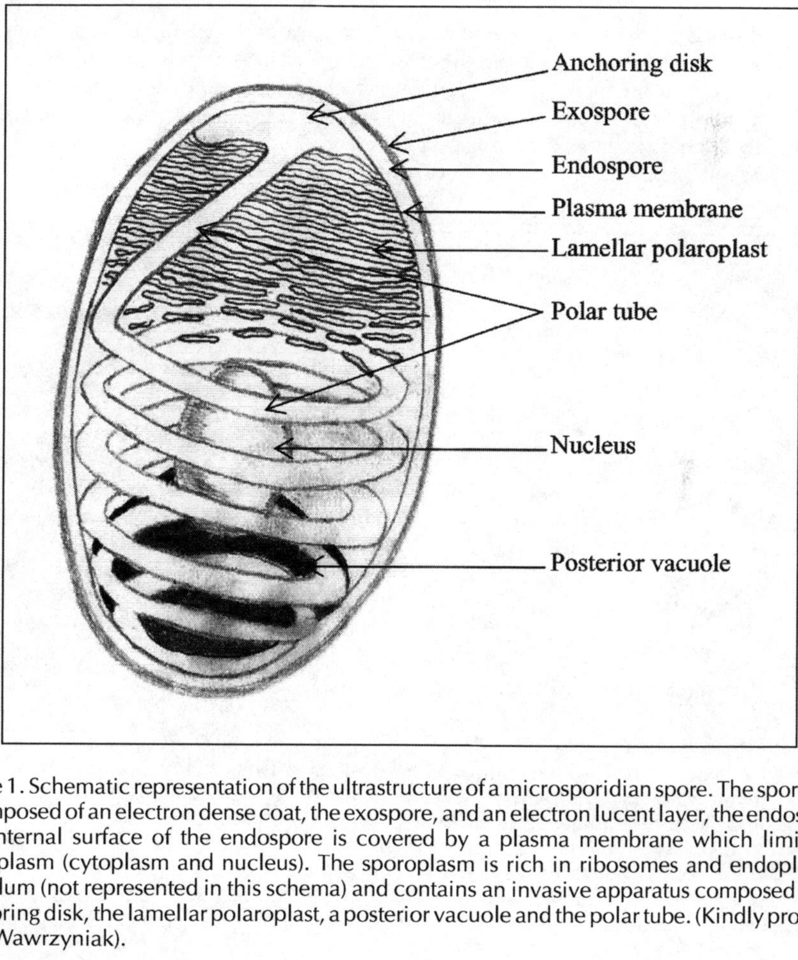

Figure 1. Schematic representation of the ultrastructure of a microsporidian spore. The spore wall is composed of an electron dense coat, the exospore, and an electron lucent layer, the endospore. The internal surface of the endospore is covered by a plasma membrane which limits the sporoplasm (cytoplasm and nucleus). The sporoplasm is rich in ribosomes and endoplasmic reticulum (not represented in this schema) and contains an invasive apparatus composed of the anchoring disk, the lamellar polaroplast, a posterior vacuole and the polar tube. (Kindly provided by I. Wawrzyniak).

pyrophosphatase activity has been identified as the Golgi apparatus.[11] The highly specialised extrusion apparatus generally comprises a very long tubular structure, the polar tube (or polar filament), coiled obliquely around the posterior region of the spore, an anchoring disk (or polar cap) and an anterior membrane-bounded organelle, termed the lamellar polaroplast. A vacuole at the posterior end of the spore is also present in some species. During invasion, the sporoplasm contents are injected into a target host cell through the hollow polar tube, suddenly extruded at the anterior pole of the spore (Fig. 2A,B) in response to ill-defined stimuli. It has been hypothesized that the lamellar polaroplast provides a new plasma membrane for discharged sporoplasms, the preexisting plasma membrane remaining in the spore.[12]

The first part of intracellular development of the Microsporidia corresponds to a proliferative phase (merogony), during which meronts divide by binary or multiple fission (Fig. 2). The second part called sporogony begins when meronts transform into sporonts, that are characterized by the deposition of an electron-dense material on the plasma membrane. Sporonts divide into sporoblasts in which the formation of the thick wall is associated with the differentiation of the invasion apparatus. In *Encephalitozoon* species, all the development takes place inside a parasitophorous vacuole (Fig. 2A,C), whereas in several other microsporidian species development occurs in direct contact with the host cell cytoplasm. After completion of the maturation of sporoblasts into spores, the rupture of host cells may lead to the release of spores in the environment.

The rigid sporal cell wall consists of two major layers surrounding a unit plasma membrane (Fig. 1): (i) an electron-dense outer layer, the exospore, which is principally proteinaceous, and (ii) an electron-lucent inner layer, the endospore, which contains chitin and protein components.[13] The endospore is thicker than the exospore, but is relatively thin at the apical site for extrusion corresponding to the anchoring disk region where the polar tube is connected. The rigidity of the wall makes the spore resistant to various environmental stresses. In addition, this

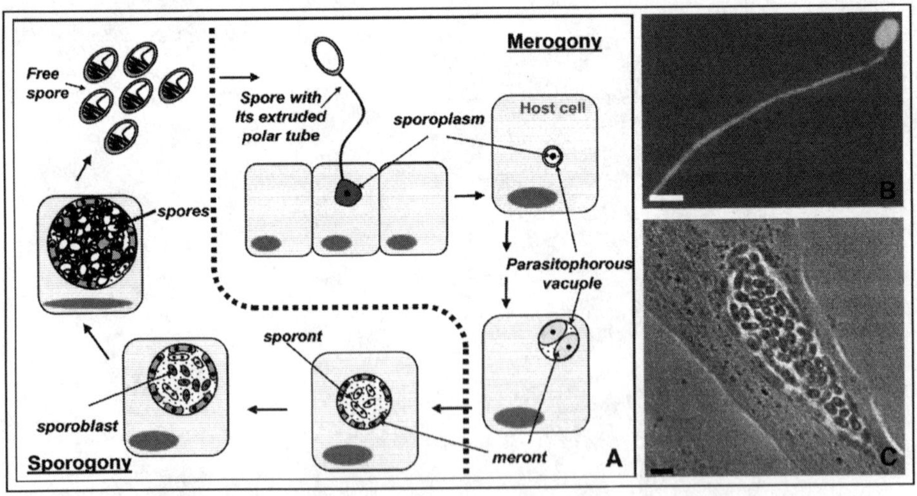

Figure 2. Representation of the asexual life cycle in the microsporidia *Encephalitozoon cuniculi*. A) The life cycle comprises three distinct phases: the infective phase characterized by polar tube extrusion and injection of the sporoplasm within the host cytoplasm, the proliferative phase or merogony and the spore-forming phase or sporogony (kindly provided by Dr D. Brosson). B) An *E. cuniculi* spore after polar tube extrusion. The spore wall is stained with Uvitex 2B, a fluorochrome specific of chitin that is the major component of the endospore. The extruded polar tube is labeled with anti-PTP1 antibodies. Secondary antibody is Alexa 488-conjugated goat anti-mouse IgG. C) *E. cuniculi* developmental stages in a parasitophorous vacuole of infected HFF (human foreskin fibroblasts) cells, as viewed in phase contrast microscopy. Bars = 2 μm.

rigidity would maintain a high intrasporal hydrostatic pressure, which is believed to be important for the process of germination. Finally, its aqueous permeability might play a role in the control of the polar tube extrusion activation.[14] It cannot be ruled out that an interaction between some component(s) of the cell wall and some receptor(s) of the host is also a prerequisite for polar tube extrusion; however, current knowledge of the composition of the microsporidian wall is limited. Only one or two spore wall proteins (SWPs) have been clearly assigned to the exospore in two *Encephalitozoon* species: SWP1 in *E. cuniculi*,[15] SWP1 and SWP2 in *E. intestinalis*.[16] Three components have then been shown to localize to the *E. cuniculi* endospore: a chitin deacetylase like-protein[17] and two antigens with unknown functions named endospore protein 1 (EnP1) and endospore protein 2 (EnP2)[18] also called SWP3 in another study.[19] More recently, a new candidate protein for a cell wall location has been described through a proteomic analysis of *E. cuniculi*.[20]

Structure, Formation and Composition of the Polar Tube

Although microsporidia differ greatly in host range and cell type specificity, they share a similar mechanism for host cell invasion. These unicellular eukaryotes have developed a specific process for invading their host, involving the rapid extrusion of the polar tube that remains attached to the anterior part of the spore and can pierce a host cell membrane for delivering an infective sporoplasm inside the host cytoplasm (for a review see ref. 21).

In the spore, the polar tube can be divided into two regions (Fig. 1): (i) an anterior straight part, called manubrium, which is surrounded by the lamellar polaroplast and is connected to the anchoring disk at the anterior end of the spore, and (ii) a posterior part which forms peripheral coils within the sporoplasm, the number of turns varying among species from four to more than 30.[22] Five to seven coils arranged in a single row are representative of *Encephalitozoon*.[23] The polar tube coiled within the sporoplasm seems to be limited by a membrane and consists of several electron-dense and electron-lucent concentric layers around a central core.[23,24] The number of layers varies when comparing the ultrastructural aspects of *Nosema algerae* spores before, during and after polar tube extrusion.[24] Three layers of different electron density were observed for extruded polar tube and in the straight part of the polar tube within the spore, whereas the coiled part showed six concentric layers. This can be interpreted as a folding of the tube onto itself in its coiled region and may explain why the polar tube is longer after extrusion than when it is still coiled within the spore. Another explanation for the increased length of the extruded polar tube is that a self-assembling protein may be incorporated at the polar tube growing tip during extrusion.[25-26] After extrusion, the polar tube extends approximately over 20-150 μm depending on the species and is about 0.1-0.15 μm wide.[27] During sporoplasm discharge, its diameter can increase until 0.4 μm.

In most microsporidian species, the polar tube emerges from a rather enigmatic reticular structure.[28] Certain vesicles associated with polar tube differentiation steps have been considered as being of Golgi nature.[29] Thus, the intracellular transport of polar tube protein components might involve an ER-Golgi classical pathway. This is consistent with the presence of signal peptides in both polar tube protein 1 (PTP1) and polar tube protein 2 (PTP2).[30] However, some observations mainly done in *Enterocytozoon bieneusi* indicate that the polar tube precursors are spherical peculiar vesicles scattered in the cytoplasm of sporogonial plasmodia which do not originate from the Golgi-like network.[31,32]

Close inter-species similarities in the structural organization and solubility properties of the polar tube suggest that some of its molecular components may have been conserved during evolution of microsporidia. Initially, a 23-kDa protein was identified and proposed as being the single component of the polar tube in *Ameson* (formerly *Nosema*) *michaelis*.[25] Although no amino acid sequence identification was obtained, this study provided evidence that the 23-kDa protein was able to assemble into structures in vitro and that the binding between polar tube protein (PTP) subunits was due to interprotein disulfide linkages. In *Glugea americanus*, four proteins that were differentially solubilized with 2% dithiothreitol (DTT) have been assumed to originate from the polar tube. This localization has been demonstrated for a 43-kDa protein purified by high-pressure liquid chromatography and shown to contain numerous proline residues, but information on the amino acid sequence was limited to a short N-terminal region.[33] All these studies indicated that disulfide linkages seem to be important in stabilizing the PTPs since the polar tube resists treatment with either detergents or acids, but dissociates in the presence of a high concentration of a reducing agent such as DTT or 2-mercaptoethanol.[25,33] Using these solubility characteristics, several PTPs were then similarly isolated from *Encephalitozoon* species, with apparent molecular sizes varying from 45 to 55 kDa.[34]

Since 1998, significant advances in the molecular characterization of the polar tube have been achieved through the identification of the entire amino acid sequences of three different proteins (PTP1, PTP2 and PTP3), with no similarity to other proteins in Genebank and other databases. These polar tube proteins appear to be members of novel protein families. The complete sequence of a polar tube protein was reported initially from *Encephalitozoon cuniculi*.[35] This protein was designated PTP1 and is an acidic proline-rich protein of 395 residues containing four tandemly arranged 26-amino acid repeats. PTP1-like protein encoding genes were subsequently identified in the two other *Encephalitozoon* species pathogenic to humans, *E. hellem* and *E. intestinalis*.[30,36] The PTP1-coding region in *E. intestinalis*, representing a 371-residue polypeptide, is shorter than those in other species (453 aa in *E. hellem* and 395 aa in *E. cuniculi*). A remarkably conserved N-terminal region of 22 residues represents a cleaved signal peptide, probably involved in the targetting of PTP1. *Encephalitozoon* PTP1s have similarities in overall amino acid composition mainly characterized by their richness in proline and glycine residues (Table 1), but have only limited identity in amino acid sequences. Sequence alignment indicates the highest homology is in the N- and C-terminal domains and not in the strongly divergent hydrophilic central region which consists of tandemly repeated sequences. Both the composition and the number of repeats are different among the three species. The repetitive character of the central core is less evident in *E. intestinalis*. Only two highly degenerated major repeats of 27 or 28 amino acids were distinguishable, contrasting with the nearly perfect repeats seen in *E. hellem* PTP1 (six 20-aa repeats) and *E. cuniculi* PTP1 (four 26-aa repeats). An intra-species variability of the repetitive central core was also demonstrated through numerous point mutations and variation in repeat number, both in *E. cuniculi* and *E. hellem*.[37,38]

PTP1s are also characterized by a high number of cysteine residues including one at the C-terminal end (Table 1). This suggests a possible formation of interprotein or intraprotein disulfide linkages consistent with the solubilization of PTPs by reducing agents and the reported in vitro conditions of assembly of the purified 23-kDa PTP of *Ameson michaelis*.[25]

Several O-glycosylation sites are predicted in PTP1 (Table 1). Cytochemical investigations indicate that the polar tube contains glycoconjugates,[39] and a recent study has provided evidence for O-linked mannosylation of *E. hellem* PTP1.[40] The modification of PTP1 by linked-carbohydrates could explain its aberrant migration in SDS-PAGE, as shown by deglycosylation experiments in *E. intestinalis*.[41] Interestingly, PTP1 appears as a good candidate antigen for serological diagnosis, given that a recombinant *E. cuniculi* PTP1 is strongly recognized by sera from *E. cuniculi*-infected patients.[42]

While PTP1 is considered to be the major component of the polar tube, accounting for at least 70% of the mass of the polar tube, biochemical and immunological data clearly support protein heterogeneity of the polar tube. Indeed, DTT-soluble materials from *Glugea americanus* revealed the presence of several potential polar tube components.[33] With the use of specific polyclonal and monoclonal antibodies, several PTPs of different size were also identified: 34, 75 and 170 kDa in *Glugea atherinae*; 28, 35, 55 and 150 kDa in *E. cuniculi*;[43] 60 and 120 kDa in *E. intestinalis*.[44]

A 35-kDa migrating protein band in *E. cuniculi* has been demonstrated to represent a 277 amino-acid polypeptide, named PTP2[30] that has no significant homology in protein databases. PTP2 is a novel structural protein with a basic lysine-rich core and an acidic tail. The N-terminal region has a characteristic signal peptide. In the *E. cuniculi* genome, the *ptp2* gene

Table 1. *Major characteristics of PTP1 (A) and PTP2 (B) from* E. cuniculi *(Ec),* E. hellem *(Eh),* E. intestinalis *(Ei),* A. locustae *(Al) and* P. grylli *(Pg)*

A Protein	Length (aa Number)	pI	Major aa (%) Pro	Gly	Cysteine Residue Number	Carboxy-Terminal Residue	Number of O-Glycosylation Potential Sites
Ec-PTP1	395	4.7	13.4	11.8	17	C	37
Ei-PTP1	371	4.5	13.7	8.3	17	C	56
Eh-PTP1	453	4.4	14.6	12.9	21	C	32
Al-PTP1	355	5.2	19.8	10.6	12	C	19
Pg-PTP1	351	5.2	21.6	14.7	12	C	19

B Protein	Length (aa Number)	pI	Major aa (%) Lys	Glu	Cysteine Residue Number	Carboxy-Terminal Residue	Number of O-Glycosylation Potential Sites
Ec-PTP2	277	8.6	12.1	9.5	8	E	3
Ei-PTP2	275	8.6	11.8	6.2	8	E	4
Eh-PTP2	272	8.6	10.8	9.5	8	E	5
Al-PTP2	287	9.1	12.3	9.3	8	G	1
Pg-PTP2	287	8.9	12.3	8.6	8	K	2

The most abundant amino acid is proline in PTP1 and lysine in PTP2. The pI, amino acid percentages and the number of O-glycosylation potential sites are deduced from the mature proteins. (Modified from: Polonais et al. Fungal Genet Biol 42:791-803; ©2005 with permission from Elsevier.[45])

was found downstream of the *ptp1* gene on the same DNA strand of chromosome VI and separated from *ptp1* by only 860 nucleotides (Fig. 3). Conserved clustering of the *ptp1-ptp2* genes facilitated the further identification of *ptp2*-like genes in *E. hellem* and *E. intestinalis.*[30] All *Encephalitozoon* PTP2s are basic proteins of similar size, close to 30 kDa. Extending throughout the entire coding region, the degree of conservation at the amino acid level is higher than for PTP1s. Three different regions can be distinguished: an N-terminal part of 50 noncharged residues, an internal basic region including a central lysine-rich hexapeptide and a short acidic C-terminal part (24 to 27 aa). The major residue is lysine and one glutamate is observed at the C-terminus of each PTP2. The eight cysteine residues are similarly located in the three PTP2 sequences.

Numerous clusters of annotated genes in *Antonospora* (*Nosema*) *locustae*, a microsporidia parasite of insects, have the same order and orientation as in *E. cuniculi*. This syntenic relationship has been successfully exploited to identify two neighboring orphan genes in *A. locustae* corresponding to divergent homologues of *Encephalitozoon ptp1* and *ptp2.*[45] The same study also provided evidence for a similar locus in *Paranosema grylli*, a parasite of crickets that is closely related to *Antonospora* species (Fig. 3). When comparing either PTP1s or PTP2s from *Encephalitozoon* and *Antonospora-Paranosema* species, no significant amino acid signatures were identified. This inter-genus protein sequence divergence contrasts with the conserved morphological characteristics of the polar tube, but is consistent with an absence of labeling of the *A. locustae* polar tube by antibodies raised against *E. cuniculi* PTPs. Similarities in overall amino acid composition, solubility and electrophoretic properties of PTPs are however evident and some higher-order structural features are likely preserved. PTP1-like features include an acidic pI, a high proline content, several internal repeats containing O-glycosylation potential sites and a C-terminal cysteine residue (Table 1). Although varying in size from 351 amino acids in *P. grylli* to 453 amino acids in *E. hellem*, all the proteins of the PTP1 family migrated in

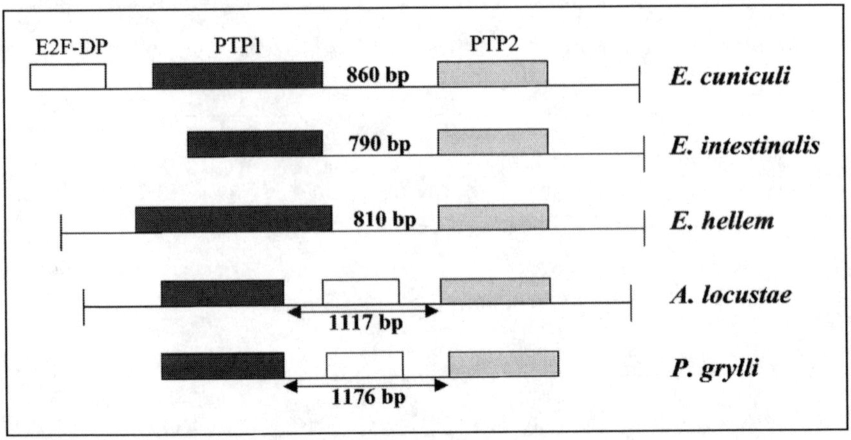

Figure 3. Conserved clustering of *ptp1* and *ptp2* genes in microsporidian species belonging to the three different genera *Encephalitozoon, Antonospora* and *Paranosema*. *ptp1* genes are represented by dark grey boxes, *ptp2* genes by clear grey boxes. White boxes represent *e2f-dp* genes encoding a transcription factor probably involved in the control of microsporidian cell cycle and differentiation. (Modified from: Polonais et al. Fungal Genet Biol 42:791-803; ©2005 with permission from Elsevier.[45])

SDS-PAGE gels as 50-55 kDa broad bands.[35,36,45] Considering that proline-rich proteins such as collagen and elastin, are characterized by a high tensile strength, Keohane et al[33] have postulated a similar property for PTP1s in relation to sporoplasm discharge and flow through the polar tube. PTP2s in *A. locustae* and *P. grylli* have calculated molecular masses close to those of *Encephalitozoon* spp (~30 kDa). Basic pI, lysine and glutamate richness and conservation of eight cysteine residues are also reminiscent of *Encephalitozoon* PTP2s (Table 1). Similar to *Encephalitozoon* spp PTP1 and PTP2, the PTP1 and PTP2 from the two insect microsporidians are solubilized in the presence of DTT and possess candidate cysteine residues for the establishment of covalent linkages that may be of primary importance for the high tensile strength and functioning of the polar tube. Whatever the considered microsporidian genus, eight cysteine residues are present in PTP2s and one cysteine is typically found at the extreme C-terminus of PTP1s. The conservation of the C-terminal cysteine in PTP1 sequences is suggestive of a critical residue involved in the branching between only PTP1 molecules or both PTP1 and PTP2 molecules.

The third known PTP (PTP3) is encoded by a single transcription unit located on chromosome XI of *E. cuniculi* and was identified by immunoscreening of a cDNA library.[46] PTP3 is predicted to be synthesized as a 1256-amino acid precursor (136 kDa) with a cleavable signal peptide. Unlike PTP1 and PTP2, the mature PTP3 protein lacks cysteine residue and can be solubilized without the need of a reducing agent. This protein comprises a large acidic core, rich in aspartate and glutamate residues, that is flanked by highly basic N- and C-terminal regions. PTP3 seems to be more conserved between microsporidia since anti-PTP3 antibodies cross react with the polar tubes of both *E. intestinalis* and *E. hellem*[46] and an homologous gene exists in *A. locustae* (F. Delbac, unpublished data).

The presence of a predicted signal peptide at the N-terminus of the three PTPs so far characterized is consistent with data indicating that the polar tube appears to form by the coalescence of vesicles that may be processed through the ER-Golgi pathway. Although *E. cuniculi* harbours an extremely reduced nuclear genome and an unstacked Golgi apparatus, gene annotation predicted several proteins that are well representative of the functioning of an ER-Golgi system in protein translocation, folding and modification.[10,47] Among these proteins, we can cite the signal recognition particle (SRP), the SRP receptor and those involved in translocon formation (Sec61, Sec62, Sec63). A signal processing peptidase of type I that is presumably involved in the removal of signal peptides is present. Genes coding for subunits of the COPI and COPII coats and for some GTPases as well as those coding for components of the SNARE complexes are also found in the genome. However, a remarkable feature is the lack of putative enzymes responsible for N-linked glycosylations, suggesting that *E. cuniculi* should be unable to perform this ubiquitous pathway of protein modification, a unique case among eukaryotes.

Using chemical cross-linkers, a large multimeric complex was formed and shown to contain PTP1, PTP2 and PTP3 with a few other proteins.[46] The two major antigenic polar tube proteins PTP1 and PTP2 are predicted to contribute to the high tensile strength of the polar tube via an assembly process dependent on disulfide linkages, whereas PTP3 is hypothesized to play a role in the control of the polar tube extrusion as part of a specific response to ionic stimuli.[46] However, the precise interactions between these PTPs during the formation and functioning of the polar tube remain to be determined.

Polar Tube Extrusion and Sporoplasm Discharge

The invasion strategy used by microsporidia is primarily related to spore germination, a unique motile process during which these fungi-related parasites inject their contents into target cells through the lumen of a rapidly extruded polar tube, as a prerequisite to obligate intracellular development. Analyses by microscopy and videomicroscopy have revealed that the process of tube extrusion and transfer of genetic material is completed in less than 2 s[27] and can be divided in 4 different steps: (1) activation of spore discharge by appropriate stimuli, (2) increase of the intrasporal osmotic pressure, (3) polar tube extrusion by eversion and (4) sporoplasm discharge through the hollow tube. Different conditions have been shown to activate spore germination in vitro, including variations of pH, addition of various cations or anions, H_2O_2 or low dose UV radiation, depending on the species.[48-51] This probably reflects their adaptation to their host and external environment since microsporidia are able to infect a wide range of invertebrate and vertebrate hosts and a large number of tissue and cell types. The existence of specific conditions for initiation of the germination process could also prevent accidental discharge in the environment.[22] It seems that calcium, that is probably sequestered in the polaroplast compartment, may be a key ion in this process. Addition of the calcium ionophore A 23187 stimulates spore discharge[52] whereas calcium chloride can either induce or block spore discharge according concentration.[53] Furthermore, it has been suggested that Ca^{2+} and its interaction with cytoskeletal components influenced spore discharge because treatment with the microtubule inhibitor demecolcine, the filamentous actin disruptor cytochalasin D, or the Ca^{2+} channel blocker nifedipine, inhibited *E. hellem* polar tube discharge.[50] Activation of polar tube extrusion could be also the result of an adherence of the microsporidian spore to host cell sulfated glycosaminoglycans.[54] Southern et al[55] recently demonstrated that spore adherence is augmented by manganese (Mn^{2+}) and magnesium (Mg^{2+}) and leads to higher host cell infection efficiency.

Whatever the stimuli required for germination activation, a common early feature, documented in aquatic microsporidia, has been a strong increase of the intrasporal osmotic pressure.[52,56] An influx of water into the spore leads to the swelling of both the polaroplast and the posterior vacuole, thereby producing sufficient force for polar tube extrusion and sporoplasm delivering. It has been proposed that an integral membrane channel of the aquaporin family, recently described in *E. cuniculi*, might be involved in the water influx during germination.[57] A possible cause of the osmotic pressure increase could be hydrolysis of trehalose, a nonreducing disaccharide, each trehalose molecule being converted into two glucose molecules.[58]

The first sign of spore discharge is the appearance of a protrusion at the anterior end of the spore at the anchoring disk. This is followed by the rapid emergence of the polar tube in a helicoidal fashion.[27,56] Some studies indicate that this process occurs by eversion of the polar tube that resembles an inner tube sliding through an outer tube, similarly to an everting glove finger. When the polar tube is fully extruded, the sporoplasm including the nucleus flows through the hollow tube and appears after 15-500 ms as a droplet at its distal end.[26,27] The new plasma membrane for delivered sporoplasm may be provided by the lamellar polaroplast.[12]

Various Pathways of Host Cell Invasion by Microsporidia

The polar tube discharges through the thin anterior end of the spore and provides a duct through which the infective sporoplasm can flow to be finally transferred inside the cytoplasm of a new host cell.[21] This invasion mechanism is an active process, the extruded polar tube appearing to be able to pierce the host cell plasma membrane acting like an hypodermic needle (Fig. 4B).

The likelihood of a successful invasion may be increased when parasites are in close proximity to target cells and microsporidian spore adherence to the host cell surface may be the initial event that signals the spore to activate (Fig. 4C). Some electron microscopic studies indicate that the microsporidian spore can attach to the host cell membrane prior to the invasion. Sulfated glycosaminoglycans (GAGs) at the host cell surface have been recently shown to play an important role in the attachment of *E. intestinalis* spores.[54] This spore adherence is presumably a precursor to spore activation and polar tube extrusion since host cell infection is reduced when spore adherence is inhibited; however, the spore surface ligand(s) involved in

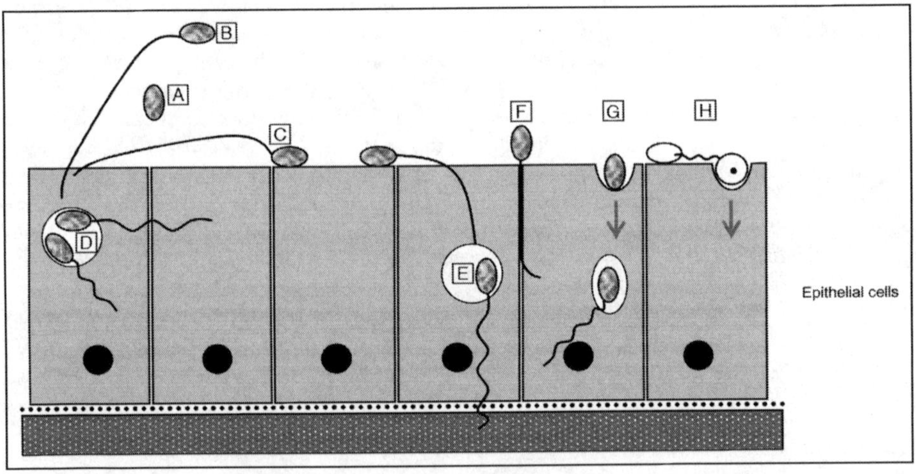

Figure 4. Different modes of host cell invasion by microsporidia. A) Free spore before devagination. B) Spore with its extruded polar tube that pierces the host cell membrane and discharges the sporoplasm inside the host cytoplasm. C) Attachment of the spore to host cell receptor(s) followed by polar tube extrusion and discharge of the sporoplasm inside the host cytoplasm. D) Secondary infestation: spores produced within a parasitophorous vacuole can germinate thereby injecting their sporoplasm in the same cell or towards an adjacent cell. E) Secondary infestation of an underlying cell or another tissue. F) Basolateral penetration involving a passage through intercellular junctions. G) Phagocytosis of a spore followed by polar tube extrusion and discharge of the sporoplasm inside the host cytoplasm. The germination after phagocytosis allowed the spore to escape from the phagosomes that mature into lysosomes. The membrane of the vacuole surrounding the spore internalized by phagocytosis (G) is thus different from that corresponding to the parasitophorous vacuole (D and E). H) Phagocytosis of the extracellular discharged sporoplasm. (Modified after ref. 66).

the interaction with host cell GAGs remain(s) to be determined. Some secondary infections may also occur when some mature spores produced in the initially infested host cell germinate, thereby infecting the same host cell or an adjacent cell (Fig. 4D). Infections become disseminated when the polar tube injects the spore contents within another tissue (Fig. 4E).

Some observations support the possibility of additional modes of microsporidian entry and dissemination into host cells (Fig. 4G-H). It has been demonstrated that microsporidian spores can be ingested by professional and non professional phagocytes via an host actin-based mechanism.[59,60,61] After contact of the spore apex with the host cell membrane, the polar tube is extruded within an invagination of the host cell membrane that extends inside a pseudopod containing filamentous actin, suggesting that microsporidia can induce host cell alterations facilitating the invasion.[59] Confocal and electron microscopy observations also suggest that after initial contacts through the posterior pole of the microsporidian spore, the basolateral surface of Caco-2 cells may be the portal of entry for *E. intestinalis* sporoplasm.[61] A recent study suggested that relatively few cells were infected through the traditional penetration of the polar tube from outside and that alternative mechanisms may be very important in the spread of microsporidia within their hosts.[62] In this study, phagocytosis of spores occurred at least 10 times more frequently than injection of sporoplasms. Although spores that remained inside these compartments disappeared within 3 days, some microsporidia extruded their polar tube inside the cells following phagocytosis, thereby escaping from the phagosomes that mature into lysosomes and causing infection (Fig. 4G). In contrast, another recent study provided evidence that phagocytosis is not a significant mode of infection in differentiated intestinal epithelial cell.[63] As shown in Figure 4, microsporidia of the genus *Encephalitozoon* develop inside a parasitophorous vacuole. The membrane of this vacuole is of unknown origin and does not possess any markers of late endosomal or lysosomal differentiation.[21,64] Therefore, it is important to differentiate between phagosomes containing internalized spores (Fig. 4G) that will mature to lysosomes and parasitophorous vacuoles (Fig. 4D,E) in which the development of the parasites will take place. Extracellular discharged sporoplasms tightly abutted to the host plasmalemma, may be also incorporated into the host cytoplasm by induced phagocytosis (Takvorian et al[65]; Fig. 4H). In this case, the origin of the parasitophorous vacuole membrane would be the invaginated host cell cytoplasm membrane. However, Fasshauer et al[64] showed that the parasitophorous vacuole membrane lacks host cell membrane proteins immediately after invasion. This suggests that the host cell transmembrane proteins and the transferrin receptor as well as fusion-mediating proteins must be removed very rapidly from the emerging parasitophorous vacuole membrane.

The invasion process may also require a close association between polar tube component(s) and host cell surface receptor(s), and it has been suggested that PTP-linked oligosaccharides might be essential for these early interactions. Two mannose-binding lectins (GNA and ConA) have been shown to bind both the polar tube and its major component PTP1 in *A. locustae* and *P. grylli*.[45] Moreover, *A. locustae* PTP1 was retained on ConA affinity chromatography columns. Another biochemical study has revealed that the purified PTP1 of *E. hellem* can bind ConA and is modified by O-linked mannosylation.[40] This is in perfect agreement with the high number of predicted O-glycosylation sites, mainly distributed through the repeat-containing central region of PTP1 (Table 1), and with previous cytochemical evidence for some glycoconjugates in the polar tube.[39] The *E. cuniculi* genome sequence indeed contains a minimal set of genes required for O-mannosylation pathway but those characteristic for N-glycosylation have not been identified.[10,47] Deglycosylation experiments conducted with *E. hellem* PTP1 did not demonstrate any evidence for N-glycosylation.[40] PTP1 mannosylation may be essential to protect the microsporidian polar tube against proteolytic degradation and/or to interact with host cell surface. Interactions between mannosylated PTP1 and some unknown host cell mannose-binding molecules are also supported by the decreased level of infection by *E. hellem* in mannose-pretreated RK13 cells.[40]

Conclusion

Much remains to be studied about the molecular architecture of the microsporidian polar tube, including identification of other PTPs and understanding the interactions between these components which are required for polar tube assembly. Further studies are also needed to precise the origin of the parasitophorous vacuole. The development of tools for genetically modifying microsporidia would be useful for elucidating the events in the rupture of the anterior attachment complex, eversion of the polar tube as well as the mechanism of host cell attachment and penetration.

References

1. Weiss LM. Microsporidia: Emerging pathogenic protists. Acta Trop 2001; 78:89-102.
2. Didier ES. Microsporidiosis: An emerging and opportunistic infection in humans and animals. Acta Trop 2005; 94:61-76.
3. Matsubayashi H, Koike T, Mikata I et al. A case of Encephalitozoon-like body infection in man. AMA Arch Pathol 1959; 67:181-187.
4. van Gool T, Vetter JC, Weinmayr B et al. High seroprevalence of Encephalitozoon species in immunocompetent subjects. J Infect Dis 1997; 175:1020-1024.
5. Mathis A, Weber R, Deplazes P. Zoonotic potential of the microsporidia. Clin Microbiol Rev 2005; 18:423-445.
6. Van de Peer Y, Ben Ali A, Meyer A. Microsporidia: Accumulating molecular evidence that a group of amitochondriate and suspectedly primitive eukaryotes are just curious fungi. Gene 2000; 246:1-8.
7. Keeling PJ. Congruent evidence from alpha-tubulin and beta-tubulin gene phylogenies for a zygomycete origin of microsporidia. Fungal Genet Biol 2003; 38:298-309.
8. Thomarat F, Vivares CP, Gouy M. Phylogenetic analysis of the complete genome sequence of Encephalitozoon cuniculi supports the fungal origin of microsporidia and reveals a high frequency of fast-evolving genes. J Mol Evol 2004; 59:780-791.
9. Williams BA, Hirt RP, Lucocq JM et al. A mitochondrial remnant in the microsporidian Trachipleistophora hominis. Nature 2002; 418:865-869.
10. Katinka MD, Duprat S, Cornillot E et al. Genome sequence and gene compaction of the eukaryote parasite Encephalitozoon cuniculi. Nature 2001; 414:450-453.
11. Takvorian PM, Cali A. Enzyme histochemical identification of the Golgi apparatus in the microsporidian, Glugea stephani. J Eukaryot Microbiol 1994; 41:63S-64S.
12. Weidner E, Byrd W, Scarborough A et al. Microsporidian spore discharge and the transfer of polaroplast organelle membrane into plasma membrane. J Protozool 1984; 31:195-198.
13. Bigliardi E, Sacchi L. Cell biology and invasion of the microsporidia. Microbes Infect 2001; 3:373-379.
14. Frixione E, Ruiz L, Cerbon J et al. Germination of Nosema algerae (Microspora) spores: Conditional inhibition by D2O, ethanol and Hg2+ suggests dependence of water influxupon membrane hydration and specific transmembrane pathways. J Eukaryot Microbiol 1997; 44:109-116.
15. Bohne W, Ferguson DJP, Kohler K et al. Developmental expression of tandemly repeated, glycine- and serine-rich spore wall protein in the microsporidian pathogen Encephalitozoon cuniculi. Infect Immun 2000; 68:2268-2275.
16. Hayman JR, Hayes SF, Amon J et al. Developmental expression of two spore wall proteins during maturation of the microsporidian Encephalitozoon intestinalis. Infect Immun 2001; 69:7057-7066.
17. Brosson D, Kuhn L, Prensier G et al. The putative chitin deacetylase of Encephalitozoon cuniculi: A surface protein implicated in microsporidian spore-wall formation. FEMS Microbiol Lett 2005; 247:81-90.
18. Peuvel-Fanget I, Polonais V, Brosson D et al. EnP1 and EnP2, two proteins associated with the Encephalitozoon cuniculi endospore, the chitin-rich inner layer of the microsporidian spore wall. Int J Parasitol 2006; 36:309-318.
19. Xu Y, Takvorian P, Cali A et al. Identification of a new spore wall protein from Encephalitozoon cuniculi. Infect Immun 2006; 74:239-247.
20. Brosson D, Kuhn L, Delbac F et al. Proteomic analysis of the eukaryotic parasite Encephalitozoon cuniculi (Microsporidia): A reference map for proteins expressed in late sporogonial stages. Proteomics 2006; 6:3625-3635.
21. Franzen C. Microsporidia: How can they invade other cells? Trends in Parasitol 2004; 20:275-276.
22. Xu Y, Weiss LM. The microsporidian polar tube: A highly specialised invasion organelle. Int J Parasitol 2005; 35:941-953.

23. Canning EU, Curry A, Lacey CJN et al. Ultrastructure of Encephalitozoon sp. Infecting the conjunctival, corneal and nasal epithelia of a patient with AIDS. Europ J Protistol 1992; 28:226-237.
24. Chioralia G, Trammer T, Maier WA et al. Morphologic changes in Nosema algerae (Microspora) during extrusion. Parasitol Res 1998; 84:123-131.
25. Weidner E. The microsporidian spore invasion tube. The ultrastructure, isolation, and characterization of the protein comprising the tube. J Cell Biol 1976; 71:23-34.
26. Weidner E. The microsporidian spore invasion tube. III. Tube extrusion and assembly. J Cell Biol 1982; 93:976-979.
27. Frixione E, Ruiz L, Santillan M et al. Dynamics of polar filament discharge and sporoplasm expulsion by microsporidian spores. Cell Motil Cytoskel 1992; 22:38-50.
28. Desportes I. Ultrastructure de Stempellia mutabilis Leger et Hesse, microsporidie parasite de l'éphémère Ephemera vulgata. Protistologica 1976; 12:121-150.
29. Takvorian PM, Cali A. Polar tube formation and nucleoside diphosphatase activity in the microsporidian, Glugea stephani. J Eukaryot Microbiol 1996; 43:102S-103S.
30. Delbac F, Peuvel I, Méténier G et al. Microsporidian invasion apparatus : Identification of a novel Polar Tube Protein and evidence for clustering of ptp1 and ptp2 genes in three Encephalitozoon species. Infect Immun 2001; 69:1016-1024.
31. Desportes-Livage I, Hilmarsdottir I, Romana C et al. The characteristics of the microsporidian Enterocytozoon bieneusi: A consequence of its development within the short living-enterocytes. J Protozool 1991; 38:111S-113S.
32. Hilmarsdottir I, Desportes-Livage I, Datry A et al. Morphogenesis of the polaroplast in Enterocytozoon bieneusi Desportes et al, 1985, a microsporidian parasite of HIV infected patients. Europ J Protistol 1993; 29:88-97.
33. Keohane EM, Orr GA, Takvorian PM et al. Purification and characterization of a microsporidian polar tube protein. Mol Biochem Parasitol 1996; 79:255-259.
34. Keohane EM, Orr GA, Takvorian PM et al. Polar tube proteins of microsporidia of the family Encephalitozoonidae. J Eukaryot Microbiol 1999; 46:1-5.
35. Delbac F, Peyret P, Méténier G et al. On proteins of the microsporidian invasive apparatus: Complete sequence of a Polar Tube Protein of Encephalitozoon cuniculi. Mol Microbiol 1998; 29:825-834.
36. Keohane EM, Orr GA, Zhang HS et al. The molecular characterization of the major polar tube protein gene from Encephalitozoon hellem, a microsporidian parasite of humans. Mol Biochem Parasitol 1998; 94:227-236.
37. Peuvel I, Delbac F, Metenier G et al. Polymorphism of the gene encoding a major polar tube protein PTP1 in two microsporidia of the genus Encephalitozoon. Parasitology 2000; 121:581-587.
38. Xiao L, Li L, Moura H et al. Genotyping Encephalitozoon hellem isolates by analysis of the polar tube protein gene. J Clin Microbiol 2001; 39:2191-2196.
39. Vavra J. Detection of polysaccharides in microsporidian spores by means of the periodic acid-thio semicarbazide-silver proteinate test. J Microscopie 1972; 14:357-360.
40. Xu Y, Takvorian PM, Cali A et al. Glycosylation of the major polar tube protein of Encephalitozoon hellem, a microsporidian parasite that infects humans. Infect Immun 2004; 72:6341-6350.
41. Peek R, Delbac F, Speijer D et al. Carbohydrate moieties of microsporidian polar tube proteins are targeted by immunoglobulin G in immunocompetent individuals. Infect Immun 2005; 73:7906-7913.
42. van Gool T, Biderre C, Delbac F et al. Serodiagnostic studies in an immunocompetent individual infected with Encephalitozoon cuniculi. J Infect Dis 2004; 189:2243-2249.
43. Delbac F, Duffieux F, David D et al. Immunocytochemical identification of spore proteins in two microsporidia, with emphasis on extrusion apparatus. J Eukaryot Microbiol 1998; 45:224-231.
44. Beckers PJ, Derks GJ, Gool T et al. Encephalytozoon intestinalis-specific monoclonal antibodies for laboratory diagnosis of microsporidiosis. J Clin Microbiol 1996; 34:282-285.
45. Polonais V, Prensier G, Méténier G et al. Microsporidian polar tube proteins: Highly divergent but closely linked genes encode PTP1 and PTP2 in members of the evolutionarily distant Antonospora and Encephalitozoon groups. Fungal Genet Biol 2005; 42:791-803.
46. Peuvel I, Peyret P, Méténier G et al. The microsporidian polar tube: Evidence for a third polar tube protein (PTP3) in Encephalitozoon cuniculi. Mol Biochem Parasitol 2002; 122:69-80.
47. Vivarès CP, Méténier G. The microsporidia genome: Living with minimal genes as an intracellular eukaryote. In: Lindsay DS, Weiss LM, eds. Opportunistic Infections: Toxoplasma, Sarcocystis and Microsporidia, World Class Parasites. Vol. 9. Boston: Kluwer Acad Pub, 2004:215-242.
48. Undeen AH, Avery SW. Effect of anions on the germination of Nosema algerae (Microspora: Nosematidae) spores. J Invertebr Pathol 1988; 52:84-89.

49. Undeen AH, Vander Meer RK. The effect of ultraviolet radiation on the germination of Nosema algerae Vavra and Undeen (Microsporida: Nosematidae) spores. J Protozool 1990; 37:194-199.
50. Leitch GJ, He Q, Wallace S et al. Inhibition of the spore polar filament extrusion of the microsporidium, Encephalitozoon hellem, isolated from an AIDS patient. J Eukaryot Microbiol 1993; 40:711-717.
51. Frixione E, Ruiz L, Undeen AH. Monovalent cations induce microsporidian spore germination in vitro. J Eukaryot Microbiol 1994; 41:464-468.
52. Weidner E, Byrd W. The microsporidian spore invasion tube. II. Role of calcium in the activation of invasion tube discharge. J Cell Biol 1982; 93:970-975.
53. Pleshinger J, Weidner E. The microsporidian spore invasion tube. IV. Discharge activation begins with pH-triggered Ca^{2+} influx. J Cell Biol 1985; 100:1834-1838.
54. Hayman JR, Southern TR, Nash TE. Role of sulfated glycans in adherence of the microsporidian Encephalitozoon intestinalis to host cells in vitro. Infect Immun 2005; 73:841-848.
55. Southern TR, Jolly CE, Russell Hayman J. Augmentation of microsporidia adherence and host cell infection by divalent cations. FEMS Microbiol Lett 2006; 260:143-149.
56. Lom J, Vavra J. The mode of sporoplasm extrusion in microsporidian spores. Acta Protozol 1963; 1:81-92.
57. Ghosh K, Cappiello CD, McBride SM et al. Functional characterization of a putative aquaporin from Encephalitozoon cuniculi, a microsporidia pathogenic to humans. Int J Parasitol 2006; 36:57-62.
58. Undeen AH, Vander Meer RK. Conversion of intrasporal trehalose into reducing sugars during germination of Nosema algerae (Protista: Microspora) spores: A quantitative study. J Eukaryot Microbiol 1994; 41:129-132.
59. Magaud A, Achbarou A, Desportes-Livage I. Cell invasion by the microsporidium Encephalitozoon intestinalis. J Eukaryot Microbiol 1997; 44:81S.
60. Couzinet S, Cejas E, Schittny J et al. Phagocytic uptake of Encephalitozoon cuniculi by nonprofessional phagocytes. Infect Immun 2000; 68:6939-6945.
61. Foucault C, Drancourt M. Actin mediates Encephalitozoon intestinalis entry into the human enterocyte-like cell line, Caco-2. Microb Pathog 2000; 28:51-58.
62. Franzen C, Muller A, Hartmann P et al. Cell invasion and intracellular fate of Encephalitozoon cuniculi (Microsporidia). Parasitology 2005; 130:285-292.
63. Leitch GJ, Ward TL, Shaw AP et al. Apical spore phagocytosis is not a significant route of infection of differentiated enterocytes by Encephalitozoon intestinalis. Infect Immun 2005; 73:7697-7704.
64. Fasshauer V, Gross U, Bohne W. The parasitophorous vacuole membrane of Encephalitozoon cuniculi lacks host cell membrane proteins immediately after invasion. Eukaryot Cell 2005; 4:221-224.
65. Takvorian PM, Weiss LM, Cali A. The early events of Brachiola algerae (Microsporidia) infection: Spore germination, sporoplasm structure, and development within host cells. Folia Parasitol 2005; 52:118-129.
66. Vivares CP, Metenier G. The microsporidian Encephalitozoon. Bioessays 2001; 23:194-202.

Chapter 18

Intestinal Invasion by *Entamoeba histolytica*

Shahram Solaymani-Mohammadi and William A. Petri, Jr.*

Introduction

Entamoeba histolytica is a protozoan parasite that infects humans and causes the disease amebiasis. The spectrum of intestinal amebiasis varies from colonization without symptoms to fulminating diarrhea and intestinal hemorrhage. The dissemination of the parasite via invasion of the intestinal epithelium allows the trophozoites to invade extra-intestinal sites, most usually the liver. Without treatment, the amebic liver abscesses may continue to enlarge and, if ruptured, cause mortality owing to acute peritonitis. Cases of clinical amebiasis have been reported worldwide, in particular in under-developed and developing counties in Africa, South America, the Indian subcontinent, and Mexico. It has been estimated that approximately 50 million individuals are infected with *E. histolytica* and about 100,000 people die of invasive amebiasis annually, making it the third leading parasitic cause of death, after malaria and schistosomiasis.[1] The host-parasite interaction in human amebiasis is very complicated, and different aspects of innate immunity of the human host against the parasite still are unknown. New insights into the pathogenesis of amebic infections have come from development of in vitro and in vivo models of disease, new molecular and genetic approaches, the identification of key factors in *E. histolytica* pathogenesis, recognition of the mechanisms of evasion from the host's harmful responses, and detection of crucial elements of the host immune responses both innate and acquired. In this chapter, we discuss the innate immunity of human hosts against the parasite and the most important parasite virulence factors and survival strategies that are implicated in pathogenesis.

Innate Mechanisms of Host Resistance to *E. histolytica*

The invasive form of the parasite is the trophozoite (Fig. 1). Amebas encounter natural "barriers" in both the intestine and systemic circulation after extra-intestinal invasion. Although the exact role of these "natural barriers" and their potential roles are still undefined, in the gut, for example, these innate "barriers" prevent potential pathogens and antigens from gaining access to the underlying epithelium, a process called nonimmune exclusion.

Intestinal mucins are highly glycosylated molecules and consist of a core protein (apomucin) joined to oligosaccharides. Mucin glycoproteins line the surface epithelium of respiratory, urogenital, and digestive tracts from the nasal cavity/oropharynx to rectum.[2] Member of the mucin family can differ considerably in size; some are small, whereas others contain several thousands of residues and are among the largest known.[3] However, gastrointestinal mucins consist solely of high molecule mass glycoproteins and their size varies from 0.5×10^6 to 25×10^6 Dalton.[4] Mucins consist of a peptide backbone containing alternating glycosylated and nonglycosylated domains, with O-linked glycosylated regions comprising 70-80% of

*Corresponding Author: William A. Petri, Jr.—Division of Infectious Diseases and International Health, University of Virginia Health System, Charlottesville, Virginia, USA.
Email: wap3g@virginia.edu

Molecular Mechanisms of Parasite Invasion, edited by Barbara A. Burleigh and Dominique Soldati-Favre.

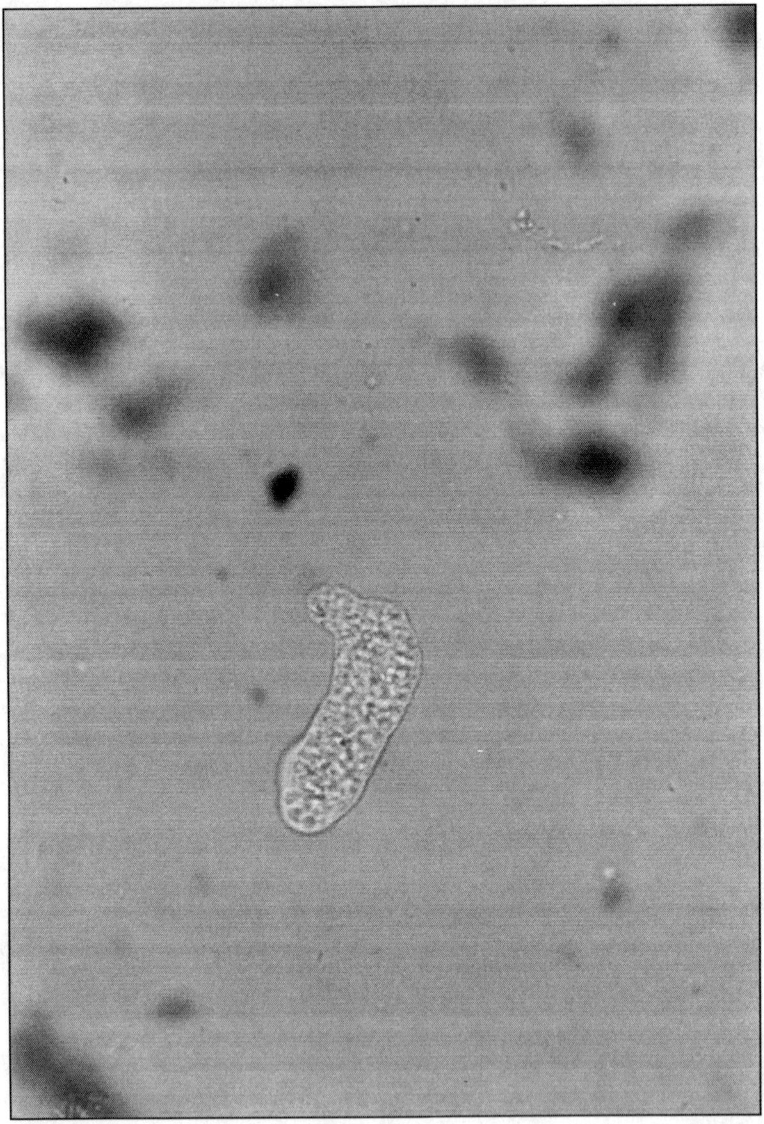

Figure 1. *Entamoeba histolytica* trophozoite in xenic culture (Robinson's medium) (x 400).

the polymer. N-Acetylglucosamine, N-acetylgalactosamine, fucose, and galactose are the 4 primary mucin oligosaccharides.[5,6] Mucin oligosaccharide chains are often terminated with sialic acid or sulfate groups, which account for the polyanionic nature of mucins at a neutral pH.[5] Gastrointestinal mucins are the first line of host defense against enteric pathogens, including *E. histolytica*. Binding sites of mucins have been shown to compete with those of underlying epithelium, preventing attachment of pathogens to the intestinal wall. To establish colonization, *E. histolytica* needs to bind to colonic mucin oligosaccharides via the 170kDa heavy subunit of the parasite's Gal/GalNAc lectin. The Gal/GalNAc lectin binds to rat and human colonic mucins with a very high affinity.[7] The ability of rat and human colonic mucins to inhibit amebic attachment and cytolysis of target epithelial cells demonstrate the protective role of mucin against *E. histolytica.* In the gerbil model of amebic colitis, it has been shown that goblet cell mucin stores have been depleted before invasion of *E. histolytica* to colonic epithelium.[8] Although the cause of such depletion is still unknown, it is speculated that parasite-derived secretagogues are responsible.[9] In addition to the direct protective role of intestinal mucins, intestinal bacterial flora compete for attachment to mucin and may prevent amebic lodgment.[10] The protective key role of mucus blanket in natural immunity against *E. histolytica* infections has been verified both in vitro and in vivo. For example, the mucus-producing colonic cell line, LS174T, inhibits amebic adherence to Chinese hamster ovary (CHO) cells, showing the ability of these mucus-producing cells to compete with attachment sites that are of the utmost importance for *E. histolytica* colonization.[11] Also, the protective role of the mucus layer has been detected in vitro by the finding that *E. histolytica* trophozoites obliterate epithelial cell monolayers without a mucus barrier more quickly and easily than those protected by a mucus cover.[11] Additionally, studies by using the mouse model of intestinal amebiasis showed *E. histolytica* induced the expression of cyclooxygenase-2 in epithelial cells and macrophages, and the resultant prostaglandins enhance epithelial permeability, mediating neutrophil responses.[12]

The human complement system is an important early host defense against amebic infection. *E. histolytica* that disseminates from the bowl through the blood stream is exposed to complement, another component of innate mechanism of host resistance against invading pathogens. Earlier, it had been demonstrated that nonimmune sera lysed axenic strains of *E. histolytica,* implying a role of complement against the parasite in blood circulation.[13] Axenic *E. histolytica* isolates activate both the alternative and classical pathways.[14,15] Additionally, this event has been shown to occur in patients with ALA by finding the high concentrations of serum or plasma concentrations of components of the classical (C1q, C4) and alternative (C3, factor B) pathways, regulatory protein factor H, and one of the C3 products of degradation, C3d, in patients with amebic liver abscess.[16] It has been shown that the major *E. histolytica* extracellular proteinase, a 56kDa neutral cysteine proteinase, activates complement by *E. histolytica* in the fluid phase.[17] It is generally believed that both pathogenic *E. histolytica* and nonpathogenic *E. dispar* are susceptible to human complement. It was shown that more than 90% of *E. histolytica* trophozoites were lysed after exposure to the alternative pathway components.[18] However, even though human complement-mediated cytolysis of *E. histolytica* has an effective amebicidal activity in vitro, a considerable number of ameba can survive and escape human complement activity in vivo (see below). Reed et al,[19] for the first time, showed that both serum-sensitive and serum-resistant stains of the parasite could activate complement. The increased frequency, rate, and severity of amebic liver abscess (ALA) in complement-depleted hamsters treated with cobra venom factor (CoF) reinforced the fact that complement components may play an important role in innate immunity against amebic infections.[20] The Gal/GalNAc lectin molecule of *E. histolytica* has sequence homology and antigenic cross-reactivity with CD59, a membrane inhibitor of C5b-9 in human erythrocytes, suggesting the *E. histolytica* adhesin has both molecular mimicry and shared complement-inhibitory functions.[21] The Gal/GalNAc lectin bound C8 and C9 and efficiently, preventing membrane attack complex (C5b-9) formation and subsequent cell lysis. A definite role for the Gal/GalNAc lectin was confirmed by abrogation of amebic complement resistance following treatment with a monoclonal antibody to Gal/GalNAc lectin molecule.[21]

Different Mechanisms of Pathogenesis in *E. histolytica*

Several factors contribute to the pathogenicity of *E. histolytica*, and some may still await identification. However, three pathogenic factors of the parasite have been investigated extensively and characterized at molecular levels. These three virulence factors are: (1) the Gal/GalNAc adhesin, mediating adherence to host cells and contributing to amebic resistance to complement, (2) the amoebapores, small peptides that produce pores in target cell membranes, and (3) the cysteine proteinases that play a key role in *E. histolytica* tissue invasion, evasion of host defenses, and parasite induction of gut inflammation.

Gal/GalNAc Lectin

The Gal/GalNAc lectin is a novel multifunctional virulence factor of *E. histolytica*, participating in adherence, cytolysis, invasion, resistance to human complement, and also perhaps encystation.[22] Perhaps, the most important part in amebic pathogenesis and pathology is to adhere to the colonic wall. Adhesion of the parasite occurs mainly through the Gal/GalNAc lectin, which binds to exposed terminal Gal/GalNAc residues of target cell glycoproteins.[23] Other molecules thought to be involved in part in adhesion of the parasite are: a 220-kDa lectin, a 112-kDa adhesin, and a surface lipophosphoglycan.[24]

The Gal/GalNAc adhesin is a novel multifunctional protein composed of a heterodimer of heavy (170-kDa), light (35/31 kDa),[23] with a noncovalently-linked intermediate (150-kDa) subunit.[24] The 170-kDa heavy subunit (hgl) is a type I transmembrane protein with a small intracellular domain and a carbohydrate recognition domain (CRD) contained in its extracellular domain.[25] The 30-kDa light subunit (lgl) is covalently attached to the heavy subunit through disulfide linkages. The light subunit has been shown to have several isoforms,[23,26] although the significance of the different isoforms is not clear yet. The intermediate subunit (igl) of the Gal/GalNAc has been cloned and characterized recently, and it has been shown that it lacks a carbohydrate recognition domain (CRD).[24] The Gal/GalNAc lectin mediates adherence of trophozoites to human colonic glycoproteins, human colonic epithelium, human neutrophils, and erythrocytes, and to certain bacteria.[27,28] Evidence for the participation of this molecule in the adhesion event of the parasite has been detected by decreased amebic adherence to target cells when the lectin is inhibited by galactose,[29] by inhibition of adherence with monoclonal antibodies (mAbs) directed against the carbohydrate recognition domain (CRD) of the lectin, and finally by the lack of amebic adherence to Chinese hamster ovary (CHO) cell mutants lacking Gal/GalNAc.[30] In addition to its role in adherence, the Gal/GalNAc lectin also participates in the cytolytic events, since contact-dependent target cell lysis is reduced in the presence of galactose as well as by a monoclonal antibody against the heavy subunit is able to inhibit cytolysis in part without blocking adherence.[31,32] Interestingly, the purified lectin, even at high concentrations, has no cytotoxic effect, suggesting that this protein may be involved in signaling of cytolysis most likely via stimulation of actin polymerization.[32,33]

Amoebapore

Once *E. histolytica* establishes contact with mammalian cells, a rapid cytolytic event takes place that result in swelling, surface blebbing, and lysis of the target cell and leaving the parasite intact. The similarity of this phenomenon to perforin-mediated lysis of target cells by cytotoxic T lymphocytes[34] suggested the possible presence of a channel-forming protein, amoebapore, in *E. histolytica*.[35,36] Amoebapores are a family of small peptides contained in cytoplasmic vesicles in the trophozoites with maximum activity at acidic pH. The amebapore of *E. histolytica* is a channel-forming peptide of 77 amino acid residues; these proteins have now been purified, sequenced, and the relevant genes have been cloned.[37] Three amoebapore isoforms, A, B, and C, at a ratio of 35:10:1, respectively, have been characterized; these peptides showing 35 to 57% deduced amino acids sequence identity and are encoded by a family of three genes.[37,38] In addition to these three different amoebapores, a homologue of haemolysin III has been identified, suggesting that, in addition to amoebapores, haemolysins may have a role in host cell lysis.[39]

All of these three peptides have a common six cysteine residues at identical positions and also a histidine residue near the C terminus. Significant similarities have been determined between these peptides in both structural and functional basis and NK-lysin, pore-forming peptide occurring in natural killer (NK) cells and porcine T cells.[40] Structural modeling with the use of genetic algorithm, suggests a compact tertiary structure composed of four α-helix bundles stabilized by three disulfide bonds, a structure that is also present in NK-lysin.[41]

Amebapores are now believed to aggregate through the arrangement of their amphipathic α-helices and finally they form a channel within the plasma membrane through which water, ions, and other small molecules pass and thus the target cell lyses. Amebapores have cytolytic activity against several human cell lines including human Jurkat T cells;[41] these peptides also are effective in forming pores in gram-positive bacterial membrane.[38,41] However, causing damage to thick gram-negative bacterial shield requires high concentrations of amebapore or removal of thick wall in advance with lysosome.[41]

A homologous peptide to *E. histolytica* has been determined in nonpathogenic *E. dispar* trophozoites.[42] This peptide has significant shared structural and functional properties to that of *E. histolytica*, including highest activity at acidic pH, presence in cytoplasmic granules, and a 95% identity of primary structures. However, irrespective of these similarities, the activity of *E. dispar* amebapore is 60% lower than that from *E. histolytica*. It has been suggested that this lower activity may be related to shortened putative amino-terminal α-helix of the *E. dispar* porin.[42] Irrespective of all advances in the molecular biology and biochemistry of amebapores, their exact roles in *E. histolytica* cytolytic events have not been yet known. Amebapores are not continuingly secreted from viable trophozoites in vitro,[43] suggesting that these peptides may be secreted upon stimulus, including target cell contact, and play a role in lysis of host cells during invasion.[40] The presence of pore-forming activity in nonpathogenic *E. dispar* suggests that the primary role of these peptides is likely to destroy phagocyted bacteria, the main source of food for *Entamoeba* sp. in the gut.[40,44]

Cysteine Proteinases

Cysteine proteinases occur in a wide range of organisms including bacteria, plants, invertebrates, and vertebrates.[45] In mammals, these enzymes are involved in protein turnover within lysosomes. In addition, extracellular cysteine proteases have been implicated in various physiological and pathophysiological processes, including tumor invasion and metastasis.[46] In protozoan parasites, different types of cysteine proteases have been characterized widely and shown to have a variety of functions (including evasion from the host immune responses and roles in developmental cycle of the parasite) and cytopathic effects (including induction of apoptosis).[47]

Several cysteine proteinases with molecular weight between 16 to 96 kDa have been observed in *E. histolytica* extracts. Previously, investigators purified two distinct cysteine proteinases from *E. histolytica* and designated them as amoebapain (now know as ehcp3)[48] and histolysin (now known as ehcp1).[49] However, three different main isoforms of cysteine proteinases are produced in *E. histolytica* of about 30kDa (27-to 30-kDa) and are encoded by a family of more than 40 genes (Tannich, personal communication).

E. histolytica trophozoites show strong proteolytic activity and release large amounts of cysteine proteinases into growth media.[50] Previous studies showed that *E. histolytica* secreted 10- to 1,000-fold more cysteine proteinase activity than did *E. dispar* isolates.[51] It appears that *ehcp5* is the only cysteine proteinase known that is present on ameba surface.[52] In addition to *ehcp5,* it has been suggested that *ehcp1* is also important in *E. histolytica*-induced pathogenesis. The nonpathogenic *E. dispar* has four cysteine proteinase genes (*edcp2, edcp3, edcp4, edcp6*), with the highest expression belonging to *edcp3*. Using *ehcp* sequences as cross-hybridizing probes, it was revealed that functional orthologs corresponding to *ehcp1* and *ehcp5* are absent in *E. dispar*.[53,54] Apparently, only two of *E. dispar* cysteine proteinase genes are expressed,[53] and this may explain why *E. dispar* has low levels of cysteine proteinase activity. Additionally, *E. dispar* lacks several of the most important *E. histolytica* cysteine proteinases, and this again might explain in part its noninvasive nature.

Cysteine proteinases have been shown to be involved in host invasion by some other parasites including *Trypanosoma cruzi*, *Plasmodium falciparum*, *Cryptosporidium parvum*, and *Toxoplasma gondii*.[55] Cysteine proteinases have been implicated in the cytopathic effects of *E. histolytica* upon target cells, resulting in the release of adherent cells from monolayers, presumably by degradation of the components of the extra-cellular matrix, including fibronectin, laminin, and collagen as well as an extra-cellular matrix from vascular smooth muscles.[56] These cytopathic effects correspond with the amount of CP activity secreted and can be inhibited by some specific peptide inhibitors.[57] For example, using exogenous laminin or cysteine proteinase-specific inhibitor L-trans-epoxysuccinyl-leucylamido- (4-guanidino) butane (E-64) to neutralize cysteine proteinase reduced the formation of ALA significantly in the severe combined immunodeficient (SCID) mouse.[58] In the recent years, new generations of cysteine proteinase-specific inhibitors have been introduced which have been effective against *E. histolytica* CPs,[59] suggesting the potential of these specific inhibitors as novel antiamebic chemotherapy. By antisense inhibiting of *E. histolytica* cysteine proteinases, it more recently was shown that trophozoites with reduced cysteine protease activity were ineffective at degrading cysteine-labeled colonic mucin compared to wild-type amoebae by >60%.[60]

The Role of Signal Transduction in *E. histolytica* Pathogenicity

Interaction of *E. histolytica* trophozoites with diverse external stimuli, such as exposure to extracellular matrix (ECM) proteins, appears to activate signaling pathways through G-protein-coupled receptors. In vitro studies indicate that the parasite releases proteases and/or toxin-like molecules when in contact with several types of cultured cells.[61] Adhesion to fibronectin (FN) and its proteolytic fragments is known to induce a wide variety of cellular responses, such as expression of genes encoding proteases, secretion of proteins, activation of lymphocytes and differentiation of neural, endothelial, myoblastic and many other types of cells;[62,63] an early reaction to fibronectin binding is formation of actin adherence plates and focal contact in trophozoites. It has been shown that FN action is mainly dependent on the influx of external Ca^{2+}.[62] It has been shown that the protein kinase C (PKC) pathways are activated in amebas by information transduced as a result of trophozoite binding to FN.[63] Adherence of *E. histolytica* trophozoites to another extracellular component, collagen, is a known stimulus for parasite activation, leading to subsequent tissue destruction and invasion. In vitro interaction of *E. histolytica* with collagen induces intracellular formation and release of electron-dense granules (EDGs) and stimulation of collagenolytic activity.[64] Additionally, results of one study showed that tyrosine phosphorylation is involved in collagen signaling in amoebas and that pp125FAK and p42MAPK homologs may play an active role in turning on the genetic program that enables the parasite to invade its host.[65] There is also evidence suggesting that adhesion to collagen and activation of EDGs secretion are integrin-dependent events and the involvement of actin, vimentin, and tubulin in restructuring cytoskeleton during EDGs secretion are evident.[66] More recent experimental studies attempting to identify and characterize the gene(s) that are upregulated by the human collagen type I and Ca^{2+} interactions showed that interaction of *E. histolytica* with human collagen type I and Ca^{2+} triggers the transcriptional activation of at least two important genes(AD1 and AD2) responsible for pathogenesis of amebiasis.[67]

It was demonstrated that there were similarities between mechanisms of phagocytosis of bacteria and erythrocytes by ameba and macrophages, support the idea of coincidental selection of amebic genes encoding proteins that mediate destruction of host cells.[68] *Rho* family GTPases regulates many features of the cell behavior in eukaryotic cells and the members of the *rac* superfamily have critical role in regulating a wide range of cellular processes such as cellular growth, differentiation, vesicle transport, nuclear transport, and actin cytoskeleton regulation.[69,70] Genes for these components of signaling pathway have been identified in *E. histolytica*. More recently, a family of over 90 putative transmembrane kinases (TMKs) sub grouped into nine different families has been recognized in *E. histolytica*.[71] In addition to their possible role in antigenic variations, it seems that they are involved in signal transduction via sensing the *E. histolytica* ambient.

Parasite Evasion of Innate and Acquired Immune Responses

Degradation of Antibodies by Parasite Proteases

Amebic granules contain copious amounts of tissue-destructive substances, such as hydrolytic enzymes and strong cysteine proteinases, whose secretion contributes to the damage of host cells and tissues. These proteases, mainly cysteine proteinases, can escape the antibody-mediated humoral immune responses by degrading IgA and IgG antibodies. It has been well documented that human serum and secretory IgA were degraded completely when exposed to viable axenic trophozoite of *E. histolytica* (HM1:IMSS strain), parasite lysates, and medium that had been conditioned by incubation with viable trophozoites.[72] This phenomenon could be more important for ameba since *E. histolytica* trophozoites must conquer the destructive act of secretory IgA (sIgA) and the serum IgG antibodies during intestinal colonization and extra-intestinal dissemination.

Shedding of Immune Complexes by Capping

Interaction of *E. histolytica* trophozoites with sera from patients with invasive amebiasis results in initial clustering of bound antibodies into small patches, followed by a rapid mobilization of these patches to the posterior pole of the cell, where the uroid is formed.[73] It is believed that the uroid, the posterior appendix that forms during the movement of trophozoite, plays a role in the escape of ameba from the host immune responses. In one study, it was shown that myosin II is of the utmost importance in capping and formation of uroid in *E. histolytica* parasite, in such a way that myosin II was three times more concentrated within the uroid compared with the rest of the cell, suggesting that the release of caps may depend upon mechanical contraction driven by myosin II activity.[74] This results in the remarkable ability of *E. histolytica* to rapidly regenerate substantial amounts of plasma membrane. The properties of surface receptor redistribution (capping), liberation of caps, and plasma membrane regeneration, may contribute to the survival of the parasite in the host during infection (Fig. 2).

Anergy of T Cells and Suppression of Macrophages

Tissue invasion by *E. histolytica* has been associated with suppression of cell-mediated immunity. It has been known that *E. histolytica* exerts different modulatory effects on macrophages, and T cells especially T helper type 1(Th1).[75] T cells are important sources of macrophage-activating cytokines, and can be directly cytotoxic to amoeba, although the mechanisms involved are unknown. There is some evidence that amebic infections are associated with T cell modulation. Reduced delayed-type hypersensitivity reaction during the acute phase of the disease is one index of functional suppression of T cell during human amebiasis.[76] Similarly, in mice models, intestinal amebiasis has been associated with a cyclic suppression in immune responses, and it has been proposed that these alterations, observed at the cellular level, might facilitate invasion of the host by the parasite.[77] The cytotoxic ability of macrophages and their potential as antigen-presenting and cytokines-releasing elements are reduced during the acute phase of ALA. Exposure of human mononuclear phagocytes to the monocyte locomotion-inhibitory factor produced by *E. histolytica* led to a swift increase in the intracellular concentration of adenosine 3':5' cyclic monophosphate (cAMP) and inhibits the respiratory burst in human macrophages.[78] This study suggests that like other leukotactic inhibitors, the monocyte locomotion-inhibitory factor produced by *E. histolytica* operates through modulations of intracellular cAMP. In addition, the parasite exerts suppression that appears to be a local event mediated by direct exposure of macrophages to *E. histolytica* directly or its by-products. For example, the treatment of murine macrophages with amebic antigens in vitro reduces the production of molecules associated to the I region of the MHC induced by INF-γ.[79] This study also showed that *E. histolytica* subverted critical macrophage accessory function in part via prostaglandin E2 (PGE2) biosynthesis. Similarly, amebic liver abscess-derived macrophages produced low basal levels of TNF in response to stimulation with lipopolysaccharide (LPS),

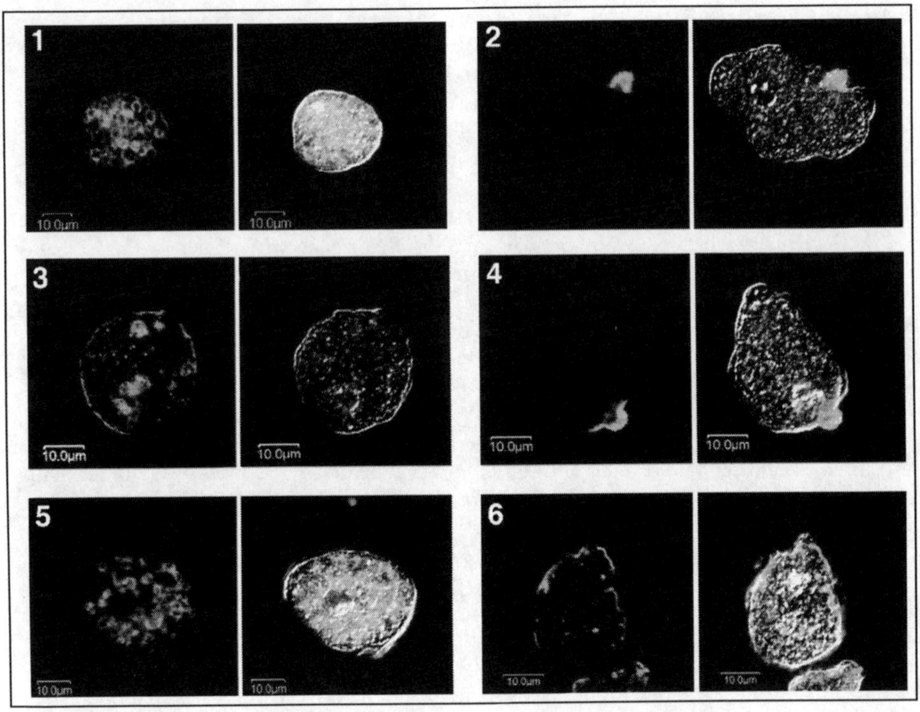

Figure 2. Induction of capping. Confocal microscopy of trophozoites after induction of capping for 30 min using antibodies against the heavy subunit of the Gal/GalNAc lectin. The fluorescence of the Gal/GalNAc lectin is shown on the left side and the fluorescent Gal/GalNAc lectin superimposed on a Nomarsky section on the right side. (Reproduced from: Katz et al. Mol Cell Biol 2002; 13:4256-65; published online before print as 10.1091/mbc.E02-06-0344; with the permission from The American Society for Cell Biology.)

whereas peritoneal and spleen macrophages as well as Kupffer cells from infected animals did not release TNF in vitro constituently.[80]

Complement Resistance

It has been shown that ameba develop complement resistance after repeated exposure to active human serum.[81] In this study, it also was shown that susceptibility to complement-dependent lysis was regained 6 weeks after serum treatments were terminated, suggesting that resistance to lysis was an acquired rather than a genetic property. It was also shown that both complement resistant and complement sensitive strains of *E. histolytica* could activate complement but only the latter could survive.[19] It has been suggested that complement resistance is necessary for *E. histolytica* tissue invasion. Braga et al[21] showed that the Gal/Gal NAc lectin molecule had sequence homology and antigenic cross reactivity with CD59, a membrane inhibitor of C5b-9 in human erythrocytes, suggesting that the Gal/Gal NAc lectin also has complement-inhibitory functions. It seems that Gal/Gal NAc lectin bound C8 and C9, preventing formation of the membrane attack complex and subsequent cell lysis.[21] Recently, it has been shown that different strain of mouse models of intestinal amebiasis and also genetically deficient mice for IL-12, IFN-γ, or inducible NO synthase are resistant against intracecal inoculation of *E. histolytica*.[82] This study also demonstrated the important role of host immune responses to resistance to infection as well as the fact that depletion of $CD4^+$ cells significantly diminished both parasite burden and host's inflammatory responses.

Antigenic Variation via Transmembrane Kinases (TMKs)

Most recently, a large family of over 80 transmembrane kinases (TMKs) in *E. histolytica* has been discovered. These proteins are CXXC-rich proteins with highly variable extracellular domains homologous to the intermediate subunit (igl) of the Gal/GalNac and variant surface proteins (VSPs) of Giardia and with cytoplasmic kinase domains. The discovery of the large family of CXXC-containing TMKs is of interest not only for their potential role in antigenic variation but also for their role in cell signaling. *E. histolytica* must respond to a wide variety of environmental stimuli as it exists into a trophozoite in the intestinal lumen and enters the host by invasion of the intestinal mucosal epithelium. Invasion involves attaching to the epithelium and responding to that attachment event through signaling events via the *E. histolytica* Gal/GalNac adherence lectin that lead to host cell killing. The changing host environment should necessitate having a variety of ways of sensing and responding to the host.[71]

In conclusion, the mechanisms involved in pathogenesis of amebiasis and immune response evasion of the parasite, however, are only partially defined, and the specific amoebic components responsible are unidentified. A more complete understanding of the mechanisms involved in amoebic downregulation of host inunumity may assist in the identification of strategies for preventing and treating amebiasis.

Refrerences

1. World Health Organization: Amebiasis. WHO Weekly Epidemiological Records 1997; 72:97-100.
2. Lamont JT. Mucus: The front line of intestinal mucosal defense. Ann NY Acad Sci 1992; 664:190-201.
3. Perez-Vilar J, Hill RL. The structure and assembly of secreted mucins. J Biol Chem 1999; 274(45):31751-31754.
4. Moncada DM, Kammanadiminti SJ, Chadee K. Mucin and Toll-like receptors in host defense against intestinal parasites. Trends Parasitol 2003; 19(7):305-311.
5. Forstner JF, Oliver MG, Sylvester FA. Production, structure, and biologic relevance of gastrointestinal mucins. In: Blaser MJ, Smith PD, Ravdin JI, Greenberg HB, Guerrant RL, eds. Infections of the gastrointestinal tract. New York: Raven Press, 1995:71-88.
6. Deplancke B, Gaskins HR. Microbial modulation of innate defense: Goblet cells and the intestinal mucus layer. Am J Clinical Nutrition 2001; 73(6):131S-141.
7. Chadee K, Petri Jr WA, Innes DJ et al. Rat and human colonic mucins bind to and inhibit adherence lectin of Entamoeba histolytica. J Clin Invest 1987; 80(5):1245-1254.
8. Chadee K, Meerovitch E. Entamoeba histolytica: Early progressive pathology in the cecum of the gerbil (Meriones unguiculatus). Am J Trop Med Hyg 1985; 34:283-291.
9. Belley A, Keller K, Gottke M et al. Intestinal mucins in colonization and host defense against pathogens. Am J Trop Med Hyg 1999; 60:10-15.
10. Ravdin JI. Immunobiology of human infection by Entamoeba histolytica. Pathol Immunopathol Res 1989; 8(3-4):179-205.
11. Belley A, Keller K, Grove J et al. Interaction of LS174T human colon cancer cell mucins with Entamoeba histolytica: An in vitro model for colonic disease. Gastroenterology 1996; 111(6):1484-1492.
12. Stenson WF, Zhang Z, Riehl T et al. Amebic infection in the human colon induces cyclooxygenase-2. Infect Immun 2001; 69(5):3382-8.
13. Huldt G, Davies P, Allison AC et al. Interactions between Entamoeba histolytica and complement. Nature 1979; 277(5693):214-216.
14. Meri S, Richaud G, Linder E. Complement activation by antigenic fractions of Entamoeba histolytica. Parasite Immunol 1985; 7(2):153-164.
15. Calderon J, Schreiber RD. Activation of the alternative and classical complement pathways by Entamoeba histolytica. Infect Immun 1985; 50:560-565.
16. Munoz LE, Salazar OG. Complement activation in patients with amebic liver abscess. J Hepatol 1987; 5(1):30-36.
17. Reed SL, Keene WE, McKerrow JH et al. Cleavage of C3 by a neutral cysteine proteinase of Entamoeba histolytica. J Immunol 1989; 143:189-195.
18. Walderich B, Weber A, Knobloch J. Sensitivity of Entamoeba histolytica and Entamoeba dispar patient isolates to human complement. Parasite Immunol 1997; 19(6):265-271.
19. Reed SL, Curd JG, Gigli I et al. Activation of complement by pathogenic and nonpathogenic Entamoeba histolytica. J Immunol 1986; 136(6):2265-2270.

20. Capin R, Capin NR, Carmona M et al. Effect of complement depletion on the induction of amebic liver abscess in the hamster. Arch Invest Med (Mex) 1980; 11(1 Suppl):173-180.
21. Braga LL, Ninomiya H, McCoy JJ et al. Inhibition of the complement membrane attack complex by the galactose-specific adhesion of Entamoeba histolytica. J Clin Invest 1992; 90(3):1131-1137.
22. Mann BJ. Structure and function of the Entamoeba histolytica Gal/GalNAc lectin. Int Rev Cytol 2002; 216:59-80.
23. Petri Jr WA, Chapman MD, Snodgrass T et al. Subunit structure of the galactose and N-acetyl-D-galactosamine-inhibitable adherence lectin of Entamoeba histolytica. J Biol Chem 1989; 264(5):3007-3012.
24. Cheng XJ, Hughes MA, Huston CD et al. Intermediate subunit of the Gal/GalNAc lectin of Entamoeba histolytica is a member of a gene family containing multiple CXXC sequence motifs. Infect Immun 2001; 69(9):5892-8.
25. Stanley Jr SL, Huizenga H, Li E. Isolation and partial characterization of a surface glycoconjugate of Entamoeba histolytica. Mol Biochem Parasitol 1992; 50(1):127-138.
26. Dodson JM, Lenkowski Jr PW, Eubanks AC et al. Infection and immunity mediated by the carbohydrate recognition domain of the Entamoeba histolytica Gal/GalNAc lectin. J Infect Dis 1999; 179(2):460-466.
27. McCoy JJ, Weaver AM, Petri Jr WA. Use of monoclonal anti-light subunit antibodies to study the structure and function of the Entamoeba histolytica Gal/GalNAc adherence lectin. Glycoconj J 1994; 11(5):432-436.
28. Ravdin JI, John JE, Johnston LI et al. Adherence of Entamoeba histolytica trophozoites to rat and human colonic mucosa. Infect Immun 1985; 48(2):292-297.
29. Bracha R, Mirelman D. Adherence and ingestion of Escherichia coli serotype 055 by trophozoites of Entamoeba histolytica. Infect Immun 1983; 40(3):882-887.
30. Petri Jr WA, Haque R, Mann BJ. The bittersweet interface of parasite and host: Lectin-carbohydrate interactions during human invasion by the parasite Entamoeba histolytica. Annu Rev Microbiol 2002; 56:39-64.
31. Petri Jr WA, Joyce MP, Broman J et al. Recognition of the galactose-or N-acetylgalactosamine-binding lectin of Entamoeba histolytica by human immune sera. Infect Immun 1987; 55(10):2327-2331.
32. Saffer LD, Petri Jr WA. Role of the galactose lectin of Entamoeba histolytica in adherence-dependent killing of mammalian cells. Infect Immun 1991; 59(12):4681-4683.
33. Mann BJ, Lockhart LA. Molecular analysis of the Gal/GalNAc adhesin of Entamoeba histolytica. J Eukaryot Microbiol 1998; 45(2):13S-16S.
34. Tschopp J, Nabholz M. Perforin-mediated target cell lysis by cytolytic T lymphocytes. Annu Rev Immunol 1990; 8:279-302.
35. Lynch EC, Rosenberg IM, Gitler C. An ion-channel forming protein produced by Entamoeba histolytica. EMBO J 1982; 1(7):801-804.
36. Rosenberg I, Gitler C. Subcellular fractionation of amoebapore and plasma membrane components of Entamoeba histolytica using self-generating Percoll gradients. Mol Biochem Parasitol 1985; 14(2):231-248.
37. Leippe M, Tannich E, Nickel R et al. Primary and secondary structure of the pore-forming peptide of pathogenic Entamoeba histolytica. EMBO J 1992; 11(10):3501-3506.
38. Leippe M, Andra J, Nickel R et al. Amoebapores, a family of membranolytic peptides from cytoplasmic granules of Entamoeba histolytica: Isolation, primary structure, and pore formation in bacterial cytoplasmic membranes. Mol Microbiol 1994; 14(5):895-904.
39. Loftus B, Anderson I, Davis R et al. The genome of the protest parasite Entamoeba histolytica. Nature 2005; 433(7028):865-868.
40. Leippe M. Amoebapores. Parasitol Today 1997; 13(5):178-183.
41. Leippe M, Andra J, Nickel R et al. Amoebapores, a family of membranolytic peptides from cytoplasmic granules of Entamoeba histolytica: Isolation, primary structure, and pore formation in bacterial cytoplasmic membranes. Mol Microbiol 1994; 14(5):895-904.
42. Leippe M, Bahr E, Tannich E et al. Comparison of pore-forming peptides from pathogenic and nonpathogenic Entamoeba histolytica. Mol Biochem Parasitol 1993; 59(1):101-109.
43. Leippe M, Sievertsen HJ, Tannich E et al. Spontaneous release of cysteine proteinases but not of pore-forming peptides by viable Entamoeba histolytica. Parasitology 1995; 111(Pt 5):569-574.
44. Tannich E. Entamoeba histolytica and E. dispar: Comparison of molecules considered important for host tissue destruction. Trans R Soc Trop Med Hyg 1998; 92(6):593-596.
45. Berti PJ, Storer AC. Alignment/phylogeny of the papain superfamily of cysteine proteases. J Mol Biol 1995; 246:273-283.
46. Sloane BF, Rozhin J, Hatfield JS et al. Plasma membrane-associated cysteine proteinases in human and animal tumors. Exp Cell Biol 1987; 55:209-224.

47. Aly AS, Matuschewski K. A malarial cysteine protease is necessary for Plasmodium sporozoite egress from oocysts. J Exp Med 2005; 202(2):225-30.
48. Scholze H, Schulte W. On the specificity of a cysteine proteinase from Entamoeba histolytica. Biomed Biochim Acta 1988; 47(2):115-123.
49. Luaces AL, Barrett AJ. Affinity purification and biochemical characterization of histolysin, the major cysteine proteinase of Entamoeba histolytica. Biochem J 1988; 250(3):903-909.
50. Keene WE, Hidalgo ME, Orozco E et al. Entamoeba histolytica: Correlation of the cytopathic effect of virulent trophozoites with secretion of a cysteine proteinase. Exp Parasitol 1990; 71(2):199-206.
51. Reed SL, Keene WE, McKerrow JH. Thiol proteinase expression and pathogenicity of Entamoeba histolytica. J Clin Microbiol 1989; 27(12):2772-2777.
52. Jacobs T, Bruchhaus I, Dandekar T et al. Isolation and molecular characterization of a surface-bound proteinase of Entamoeba histolytica. Mol Microbiol 1998; 27:269-276.
53. Bruchhaus I, Jacobs T, Leippe M et al. Entamoeba histolytica and Entamoeba dispar: Differences in numbers and expression of cysteine proteinase genes. Mol Microbiol 1996; 22(2):255-263.
54. Willhoeft U, Hamann L, Tannich E. A DNA sequence corresponding to the gene encoding cysteine proteinase 5 in Entamoeba histolytica is present and positionally conserved but highly degenerated in Entamoeba dispar. Infect Immun 1999; 67:5925-5929.
55. Forney JR, Yang S, Healy MC. Protease activity associated with excystation of Cryptosporidium parvum oocysts. J Parasitology 1996; 82(6):889-892.
56. Keene WE, Petitt MG, Allen S et al. The major neutral proteinase of Entamoeba histolytica. J Exp Med 1986; 163(3):536-549.
57. Reed S, Bouvier J, Pollack AS et al. Cloning of a virulence factor of Entamoeba histolytica: Pathogenic strains possess a unique cysteine proteinase gene. J Clin Invest 1993; 91(4):1532-1540.
58. Li E, Yang WG, Zhang T et al. Interaction of laminin with Entamoeba histolytica cysteine proteinases and its effect on amebic pathogenesis. Infect Immun 1995; 63(10):4150-4153.
59. Que X, Reed SL. Cysteine proteinases and the pathogenesis of amebiasis. Clin Microbiol Rev 2000; 13(2):196-206.
60. Moncada D, Keller K, Ankri S et al. Antisense inhibition of Entamoeba histolytica cysteine proteases inhibits colonic mucus degradation. Gastroenterology 2006; 130(3):721-30.
61. Meza I. Extracellular matrix-induced signaling in Entamoeba histolytica: Its role in invasiveness. Parasitol Today 2000; 16(1):23-28.
62. Carbajal ME, Manning-Cela R, Pina A et al. Fibronectin-induced intracellular calcium rise in Entamoeba histolytica trophozoites: Effect on adhesion and the actin cytoskeleton. Exp Parasitol 1996; 82(1):11-20.
63. Santiago A, Carbajal ME, Benitez-King G et al. Entamoeba histolytica: PKC transduction pathway activation in the trophozoite-fibronectin interaction. Exp Parasitol 1994; 79(3):436-444.
64. Munoz ML, Lamoyi E, Leon G et al. Antigens in electron-dense granules from Entamoeba histolytica as possible markers for pathogenicity. J Clin Microbiol 1990; 28(11):2418-2424.
65. Perez E, Munoz ML, Ortega A. Entamoeba histolytica: Involvement of pp125FAK in collagen-induced signal transduction. Exp Parasitol 1996; 82(2):164-170.
66. de Lourdes Munoz M, Das P, Tovar R. Entamoeba histolytica trophozoites activated by collagen type I and Ca^{2+} have a structured cytoskeleton during collagenase secretion. Cell Motil Cytoskeleton 2001; 50(1):45-54.
67. Debnath A, Akbar MA, Mazumder A et al. Entamoeba histolytica: Characterization of human collagen type I and Ca^{2+} activated differentially expressed genes. Exp Parasitol 2005; 110(3):214-219.
68. Ghosh SK, Samuelson J. Involvement of p21racA, phosphoinositide 3-kinase, and vacuolar ATPase in phagocytosis of bacteria and erythrocytes by Entamoeba histolytica: Suggestive evidence for coincidental evolution of amebic invasiveness. Infect Immun 1997; 65(10):4243-4249.
69. Aguilar-Rojas A, Almaraz-Barrera Mde J, Krzeminski M et al. Entamoeba histolytica: Inhibition of cellular functions by overexpression of EhGEF1, a novel Rho/Rac guaninenucleotide exchange factor. Exp Parasitol 2005; 109(3):150-162.
70. Lohia A, Samuelson J. Molecular cloning of a rho family gene of Entamoeba histolytica. Mol Biochem Parasitol 1993; 58(1):177-180.
71. Beck DL, Boettner DR, Dragulev B et al. Identification and gene expression analysis of a large family of transmembrane kinases related to the Gal/GalNAc lectin in Entamoeba histolytica. Eukaryot Cell 2005; 4(4):722-32.
72. Kelsall BL, Ravdin JI. Degradation of human IgA by Entamoeba histolytica. J Infect Dis 1993; 168(5):1319-1322.
73. Espinosa-Cantellano M, Martinez-Palomo A. Entamoeba histolytica: Mechanism of surface receptor capping. Exp Parasitol 1994; 79(3):424-435.

74. Arhets P, Gounon P, Sansonetti P et al. Myosin II is involved in capping and uroid formation in the human pathogen Entamoeba histolytica. Infect Immun 1995; 63(11):4358-4367.
75. Zambrano-Villa S, Rosales-Borjas D, Carrero JC et al. How protozoan parasites evade the immune response. Trends Parasitol 2002; (6):272-278.
76. Ortiz-Ortiz L, Zamacona G, Sepulveda B et al. Cell-mediated immunity in patients with amebic abscess of the liver. Clin Immunol Immunopathol 1975; 4(1):127-134.
77. Ghosh PK, Castellanos-Barba C, Ortiz-Ortiz L. Intestinal amebiasis: Cyclic suppression of the immune response. Parasitol Res 1995; 81(6):475-480.
78. Rico G, Diaz-Guerra O, Kretschmer RR. Cyclic nucleotide changes induced in human leukocytes by a product of axenically grown Entamoeba histolytica that inhibits human monocyte locomotion. Parasitol Res 1995; 81(2):158-162.
79. Wang W, Chadee K. Entamoeba histolytica suppresses gamma interferon-induced macrophage class II major histocompatibility complex Ia molecule and I-A beta mRNA expression by a prostaglandin E2-dependent mechanism. Infect Immun 1995; 63(3):1089-1094.
80. Wang W, Keller K, Chadee K. Modulation of tumor necrosis factor production by macrophages in Entamoeba histolytica infection. Infect Immun 1992; 60(8):3169-3174.
81. Calderon J, Tovar R. Loss of susceptibility to complement lysis in Entamoeba histolytica HM1 by treatment with human serum. Immunology 1986; 58(3):467-471.
82. Houpt ER, Glembocki DJ, Obrig TG et al. The mouse model of amebic colitis reveals mouse strain susceptibility to infection and exacerbation of disease by CD4+ T cells. J Immunol 2002; 169(8):4496-503.

Index

E

F

G

H

I

K

L

M

N

O

P

R

S

T

V